AF352780

Selected Papers from the 15th OpenFOAM Workshop

Selected Papers from the 15th OpenFOAM Workshop

Editor

Eric G. Paterson

Preface by Jonathan S. Pitt

MDPI • Basel • Beijing • Wuhan • Barcelona • Belgrade • Manchester • Tokyo • Cluj • Tianjin

Editor
Eric G. Paterson
Virginia Tech
USA

Editorial Office
MDPI
St. Alban-Anlage 66
4052 Basel, Switzerland

This is a reprint of articles from the Special Issue published online in the open access journal *Fluids* (ISSN 2311-5521) (available at: https://www.mdpi.com/journal/fluids/special_issues/OpenFOAM).

For citation purposes, cite each article independently as indicated on the article page online and as indicated below:

LastName, A.A.; LastName, B.B.; LastName, C.C. Article Title. *Journal Name* **Year**, *Volume Number*, Page Range.

ISBN 978-3-0365-1996-8 (Hbk)
ISBN 978-3-0365-1997-5 (PDF)

Contents

About the Editors

Eric G. Paterson is a Rolls-Royce Commonwealth Professor of Marine Propulsion, Head of the Kevin T. Crofton Department of Aerospace and Ocean Engineering, and Executive Director of the Ted and Karyn Hume Center for National Security Technology.

Professor Paterson has over 30 years' experience in fluid dynamics research and education. With students and colleagues, he has developed computational tools for the design and analysis of ocean vehicles, terrestrial and offshore wind turbines, hydroturbines and marine hydro-kinetic devices, thermal analysis of large-scale deployable space systems, artificial heart pumps and cardiovascular devices, and biomimetically inspired trace detectors. His current research is focused on the intersection of ship hydrodynamics and physical oceanography, ocean in situ and remote sensing, data assimilation from autonomous systems for numerical weather prediction, and polyfunctional design of active drag-reduction systems. Paterson is a Fellow of the Society of Naval Architects and Marine Engineers and an Associate Fellow of the American Institute of Aeronautics and Astronautics. He completed his PhD, MS, and BS degrees at The University of Iowa.

Preface to "Selected Papers from the 15th OpenFOAM Workshop"

The 15th OpenFOAM Workshop—A Virtual First, Jonathan S. Pitt, Chair

The 15th OpenFOAM Workshop (OFW15) was held virtually from 22–26 June 2020, and was hosted by Virginia Tech, from Arlington, Virginia. While the workshop was overall considered a success, given the events of 2020, there were moments when such an outcome was in doubt. Nevertheless, through agile adaptation to a completely online format by both the organizers and the attendees, OFW15 was able to not only continue the tradition of grass-roots information sharing among OpenFOAM[1] users worldwide, but was also able to set several firsts that are likely to continue in future workshops. In this article, the experiences of the organizers, and the outcomes (some unexpected) of moving a 200+ attendee workshop online, just three months before the events are detailed. Before continuing, it is imperative to acknowledge all those whose contributions were critical to the success of OFW15: The OpenFOAM Workshop Committee, members of the local organizing committee, Virginia Tech Continuing and Professional Education, the keynote speakers and trainers who volunteered their time, and the workshop's sponsors, ENGYS[2] and OpenCFD.

A Change of Trajectory—A Virtual Format

Originally planned to be held in-person in Arlington, VA, from 22–25 June 2020, Virginia Tech was selected to host the 15th OpenFOAM Workshop at the previous workshop (OFW14), in July of 2019. Expected to draw 200–250 attendees, the local organizing committee began the usual process of securing space, speakers, and developing budgets. However, in March of 2020, just three months before the planned start of the event, the COVID-19 pandemic gripped the world, and the reality of an in-person event quickly dissolved. Faced with the decision to outright cancel the event, the organizers instead saw this as an opportunity for a computer-savvy audience to set an example for a virtual conference, and the OpenFOAM community responded with enthusiasm.

OFW15 was quickly restructured to be the first virtually conducted OpenFOAM Workshop, and was planned to be a synchronous event delivered via the Zoom[3] platform. The decision to host the event synchronously was driven by the desire to enable live audience participation during technical presentations; however, it came with the adverse effect of conducting the workshop at more favorable or adverse local times worldwide. The general nature of the program remained the same: keynote talks, training sessions, and contributed individual technical presentations and posters remained on the program.

---

[1]   OpenFOAM® is a registered trademark of OpenCFD Limited, producer and distributor of the OpenFOAM software via https://www.openfoam.com.

[2]   http://www.engys.com

[3]   Copyright ©2021 Zoom Video Communications, Inc. All rights reserved. https://www.zoom.us

In addition to Zoom platform for the delivery of technical material, the WHOVA [4] application for conference delivery was engaged by the organizers. This provided a platform for the real-time delivery of schedule updates and announcements, but most importantly, provided a platform for real-time communication between attendees. Such a feature was greatly received by participants, with many discussion groups organically occurring during the workshop, enhancing the overall attendee experience.

OFW15 by the Numbers

The 15th OpenFOAM Workshop recorded the highest level of participation of any previous iteration of the meeting. Figure 1 displays the final tallies of registered participants, number of technical contributions by category, and a breakdown by continent of the self-reported country home of each of the technical contributors. At 336 registered participants, this was certainly the largest audience to attend an OFW. The number of technical contributions and training sessions, a key component of the workshop, were similar to past workshops. The organizers consider this to be an exceptional attendance, especially given the change in venue and delivery method so close to the planned start of the workshop. Figure 2 displays a visual map of the 39 countries represented by the registered attendees at the 15th OpenFOAM Workshop.

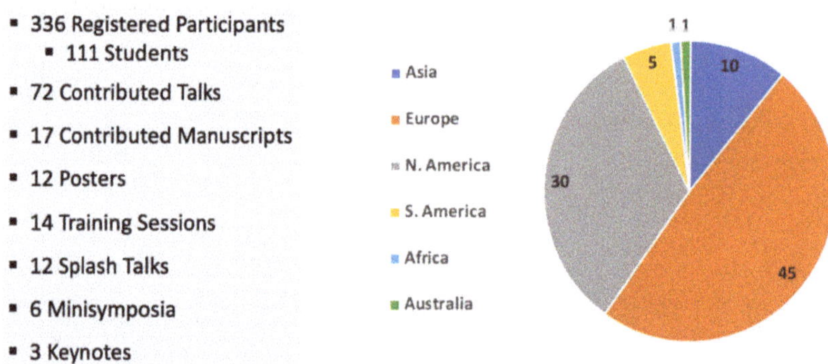

Figure 1. Attendance and participation number for OFW15 (left) and a breakdown by continent of self-reported country by technical contributors (right).

4   https://www.whova.com/

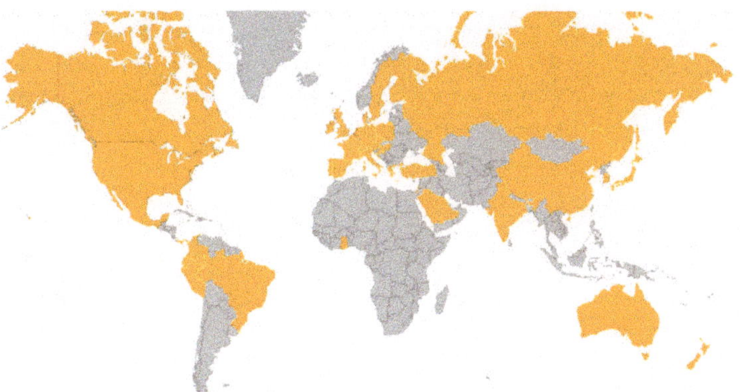

Figure 2. Visual map of the 39 countries represented by registered attendees at OFW15.

Unexpected Outcomes and Lessons Learned

Conducting the 15th OpenFOAM Workshop as a virtual event resulted in a number of lessons learned along the way, and resulted in some very positive unexpected outcomes. We list here a number of observations, supported by attendance figures above and feedback from the attendee survey:

- The virtual format and subsequent reduced attendance fee greatly expanded the aperture of individuals who were able to attend OFW15. The immediate effect of this was a wider international audience for the dissemination of OpenFOAM-related information and experiences, no doubt increasing the use of open-source CFD software worldwide.
- Participation during technical presentations, particularly in the form of asking questions, was observed to be significantly greater than during in-person events. The ability of individuals to ask questions via the Zoom chat feature was essential to interaction during OFW15. This was an unexpected, but welcomed, outcome from the virtual format of the workshop.
- The decision to host has, even synchronously, led to increased interactions among attendees; however, it may have significantly influenced participation from regions where the local time was favorable. This may be a reason for over-representation from North America and Europe in the contributed technical material; however, these two regions also represent some of the historically higher locations of OpenFOAM usage.
- While the traditional face-to-face informal conversations that are often critical to the workshop or conference experience were not possible, the inclusion of the WHOVA platform as part of the workshop allowed for individual attendees to connect with one another.

Reflections and Looking Forward

The decision to re-organize the 15th OpenFOAM Workshop into the first ever virtual instance of the event stimulated an even greater level of participation in the event than expected, and enabled access to the event for many individuals who would have otherwise been unable to attend. The net effect was to reduce the barrier to entry to learn more about the OpenFOAM platform and engage new users worldwide—precisely the goal of the OpenFOAM Workshop from its beginnings. Thinking back to a decision to possibly cancel the workshop now seems anathematic to our way of life, and the continuation of scientific pursuits.

Looking forward to future workshops, the role of virtual attendance options cannot be overlooked. The benefit of the engagement of groups that may otherwise be unable to participate is significant; however, the logistics of providing hybrid in-person and virtual workshop experiences are still in the experimentation phase. There is no doubt that future OpenFOAM Workshops will continue this upward trajectory, and set their own firsts as well.

Jonathan S. Pitt
Chair, 15th OpenFOAM Workshop

Article

OpenFOAM Simulations of Late Stage Container Draining in Microgravity

Joshua McCraney [1,*,†], **Mark Weislogel** [2] **and Paul Steen** [1,3]

[1] School of Mechanical and Aerospace Engineering, Cornell University, Ithaca, NY 14853, USA; phs7@cornell.edu
[2] School of Mechanical Engineering, Portland State University, Portland, OR 97207, USA; weisloge@pdx.edu
[3] School of Chemical and Biomolecular Engineering, Cornell University, Ithaca, NY 14853, USA
*   Correspondence: jm2555@cornell.edu
†   This paper is an extended version of our paper published in 15th OpenFOAM Workshop.

Received: 12 October 2020; Accepted: 5 November 2020; Published: 11 November 2020

Abstract: In the reduced acceleration environment aboard orbiting spacecraft, capillary forces are often exploited to access and control the location and stability of fuels, propellants, coolants, and biological liquids in containers (tanks) for life support. To access the 'far reaches' of such tanks, the passive capillary pumping mechanism of interior corner networks can be employed to achieve high levels of draining. With knowledge of maximal corner drain rates, gas ingestion can be avoided and accurate drain transients predicted. In this paper, we benchmark a numerical method for the symmetric draining of capillary liquids in simple interior corners. The free surface is modeled through a volume of fluid (VOF) algorithm via interFoam, a native OpenFOAM solver. The simulations are compared with rare space experiments conducted on the International Space Station. The results are also buttressed by simplified analytical predictions where practicable. The fact that the numerical model does well in all cases is encouraging for further spacecraft tank draining applications of significantly increased geometric complexity and fluid inertia.

Keywords: capillary flow; container draining; CFD; interior corner; capillary fluidics; contact angle; tankage

---

1. Introduction

Capillary draining dictates the fluid withdrawal rate of precious fuels, propellants, coolants, and aqueous solutions prevalent on spacecraft. For the specific example of liquid fuels aboard orbiting spacecraft, capillary draining can serve as a limit to the life of the spacecraft if and when the residual fuel in the tank is or becomes inaccessible. It is essential to establish the maximum capillary flow rate at which a container in a microgravity environment can be drained. Interior corner devices constructed of 'vanes' provide a geometric family of propellant management devices (PMDs) [1]. Such constructs have also been studied for the practical purposes of enhancing bubble coalescence and/or breakup [2,3], stabilization of liquid columns from 'g-jitter' induced by orbital maneuvers, docking and crew activity [4], and inhibiting choked flows [5]. Interior corners can be used to both enhance and hinder flows [6]. Herein, we study simple interior linear corners and corners with increased geometric complexity.

A drain application is sketched in Figure 1, where a right circular cylinder is drained in low-g. As liquid drains, its topology changes, and a dry region forms at the base. It is this late stage draining (Figure 1, far right) we consider, as such transients can be critical for fluid withdrawal. For sufficiently small stream-wise curvature, one can imagine a 'cut' along the single drain port, the unraveling of which yields a linear corner (cross section shown bottom right of Figure 1) with a drain at each end: hereafter referred to as a double-drain (Figure 2). In this work, we simulate double-drain problems. We quantitatively assess meniscus evolution profiles, maximal interfacial height, and volumetric drain rates. Simulations are validated against a simplified analytical model formulated and experimentally analyzed by Weislogel and McCraney [7] (ICF-1 test cell below). The simulations are then extended to model new, more complex geometries (ICF-8 test cell below) also conducted experimentally aboard the International Space Station (ISS).

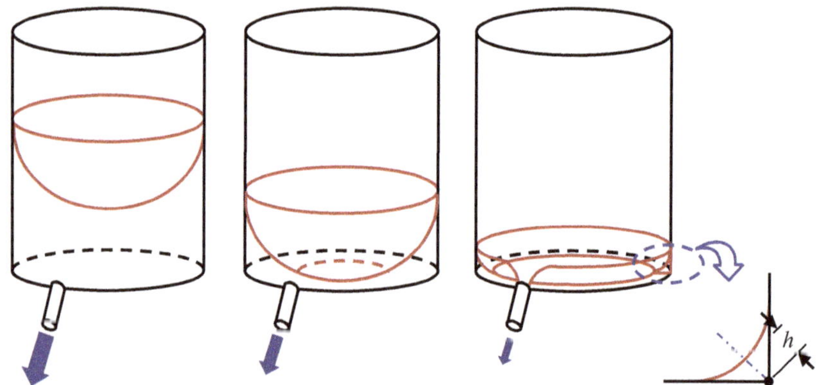

Figure 1. Simple example of capillary draining in a zero-gravity environment. The late-stage draining condition (**right**) yields a thin visco-capillary flow that is approximately linear, with approximately 2D-Cartesian cross-section sketched (**far right**).

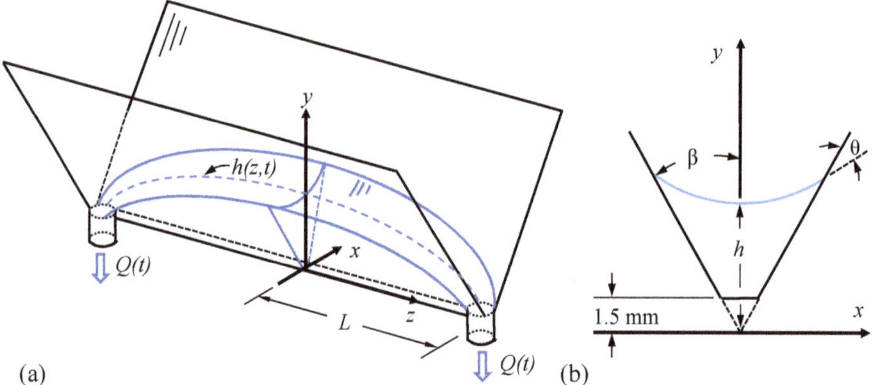

Figure 2. Schematic of double-drain flow analytical and computational domain. Drain ports are symmetric at $z = \pm L$, where volumetric sink flow rates $Q(t)$ are specified: (**a**) perspective view and (**b**) cross-sectional view, dashed end cropped from computational domain.

2. Background

The capillary driven flows of this investigation are created by underpressure gradients in the liquid caused by positive streamwise curvature gradients in the liquid free surface. Such gradients result from forced flows or in many cases spontaneous wicking flows determined by initial conditions. All such curvature gradients arise as the liquid seeks to establish contact angle boundary conditions at the liquid–solid–gas line of contact (i.e., the wall of the container). The 'contact line' is thus a critical boundary condition. Unfortunately, for the contact line to move, it must appear to violate the no slip condition. This quandary has received significant attention over many decades from theoretical [8,9], experimental [10,11], and numerical [12,13] perspectives. The concern is exaggerated in low-g environments where uncertainties in the essentially nanoscale region of the contact line can influence literally tons of liquids at the $O(1\text{m})$ scale. Practical numerical models of such flows treat shortcomings in our physical understanding of the moving contact line with a variety of simplifications in an attempt to discretize competing mathematical models of the phenomena. We highlight a limited selection of recent research that applies most directly to our numerical method.

For example, while high Capillary number oscillatory flows require dynamic contact angle models [3,14], recent studies in similar capillary regimes to those studied herein report encouraging, if not surprising, constant contact angle simulation results. Gurumurthy et al. [15] investigate the spontaneous rise of a liquid in an array of open rectangular channels, where OpenFOAM simulations agree well with power-law theoretical predictions. Malekzadeh and Roohi [16] study flow behavior and droplet formation in T-junction micro-channels, where OpenFOAM simulations agree well with micro-channel experiments. Yong-Qiang et al. [17] simulate capillary rise in fan-shaped interior corners, such that each wall of the corner has a specified constant contact angle not necessarily that of its neighbor, where simulations carried out in FLOW-3D, compare well to drop tower experiments. Arias and Montlaur [18] analyze bubble generation in a capillary T-junction geometry, and report that, while contact angle is a sensitive parameter for low Capillary number $O(10^{-3})$, for large Reynolds number $O(10^2 - 10^3)$ flows, less than 4% error is maintained for all measured quantities between ANSYS VOF simulations and micro-channel experiments. Klatte et al. [19] simulate parabolic flight and drop tower capillary drain-fill wedge geometries in microgravity using a pressure potential field within the static equilibrium Surface Evolver algorithm [20], reporting agreement within 2% for all measured quantities. Such simulation agreement with experiments and theory for a variety of capillary-dominated flow regimes gives confidence to the simple use of the static contact angle model. As such, the simulations reported herein assume a constant contact angle.

Regarding our numerical approach, the VOF method implemented in OpenFOAM's interFoam flow solver produces an interface stretching over a few computational cells [21]. This non-sharp volume fraction can lead to curvature errors, inducing non-physical spurious currents [22]. These stem from the inability of the surface tension algorithm to evaluate a constant interface curvature, thereby generating non-physical capillary waves [23]. While level set methods can be shown to run faster and more accurately calculate curvature, mass-conservative schemes remain challenging [24]. A recent comprehensive review of spurious currents in VOF and level set methods was conducted by Popinet [25]. Soh et al. [26] study droplet formation in T-junction micro-channels, where smoothing operations applied to OpenFOAM's VOF model minimize spurious velocities, improving simulation-experimental agreement. Sontti et al. [27] employ a coupled level set and VOF method, reporting reduced spurious currents compared to the sole VOF counterpart, thereby in better agreement with the micro-channel experiments. Guo et al. [28] replace the ANSYS continuum surface force model with a modified height function method, resulting in increased accuracy for micro-channel annular flows. While the state of the art is ripe with techniques to minimize spurious currents, prior to simulating transient drains, we first analyze a static liquid supported in a wedge geometry in a zero-gravity environment where we find negligible spurious velocities, adding confidence to the reported simulations.

3. Experiments and Data Reduction

The Capillary Flow Experiments (CFE) conducted aboard the ISS were a series of handheld, large length scale (<20 cm) experiments. CFE was pursued to provide data for analytical and numerical model development for capillary flow phenomena relating to moving contact line boundary conditions, critical geometric wetting, and interior corner flows. The latter were pursued via 9 Interior Corner Flow (ICF) test vessels (ICF-1, ICF-2, ..., ICF-9), each representing a geometry of practical concern. A video archive for the ICF test suites is publicly available through login at the mainpage https://psi.nasa.gov/, mission narrative explained at https://www.nasa.gov/sites/default/files/atoms/files/psi_researchers_guide-tagged.pdf. The double-drain tests analyzed herein were conducted in late 2016 and early 2017. Figures 3 and 4 provide an annotated image, solid model, and wire model for the ICF-1 and ICF-8 test vessels respectively. For each test, a given ICF vessel is placed on an ISS workbench, back-lit by a diffuse light screen via cabin lighting, and filmed via an HD Canon XF305 video camcorder fabricated in Tokyo, Japan. During the approximately 3 h of manual crew interaction with each ICF vessel on the ISS, a suite of capillary drain tests were performed. The astronauts drain liquid from the Tapered Section into the Reservoir by turning the Piston Dial counterclockwise. While draining, capillarity wicks fluid into the Interior Corner, the central geometric drain element spanning the vessel length. The drain tests reported herein were performed with Control Valve 2 fully open while Control Valve 1 was adjusted to approximately balance draining resistances in both valves. In this manner, astronauts evenly drained both sides of the Interior Corner, minimizing meniscus height at the two vessel outlets without ingesting gas and emulating a double-drain similar to Figure 1 (right), shown schematically in Figure 2.

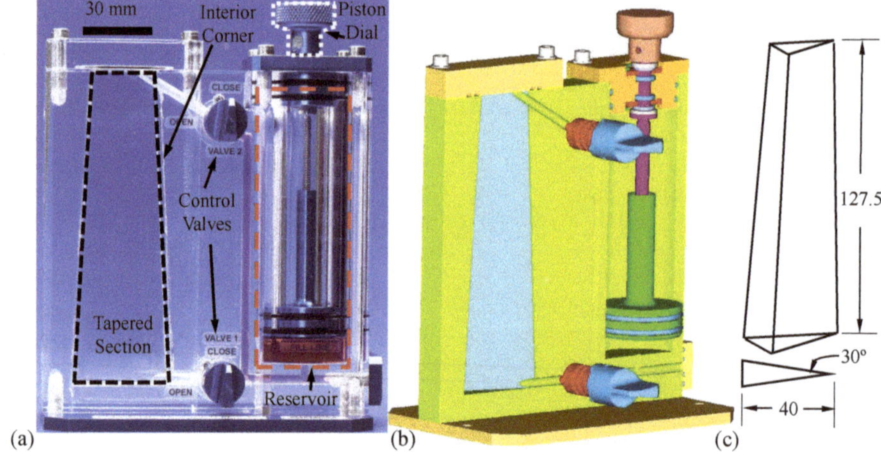

Figure 3. Annotated ICF-1 test Vessel: (**a**) apparatus, (**b**) solid model, and (**c**) wire model with dimensions in mm, where large (**bottom**) to small (**top**) cross-sectional isosceles triangles are congruent through a 20:13 ratio.

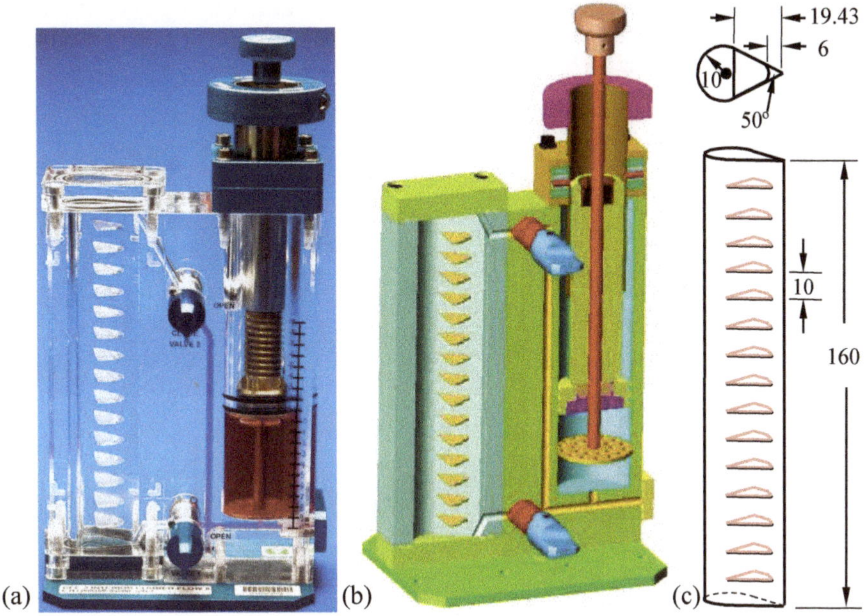

Figure 4. Annotated ICF-8 test Vessel: (**a**) apparatus, (**b**) solid model, and (**c**) wire model with dimensions in mm.

Digitized time-dependent interface profiles are readily collected from the ICF data with which comparisons to analytical and numerical predictions are straightforward. We pursue this tack herein following a brief review of a simple analysis applied to the subject double-drained interior corner. We eventually increase the geometric complexity of the corner such that the simple analytical model below no longer applies, but the experimental and numerical comparisons remain immediate.

Further details of the experimental data reduction process are briefly discussed here. The primary objective is to digitize the experimental fluid interfaces h_e to determine and compute meniscus profiles and instantaneous drain rates. In-house interfacial tracking algorithms were developed to extract the interface position from the images. The experimental volume V_e remaining in the Tapered Section was then computed via a meniscus integration over h_e using

$$V_e = F_A \int_{-L}^{L} h_e^2 \, dz, \tag{1}$$

where F_A is a geometric constant reported in Table 1, L and z shown in Figure 2. There are two ways to establish the experimental drain rate Q_e: (1) time rate of change of V_e, and (2) transient piston position during the drain process, Figure 3a. The latter is not sufficiently accurate to determine V_e: the image pixel resolution combined with the large piston area can yield measurement errors several times larger than the measured quantity, Figure 5a. Figure 5b presents a sample volume integration method for Q_e. The volumetric drain rates at experimental frame times were smoothed via trapezoidal interpolation, maintaining the overall volume-time integral average, but smoothing the drain rates. Figure 5b plots raw and smoothed values for Q_e. In this manner, draining appears continuous despite the nearly peristaltic method applied by the astronauts (i.e., nearly continuous counterclockwise hand turn of the Piston Dial). Total drain time is 70 s for ICF-1 and 69 s for ICF-8. We note h_e was not tracked near the container edge due to light reflection impeding the interfacial tracking algorithm.

Table 1. ICF-1 fluid properties, scales, and constraints. Dual values listed as (experiment, simulation) when different from each other.

Property	Units	ICF-1	ICF-8
Density, ρ	kg m^{-3}	950	910
Viscosity, μ	kg m^{-1} s^{-1}	0.0190	0.00455
Surface tension, σ	N m^{-1}	0.0206	0.0197
Contact angle, θ	deg	0°	0°
Scales	**Units**	**ICF-1**	**ICF-8**
Half angle, β	deg	15°	25°
Flow length, L	mm	63.5	80
Initial half volume, V_i	mm^3	1468, 1484	10,096, 10,549
Height, $H = \sqrt{V_i/F_A L}$	mm	8.8, 8.9	15.3, 15.6
Velocity, $W = \sigma \epsilon F_i \sin^2 \beta / \mu f$	mm s^{-1}	4.5	31.5, 32.2
Flow rate, $Q \sim W F_A H^2$	mm^3 s^{-1}	105.0, 106.2	3970, 4241
Time, $t \sim L/W$	s	14.0	2.54, 2.49
Lubrication Assumptions	**Constraint**	**ICF-1**	**ICF-8**
Slender geometry, $\epsilon = H/L$	$\epsilon^2 \ll 1$	0.0194, 0.0196	0.0364, 0.0381
Low streamwise curvature	$\epsilon^2 f \ll 1$	0.0068	0.0267, 0.0279
Capillary dominance	$Bo \ll 1$	$\sim 10^{-4}, 0$	$\sim 10^{-4}, 0$
Low intertia	$\epsilon^2 Su F_i^2 \sin^4 \beta / f \ll 1$	0.0029, 0.0030	0.5111, 0.5340
Concus-Finn wetting	$\theta < 90° - 2\beta$	satisfied	satisfied

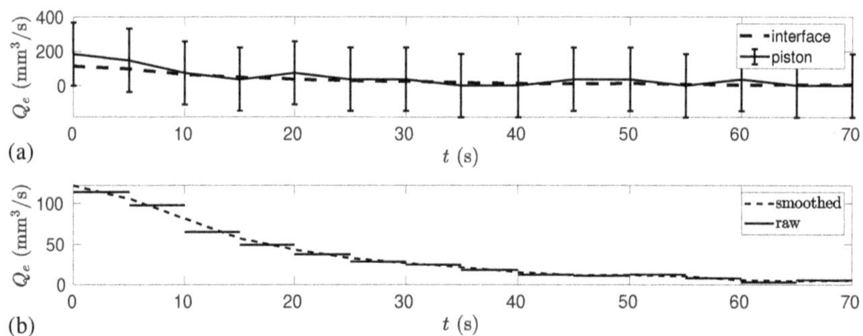

Figure 5. ICF-1 Q_e (**a**) measurement error for piston (dashed, box-whisker) and interfacial (dashed) tracking, where whisker length and dashed line thickness indicate measurement error for each respectively and (**b**) trapezoidal-integrated smooth (dashed) alongside raw (discontinuous black, solid).

4. Numerical Model Review

The ICF test cell sections were stenciled in Fusion 360, a CAD Autodesk tool. The stencil was meshed in snappyHexMesh, refining hexagonal cell layers near the surface geometry. The initial cell count was 42,571 cells for ICF-1 and 121,860 cells for ICF-8, with a one layer dynamic mesh refinement imposed at the interface (8 cell sub-refinement per interfacial cell). The simulation run time was approximately 20 and

80 h for ICF-1 and ICF-8, respectively on an AMD EPYC 7281 Hexadeca-core (16 core) processor running parallel on all 16 cores. Initial conditions were specified via swak4foam, detailed below. Transverse and axial symmetry boundary conditions were implemented in the $z = 0$ and $x = 0$ planes, respectively (Figure 2), reducing the computational domain to 1/4 the Tapered Section. The vessel walls impose no slip and a constant contact angle. Since lubrication approximations are satisfied (Table 1), and by choice of scales that absorb many dynamic contact angle geometric effects [29], a constant contact angle model is qualified [30,31]. The exit port imposes a uniform velocity that linearly interpolates specified velocities temporally. To conserve mass, an atmospheric boundary condition is imposed on a rectangular portion of the vessel roof. The sharp lower interior corner $y \in [0, 1.5]$ mm of the Tapered Section was removed for efficient meshing and run times (Figure 2b). For ICF-1, the upper portion ($y > 18$ mm) of the section was removed for computational efficiency. These assumptions had negligible effects on the numerical results.

The interFoam solver is chosen for this laminar, incompressible, two-phase fluid flow. A VOF technique at the free surface is prescribed with scalar indicator function $\alpha \in [0, 1]$ such that $\alpha = 0$ implies the computational cell is gas and $\alpha = 1$ implies the computational cell is liquid. All cells average fluid properties based on the volume fraction fraction of liquid, using density ρ as an example:

$$\rho = \alpha \rho_l + (1 - \alpha)\rho_g \tag{2}$$

where subscripts l and g denote liquid and gas, respectively. Incompressiblity implies α satisfies the advection equation:

$$\partial_t \alpha + \nabla \cdot (\alpha \boldsymbol{u}) = 0 \tag{3}$$

where $\boldsymbol{u} = \langle u_x, u_y, u_z \rangle$ is the fluid velocity. Continuity and momentum equations governing the flow are

$$\nabla \cdot \boldsymbol{u} = 0, \tag{4}$$

$$\partial_t(\rho \boldsymbol{u}) + \nabla \cdot (\rho \boldsymbol{uu}) = -\nabla P + \nabla \cdot \left(\mu \left(\nabla \boldsymbol{u} + \nabla \boldsymbol{u}^T \right) \right) + \boldsymbol{F}_b, \tag{5}$$

for pressure P and dynamic viscosity μ. Force $\boldsymbol{F}_b = \sigma \kappa \nabla \alpha$ is the surface tension body force vector modeled by the continuum surface force method of Brackbill et al. [32], where σ is the liquid–gas surface tension and $\kappa = -\nabla \cdot (\nabla \alpha / |\nabla \alpha|)$ the liquid–gas interfacial curvature. The aforementioned equations of motion are well-posed once an initial condition is specified. ICF-1 imposed a constant, initial height $h_s = 9.3$ mm, determined to match the initial experimental volume. ICF-8 imposed a constant initial height $h_s = 17.2$ mm, where draining was suspended for the next 3 s to allow the liquid to fill the vanes and establish a static equilibrium where experimental and simulation heights agree, at which point draining begun. Time $t = 0$ corresponds to the time draining ensues. The PIMPLE algorithm is applied for pressure–velocity coupling with increased stability (nCorrectors set to 3). Time integration is performed via first order Euler scheme, allowing a dynamic time step bounded by maximum Courant number 0.2 [3,14–16]. A first order Gauss linear scheme discretizes the gradient terms. Second order Gauss linear, upwind, and vanLeer schemes discretize the divergence terms.

A mesh-independence study was conducted to demonstrate spacial convergence and justify cell refinement. Figure 6 presents time averaged simulation peak center-line interfacial height h_s and time averaged volumetric flow rate Q_s (reference Figure 2a) as percent errors against the base case 42,571 cell count ICF-1 vessel. Both quantities plotted are functions of the percent of the initial 42,571 cell count. We see that increasing the cell count by 10% has less than 1% change in reported values. Then, we conclude the reported results are mesh-independent.

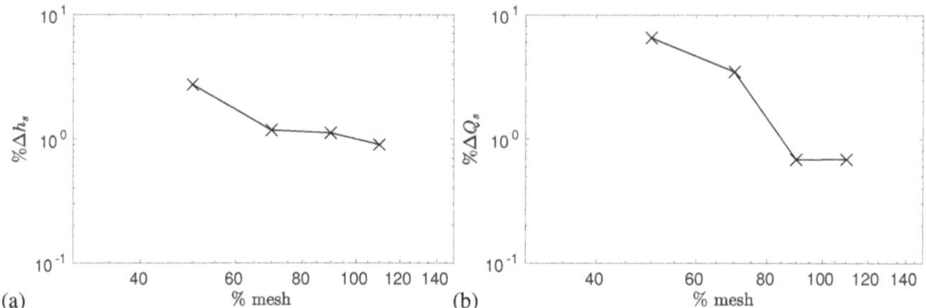

(a)   (b)

Figure 6. Convergence plots, where (**a**) percent height error $\%\Delta h_s$ and (**b**) percent volumetric flow rate error $\%\Delta Q_s$ are compared against the ICF-1 analytic study as functions of the reported percent mesh count. Values time-averaged for the first 15 s of draining.

Saufi et al. [33] study spurious currents in the interFoam solver, placing a 1 mm diameter water droplet in a medium. After 0.2 s, improper curvature calculations resulted in catastrophic droplet deformation. For the ICF-1 vessel, we conduct a similar study: draining is switched off and a slab of liquid is placed in the wedge. Sufficient time is given for the liquid to wick through the corners until static equilibrium is maintained. We then compute the Capillary number $\mu V / \sigma$ at each time step, where here $V = \|U\|_\infty$ over the entire liquid domain. For the next 20 s, we report all Capillary numbers less than 0.02. Figure 7 plots interfacial velocity vectors at their corresponding locations and the underlying liquid for several times. Clearly, the interface is stable, owing to the wedge support. Thus, while infectious to numerous flow problems, we report small spurious currents that are unlikely to significantly cloud the simulation results.

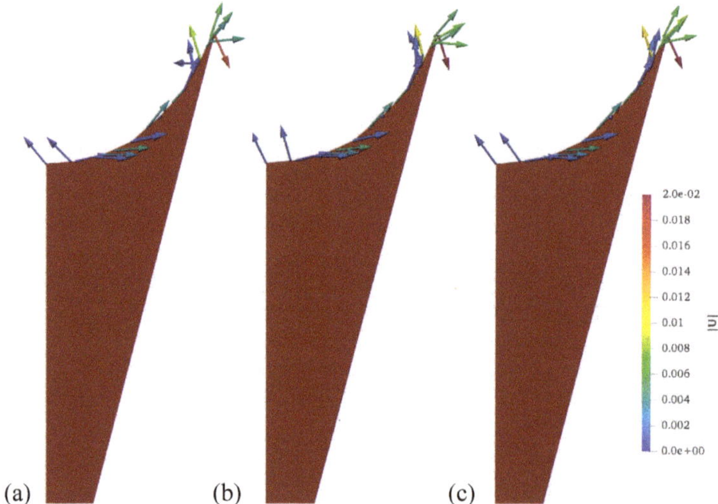

(a)   (b)   (c)

Figure 7. Part of ICF-1 cross-sectional slice with liquid (red) in equilibrium, normalized velocity $|U|$ interfacial vectors with colorbar magnitude (not pertaining to solid red liquid) at (**a**) 0 s, (**b**) 10 s, and (**c**) 20 s.

5. ICF Experiments and Simulations

5.1. Hypothesis

This study attempts to validate the working hypothesis: *OpenFOAM's non-modified interFoam solver accurately predicts large length scale (centimetric) capillary drainage in wedge geometries.* 'Accurate' is assessed via analyzing three fundamental parameters described in detail below: peak interfacial height h, volumetric flow rate Q, and meniscus evolution $h(z, t)$. We first benchmark the simulations against a simple analytical corner drain model, Section 5.2. Closed-form expressions (6)–(10) are readily compared to the simulations. We then simulate two flight experiments: smooth wedge walls ICF-1 Section 5.3, and the wedged-vane network of ICF-8, Section 5.4. An analytical model is also compared to both where appropriate. We assess the aforementioned quantities and compare simulations to experiments.

5.2. Simplified Analysis

A single symmetric double-drained interior corner is sketched in Figure 2. Symmetric flow is assumed with liquid volumetric flow rate $Q(t)$ specified at each drain port $z = \pm L$, where the meniscus height $h(\pm L, t) = 0$ is specified. If the liquid wets the corner such that the Concus–Finn corner wetting condition is satisfied [6] (Table 1), the liquid spontaneously wets into and along the interior corner. As draining at $z = \pm L$ ensues, the capillary pressure becomes increasingly negative via $P \sim -1/h(z, t)$, and liquid migrates by capillarity toward the drain ports where depth h is shallowest. Following [7], all variables are non-dimensionalized according to Table 1 values and presented as dimensionless unless otherwise specified. We invoke subscripts e, s, a to denote experimental, simulation, and analytic analysis values, respectively. For a long narrow channel satisfying the constraints of Table 1, an asymptotic lubrication prediction for $h_a(z, t)$ is

$$h_a(z, t) = \frac{F(z)}{t_i + t/t_i},\tag{6}$$

$$F(z) = \left(a_0(1 - z) - \frac{27}{20} a_0^{2/3}(1 - z)^{8/3} + \frac{243}{650} a_0^{1/3}(1 - z)^{13/3} \right)^{1/3},\tag{7}$$

where $t_i = 2.2059$ is a constant determined by the dimensionless initial half-domain volume and $a_0 = 729(10 + 7\sqrt{2})/500$ is a dimensionless constant determined by the meniscus slope at the drain port. Solutions for respective interface peak height $h_a(z = 0, t)$, liquid half-domain volume $V_a(t)$, and liquid half-domain volumetric flow rate $Q_a(t)$ are given by

$$h_a(z = 0, t) = \frac{1.178}{1 + t/t_i},\tag{8}$$

$$V_a(t) = \frac{1}{(1 + t/t_i)^2},\tag{9}$$

$$Q_a(t) = \frac{0.907}{(1 + t/t_i)^3}.\tag{10}$$

The model above assumes idealized exit conditions and geometry. Such a flow may also be simulated via OpenFoam using the geometric assumptions identified in Figure 3 for the ICF-1 vessel geometry. Excellent agreement is found for the simplified analysis and OpenFoam simulations for the first 30 s of drain process: $|h_s(z = 0) - h_a(z = 0)| < 6.5\%$ and $|Q_s - Q_a| < 4.5\%$ (Figure 8a,b). During the simulation, at approximately $t = 30$ s the drain port ingests gas, significantly retarding Q_s. The time at ingestion is shown in Figure 8b (dotted line). Figure 8c plots (7) against $h_s \cdot (t_i + t/t_i)$ at half-second intervals for the first 30 s of draining. The collapse of simulation profiles demonstrates the numerically verified self-similarity of the flow as suggested by the simple theory (6).

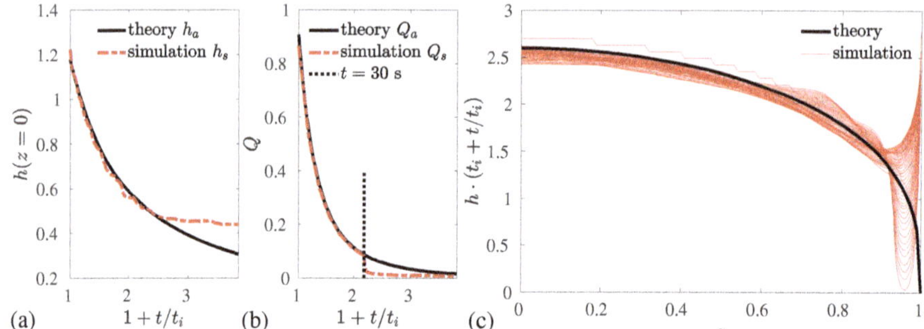

Figure 8. (**a**) h_s (dashed) and h_a (8) (solid) plotted against time, (**b**) Q_s (dashed) and Q_a (10) (solid) plotted against time, with ingestion time $t = 30$ s (dotted). (**c**) $h \cdot (t_i + t/t_i)$ (red) profiles plotted against lubrication theory (7) (black) at half second intervals for the first 30 s of draining. Simulation initial and boundary condition match lubrication prediction.

5.3. Comparison of Numerics and Analysis with Experiments for ICF-1

Figure 8 presents simulation results with the simple theoretical predictions (6)–(10). The analytical model assumes a point sink for the drain port, which excludes the possibility of gas ingestion. For drain ports with finite radii (ICF-1), a zero meniscus height specified at the drain port implies that gas ingestion is imminent. To circumvent immediate gas ingestion, the simulation initial condition incorporates a thin film ϵ_0, $h_s(z, t = 0) = F(z)/t_i + \epsilon_0$ in (7), where $\epsilon_0 = 0.5$ mm. A uniform velocity boundary condition is prescribed at the drain port $U_{vel} = V_a'(t)L/(V_i W \pi r^2)$, where $r = 2$ mm radius drain port, scales defined in Table 1. This velocity condition is consistent with the analytical model prediction, which enforces $h_s(z = L, t) = 0$. The experimental results of ICF-1 are also presented, the computational domain shown in Figure 9.

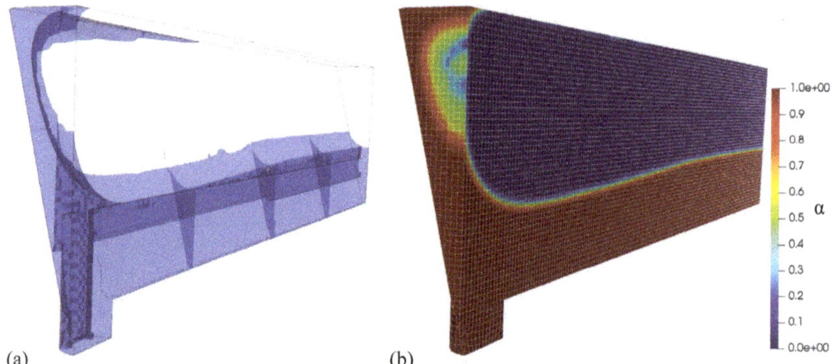

Figure 9. ICF-1 computational domain showing (**a**) liquid volume as phase fraction $\alpha < 0.6$ and (**b**) phase fraction α interpolated throughout each computational cell (outlined in white).

Figure 10 shows a still image of ICF-1 experimental draining alongside a 3D cross-sectional simulation image for qualitative comparison. Figure 11a,b presents simulation, experimental, and simple analytical results. Lack of gas ingestion is observed for the full elapsed drain time, where excellent simulation-experimental volumetric flow rate agreement is seen: $|Q_s - Q_e| < 4\%$, Figure 11b. Meniscus height h_s agrees well with h_e but coincidence is not expected: $h_s(t = 0)$ was determined to satisfy $V_a(t = 0) = V_e(t = 0)$. Additionally, the simulations are drained unevenly about $z = 0$; as such, for early times $h_e(z = -L) < h_s(z = -L)$ and $h_e(z = L) > h_s(z = L)$. Despite these discrepancies, $|h_s(z = 0) - h_e(z = 0)| < 10\%$. At $z = \pm L$ the experimental flows were observed to wick up the edge toward the upper surface of the vessel, shown in Figures 10a and 11c. This expected migration owes to the PDMS liquid perfectly wetting ($0°$ contact angle) the vessel walls. The analytical model predictions are shown to increasingly under-predict h_e at late times, as evidenced in Figure 11a. This expected result owes to the $h_a(z = \pm L) = 0$ boundary condition.

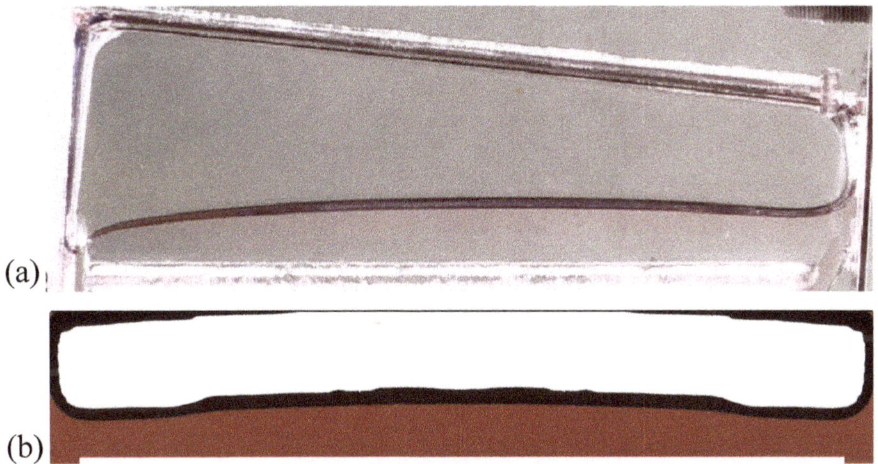

Figure 10. Profile view of the ICF-1 Tapered Section filled with liquid (red) at $t = 10$ s (**a**) during experiment and (**b**) simulation, where gray implies interfacial volume fraction $\alpha = 0.5$.

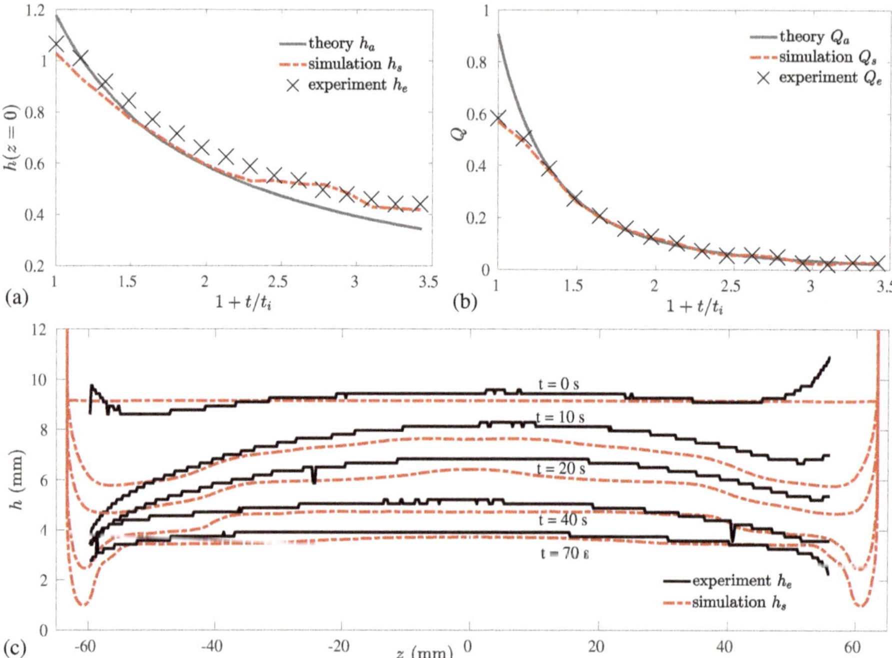

Figure 11. (a) h_s (dashed), h_a (8) (solid), and h_e (ticks) plotted against time, (b) Q_s (dashed), Q_a (10) (solid), and Q_e (ticks) plotted against time; (c) dimensional h_s (dashed) plotted against h_e (solid) at specified times.

5.4. Comparison of Numerics with Experiments for ICF-8

In a similar manner to the ICF-1 vessel, an image of ICF-8 was shown in Figure 4. In this container, a snow-cone cross-sectioned interior corner test cell is further partitioned by vane segments along the z-axis as detailed in Figure 4c. This vessel was designed to study bubble separation characteristics in such geometries, but the double-drain flow could be established and was demonstrated on the ISS. The corner flow again provides symmetric draining of the container with the vane segments adding geometric complexity by introducing a second stream-wise curvature component to the free surface. A sample draining event is shown in Figure 12. The verified OpenFoam software was re-programmed for the ICF-8 geometry. Simulation results are compared to the ICF-8 experiment data in Figure 13. The initial condition for the numerical computations of $h_s = 18$ mm was given 3 s to relax to equilibrium, as $\theta = 0°$ induces wicking along the edges and notably within the vaned segments of the container. The initial height $h_s = h_e$ was also selected resulting in the mismatched initial volumes, Table 1, which is attributable to image analysis limitations detecting the meniscus edge everywhere in the container (i.e., side walls and upper corners). To address this issue, we non-dimensionalize ICF-8 computational results with V_e. The following analysis reports draining at $t = 0$ s. The total duration of the drain time of this test is 69 s.

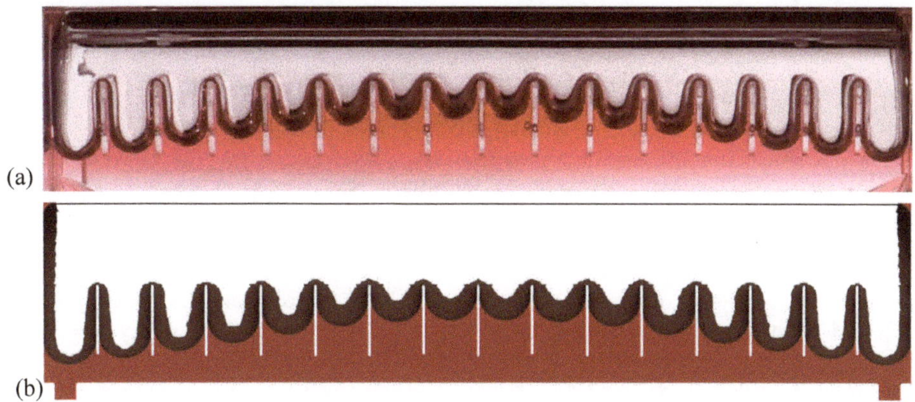

Figure 12. 3D side perspective of the ICF-8 Tapered Section filled with liquid (red) at $t = 25$ s (**a**) during experiment and (**b**) simulation, where gray implies interfacial volume fraction $\alpha = 0.5$.

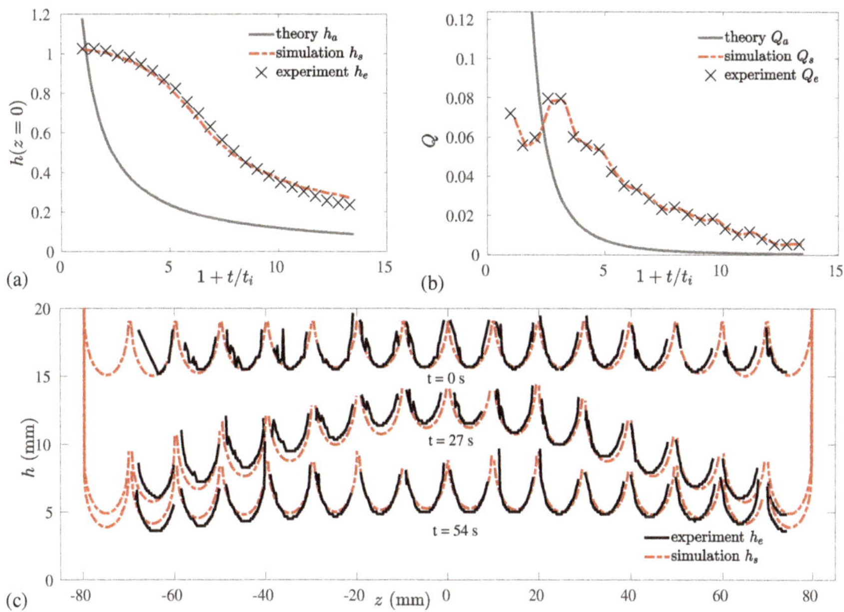

Figure 13. (**a**) h_s (dashed), h_a (8) (solid), and h_e (ticks) and (**b**) Q_s (dashed), Q_a (10) (solid), and Q_e (ticks) with time. (**c**) Dimensional h_s (dashed) against h_e (solid) at specified times.

Figure 13a,b presents comparisons of the simulations and experimental results. The absence of gas ingestion is observed for the full elapsed drain time, where excellent simulation-experimental volumetric flow rate agreement is seen: $|Q_s - Q_e| < 4.5\%$, Figure 13b. Meniscus height error is reported $|h_s - h_e| < 16\%$ for the full 69 s drain, where we note $|h_s - h_e| < 6\%$ for the first 60 s. The analytic model for a simple interior corner is also presented for token comparison despite the significant violations of model assumptions attributed to the vane-segment geometry.

6. Discussion

The Bond number $Bo \equiv \rho g H^2/\sigma$ and Suratman number $Su \equiv \rho \sigma H/\mu^2$ determine the relative strength of hydrostatic pressure to capillary pressure and capillary inertia to viscous resistance, respectively. Values for these critical dimensionless groups are listed in Table 1. The dimensionless parameters for viscous resistance $F_i = 0.1560$, section area $F_A = 0.2955$, and surface curvature $f = 0.3492$ depend on system geometery only through corner half-angle β and liquid contact angle θ. Further details are provided in [7].

In reference to Figure 11b, at small amounts of time, ICF-1 Q_a nearly doubles Q_e. Defining time $t = 0$ s as the moment the astronaut began draining the test cell, the video shows several seconds before approximate zero height is realized: the astronaut drained the container much slower than the simple analytical model prediction assumes. For large amounts of time, self-similarity is achieved, as Q_a closely predicts Q_e. Though disagreement is exaggerated for the analytical model at early times, the OpenFOAM simulations accurately capture Q_e for all time. In late stage draining, h_a under-predicts h_e.

The analytic model works well to predict the draining flow in ICF-1, but predictable errors are observed when the model is inappropriately extended to the draining flow of ICF-8. In general, the simple analytical model assumes negligible inertia. Table 1 lists the order of magnitude expectations of inertia for both ICF-1 and ICF-8 flows, where it is observed inertia is non-negligible for ICF-8. Figure 13b corroborates this, as $Q_e \gg Q_a$. The simulations account well for finite inertia, accurately predicting h_e and Q_e with near coincident menisci observed over a range of drain times, Figure 13c. While it is clear that the linear analytical model does well for simple interior corners, the OpenFOAM simulations are extendable to channel networks of greater geometric complexity and higher flow inertia.

7. Conclusions

OpenFOAM capillary drain simulations were validated against spacecraft flight experiments conducted as part of the Capillary Flow Experiments on the International Space Station. The flow scenario was that of the symmetric late stage draining in interior corner networks of varying geometric complexity including a straight interior corner (ICF-1) and one segmented by a network of vane elements (ICF-8). A simple analytical model for the flow is also used to benchmark the numerical method, where better than 6.5% agreement is found for all measured quantities prior to gas ingestion, occurring approximately 30 s into draining. Experimental meniscus heights h_e and liquid volumetric flow rates Q_e agree with the OpenFOAM simulations to within 10% and 5% respectively for the first 60 s of draining. The simple analytical model is found to under-predict the meniscus height and volumetric flow rate, and, while providing reliable height and drain rate predictions for simplified geometries, serves better as a lower-bound for complex vane networks. To this end, OpenFoam's interFoam solver is found to serve as an excellent tool for analytical model verification as well as quantitative drain rate assessment particularly for flows of increased geometric complexity and inertia.

Author Contributions: Conceptualization, J.M.; Data curation, J.M.; Formal analysis, J.M.; Investigation, J.M.; Supervision, P.S.; Writing—original draft, J.M.; Writing—review & editing, M.W. All authors have read and agreed to the published version of the manuscript.

Funding: This research was funded primarily by NASA under NNH17ZTT001N-17PSI D. M.W. is supported in part through NASA Cooperative Agreements 80NSSC18K0161 and 80NSSC18K0436.

Acknowledgments: Special thanks to JAXA astronaut K. Wakata for his patience and creativity during flight experiments. This paper is dedicated to the late Paul Steen, 6/22/1952–9/4/2020.

Conflicts of Interest: The authors declare no conflict of interest.

References

1. Srinivasan, R. Estimating zero-g flow rates in open channels having capillary pumping vanes. *Int. J. Numer. Methods Fluids* **2003**, *41*, 389–417. [CrossRef]
2. Weislogel, M.M.; Baker, J.A.; Jenson, R.M. Quasi-steady capillarity-driven flows in slender containers with interior edges. *J. Fluid Mech.* **2011**, *685*, 271–305. [CrossRef]
3. Ludwicki, J.M.; Steen, P.H. Sweeping by Sessile Drop Coalescence. *Eur. Phys. J. Spec. Top.* **2020**, *229*, 1739–1756. [CrossRef]
4. Weislogel, M.M. Some analytical tools for fluids management in space: Isothermal capillary flows along interior corners. *Adv. Space Res.* **2003**, *32*, 163–170. [CrossRef]
5. Zhang, T.T.; Yang, W.J.; Lin, Y.F.; Cao, Y.; Wang, M.; Wang, Q.; Wei, Y.X. Numerical study on flow rate limitation of open capillary channel flow through a wedge. *Adv. Mech. Eng.* **2016**, *8*. [CrossRef]
6. Concus, P.; Finn, R. On the behavior of a capillary surface in a wedge. *Proc. Natl. Acad. Sci. USA* **1969**, *63*, 292–299. [CrossRef] [PubMed]
7. Weislogel, M.M.; McCraney, J.T. The symmetric draining of capillary liquids from containers with interior corners. *J. Fluid Mech.* **2019**, *859*, 902–920. [CrossRef]
8. Bostwick, J.B.; Steen, P.H. Stability of constrained capillary surfaces. *Annu. Rev. Fluid Mech.* **2015**, *47*, 539–568. [CrossRef]
9. Bostwick, J.B.; DIjksman, J.A.; Shearer, M. Wetting dynamics of a collapsing fluid hole. *Phys. Rev. Fluids* **2017**, *2*, 1–14. [CrossRef]
10. Snoeijer, J.H.; Andreotti, B. Moving contact lines: Scales, regimes, and dynamical transitions. *Annu. Rev. Fluid Mech.* **2013**, *45*, 269–292. [CrossRef]
11. Xia, Y.; Steen, P.H. Dissipation of oscillatory contact lines using resonant mode scanning. *NPJ Microgravity* **2020**, *6*, 1–7. [CrossRef] [PubMed]
12. Sui, Y.; Ding, H.; Spelt, P.D. Numerical simulations of flows with moving contact lines. *Annu. Rev. Fluid Mech.* **2014**, *46*, 97–119. [CrossRef]
13. Wang, S.; Desjardins, O. Numerical study of the critical drop size on a thin horizontal fiber: Effect of fiber shape and contact angle. *Chem. Eng. Sci.* **2018**, *187*, 127–133. [CrossRef]
14. Antritter, T.; Mayer, M.; Hachmann, P.; Ag, H.D.; Martin, W. Suppressing artificial equilibrium states caused by spurious currents in droplet spreading simulations with dynamic contact angle model. *Prog. Comput. Fluid Dyn. Int. Journalpr* **2020**, *20*, 59–70. [CrossRef]
15. Gurumurthy, T.V.; Roisman, I.V.; Tropea, C.; Garoff, S. Spontaneous rise in open rectangular channels under gravity. *J. Colloid Interface Sci.* **2018**, *527*, 151–158. [CrossRef] [PubMed]
16. Malekzadeh, S.; Roohi, E. Investigation of Different Droplet Formation Regimes in a T-junction Microchannel Using the VOF Technique in OpenFOAM. *Microgravity Sci. Technol.* **2015**, *27*, 231–243. [CrossRef]
17. Yong-Qiang, L.; Wen-Hui, C.; Ling, L. Numerical Simulation of Capillary Flow in Fan-Shaped Asymmetric Interior Corner Under Microgravity. *Microgravity Sci. Technol.* **2017**, *29*, 65–79. [CrossRef]
18. Arias, S.; Montlaur, A. Influence of Contact Angle Boundary Condition on CFD Simulation of T-Junction. *Microgravity Sci. Technol.* **2018**, *30*, 435–443. [CrossRef]
19. Klatte, J.; Haake, D.; Weislogel, M.M.; Dreyer, M. A fast numerical procedure for steady capillary flow in open channels. *Acta Mech.* **2008**, *201*, 269–276. [CrossRef]
20. Brakke, K.A. *Surface Evolver*; Spring: Berlin, Germany, 2015. [CrossRef]
21. Vachaparambil, K.J.; Einarsrud, K.E. Comparison of Surface Tension Models for the Volume of Fluid Method. *Processes* **2019**, *7*, 542. [CrossRef]
22. Harvie, D.J.; Davidson, M.R.; Rudman, M. An analysis of parasitic current generation in Volume of Fluid simulations. *Appl. Math. Model.* **2006**, *30*, 1056–1066. [CrossRef]

23. Magnini, M.; Pulvirenti, B.; Thome, J.R. Characterization of the velocity fields generated by flow initialization in the CFD simulation of multiphase flows. *Appl. Math. Model.* **2016**, *40*, 6811–6830. [CrossRef]

24. Chiodi, R.; Desjardins, O. A reformulation of the conservative level set reinitialization equation for accurate and robust simulation of complex multiphase flows. *J. Comput. Phys.* **2017**, *343*, 186–200. [CrossRef]

25. Popinet, S. Numerical Models of Surface Tension. *Annu. Rev. Fluid Mech.* **2018**, *50*, 49–75. [CrossRef]

26. Soh, G.Y.; Yeoh, G.H.; Timchenko, V. Numerical investigation on the velocity fields during droplet formation in a microfluidic T-junction. *Chem. Eng. Sci.* **2016**, *139*, 99–108. [CrossRef]

27. Sontti, S.G.; Pallewar, P.G.; Ghosh, A.B.; Atta, A. Understanding the Influence of Rheological Properties of Shear-Thinning Liquids on Segmented Flow in Microchannel using CLSVOF Based CFD Model. *Can. J. Chem. Eng.* **2019**, *97*, 1208–1220. [CrossRef]

28. Guo, Z.; Fletcher, D.F.; Haynes, B.S. Implementation of a height function method to alleviate spurious currents in CFD modelling of annular flow in microchannels. *Appl. Math. Model.* **2015**, *39*, 4665–4686. [CrossRef]

29. Weislogel, M.M.; Lichter, S. Capillary flow in an interior corner. *J. Fluid Mech.* **1998**, *373*, 349–378. [CrossRef]

30. Weislogel, M.M. Compound capillary rise. *J. Fluid Mech.* **2012**, *709*, 622–647. [CrossRef]

31. Ramé, E.; Weislogel, M.M. Gravity effects on capillary flows in sharp corners. *Phys. Fluids* **2009**, *21*, 1–12. [CrossRef]

32. Brackbill, J.U.; Kothe, D.B.; Zemach, C. A Continuum Method for Modeling Surface Tension. *J. Comput. Phys.* **1992**, *100*, 335–354. [CrossRef]

33. Saufi, A.E.; Desjardins, O.; Cuoci, A. An accurate methodology for surface tension modeling in OpenFOAM R. *arXiv* **2020**, arXiv:2005.02865v2.

Publisher's Note: MDPI stays neutral with regard to jurisdictional claims in published maps and institutional affiliations.

 *fluids*

Article

Modelling a Moving Propeller System in a Stratified Fluid Using OpenFOAM [†]

Christian T. Jacobs

Defence Science and Technology Laboratory (Dstl), Porton Down, Salisbury, Wiltshire SP4 0JQ, UK;
cjacobs@dstl.gov.uk
† Selected Paper from the 15th OpenFOAM Workshop, Arlington,VA, USA, 22–25 June 2020.

Received: 6 October 2020; Accepted: 17 November 2020; Published: 21 November 2020

Abstract: Moving propeller systems can introduce significant disturbances in stratified environments by mixing the surrounding fluid. Restorative buoyancy forces subsequently act on this region/patch of mixed fluid, causing it to eventually collapse vertically and spread laterally in order to recover the original stratification. This work describes the use of an OpenFOAM solver, modified using existing functionality, to simulate a moving propeller system in a stratified environment. Its application considers a rotating KCD-32 propeller in a laboratory-scale wave tank which mimics published experiments on mixed patch collapse. The numerically-predicted collapse behaviour is compared with empirical data and scaling laws. The results agree closely, both qualitatively and quantitatively, thereby representing a successful step towards the validation of the numerical model.

Keywords: computational fluid dynamics; moving meshes; propeller; stratified environments; mixed patch

1. Introduction

Stratified environments are created when a fluid's temperature and/or salinity, and therefore density, changes with respect to depth [1]. A propeller system moving through such an environment is capable of rapidly mixing the surrounding fluid across isopycnals, yielding a so-called 'mixed patch' of near-uniform density in its wake. The mixed patch is subjected to buoyancy forces which attempt to restore the stratification to its initial (unperturbed) state. This results in the patch collapsing vertically and spreading out laterally, radiating internal waves in the process [2,3].

Several important stages in the collapse process have been identified through experiments and numerical models [4–7]. At early times, the turbulent kinetic energy introduced by the sudden mixing causes the patch to grow in size. This process increases the potential energy stored in the mixed patch. After the passage of the mixing source, buoyancy forces eventually overcome inertia and the patch begins to collapse vertically in an effort to restore the stratification to its equilibrium state. At later times the flow comprises 'pancake'-like eddies characterised by vertical vorticity [8]. However, many studies have not focussed on the effects of swirl from propeller systems, which may have a significant effect on the collapse and post-collapse behaviour.

Numerical approaches to modelling the action of propeller motion on fluids include actuator line and actuator disk models [9–12]; blade element momentum theory [13–16]; and models in which the propeller blades are fully resolved by the computational mesh and dynamically rotated [17–19]. The latter approach was considered by Colley [17], for example, who modelled a 5-blade KCD-32 series propeller using OpenFOAM; an open-source fluid dynamics modelling framework [20–22]. However, the environment was not stratified. The behaviour of mixed patch collapse under buoyancy could therefore not be considered.

The work presented herein concerns the application of `buoyantPimpleFoam`, an existing transient flow solver in the OpenFOAM modelling framework which takes buoyancy effects into account. Modifications were made to the `buoyantPimpleFoam` solver in order to use the dynamic mesh (DyM) and arbitrary mesh interface (AMI) functionality readily available in OpenFOAM [23–26]. The use of DyM and AMI enabled support for rotating computational meshes (and therefore a rotating propeller) and the representation of a propeller's hydrodynamic wake in a stratified environment. A step towards validating the numerical model is achieved by simulating the collapse of a mixed patch generated by a propeller, equipped with five KCD-32 blades, in a laboratory-scale stratified wave tank. The qualitative behaviour of the mixed patch throughout time is compared with similar experimental studies by Merritt [4] and Lin & Pao [5]. In particular, the numerical predictions are validated against empirical data and scaling laws describing the height and width of the mixed patch throughout its evolution. However, one difference between the model and the experiments is the use of a rotating KCD-32 propeller rather than an oscillating grid to generate the mixed patch. Furthermore, this numerical study only considers a single value of the Brunt–Väisälä frequency and Froude number, rather than a range of values. The model, once sufficiently validated, could potentially be applied at a larger scale to understand the collapse of mixed patches generated by underwater bluff bodies [7] or marine power turbines [27], for example.

The remainder of this paper is organised as follows. Section 2 briefly describes the numerical model, including the governing equations. Details on the computer-aided design (CAD) and meshing process are provided along with a description of how the simulations were set up. The numerical results are presented in Section 3. The various stages of the mixed patch collapse process are successfully observed in the model. The numerically-predicted height and width of the mixed patch agree well with the scaling laws of Merritt [4] and Lin & Pao [5]. The paper closes with some concluding remarks in Section 4.

2. Method

2.1. Model

The OpenFOAM modelling framework [20–22] was used to conduct the numerical study. In particular, a transient flow solver capable of simulating buoyant flow, `buoyantPimpleFoam`, was modified for the purpose of modelling a moving propeller in a stratified environment. The development version of OpenFOAM, available from https://github.com/OpenFOAM/OpenFOAM-dev, was forked (copied to a local repository) from Git commit `409548cbccac` and the source code was subsequently modified to make use of the existing DyM and AMI functionality [23–26]. Similar use cases of this functionality can be found in the `rhoPimpleFoam` and `interFoam` solvers, for example. Note that, since this work was conducted, a more recent development version now includes support for moving meshes in the `buoyantPimpleFoam` solver; this support was added independently of this work (see Git commit `38fff77d3537` by Henry Weller).

The numerical model considers the Navier–Stokes equations, governing the conservation laws of mass and momentum, with the Boussinesq approximation applied [28]. The fluid under consideration is inhomogeneous with small vertical density variations arising from its stratification. However, the Boussinesq approximation assumes that such density variations are small enough to be neglected except within the buoyancy term [29]. Therefore, the governing equations reduce to their incompressible form given by

$$\frac{\partial u_i}{\partial x_i} = 0, \tag{1}$$

$$\frac{\partial u_i}{\partial t} + \frac{\partial \left(u_i u_j\right)}{\partial x_j} = -\frac{1}{\rho_0}\frac{\partial p}{\partial x_i} + \frac{\rho}{\rho_0} g_i + \frac{\partial \tau_{ij}}{\partial x_j}, \tag{2}$$

where ρ is the density of the fluid, $\rho_0 = 1{,}000$ kgm^{-3} is the reference density, $u = [u_1, u_2, u_3]$ is the flow velocity, p is pressure, $g = [0, 0, -9.81]$ ms^{-2} is the acceleration due to gravity, $x = [x_1, x_2, x_3]$ is the spatial coordinate, and t denotes time [30]. The stress tensor τ is based on the fluid's kinematic viscosity $\nu = 10^{-6}$ m^{2}s^{-1}. Summation is implied over repeated indices. Note also that x_1, x_2 and x_3 are respectively referred to as x, y and z throughout the remainder of this paper. Similarly, u_x, u_y and u_z are used to respectively refer to u_1, u_2 and u_3.

2.2. Equation of State

A linear equation of state was used to compute the density field ρ, based on the temperature T, as follows

$$\rho(T) = \rho_0 \left(1 - \alpha(T - T_0)\right), \tag{3}$$

where $\alpha = 10^{-4}$ K^{-1} is the thermal expansion coefficient [31] and $T_0 = 293.15$ K is the reference temperature.

2.3. Domain

A three-dimensional cuboid domain representing the laboratory-scale wave tank of Merritt [4] was considered, with dimensions $-0.05 \le x \le 0.05$ m, $-2.0 \le y \le 0.02$ m, and $-0.1 \le z \le 0$ m. The propeller's hub was centred at $(0, 0, -0.05)$ as shown in Figure 1. The model was implemented in a reference frame where the fluid flows through a fixed rotating propeller system (rather than a propeller travelling through quiescent fluid). Inlet and outlet boundary planes were therefore defined at $y = 0.02$ m and $y = -2$ m, respectively.

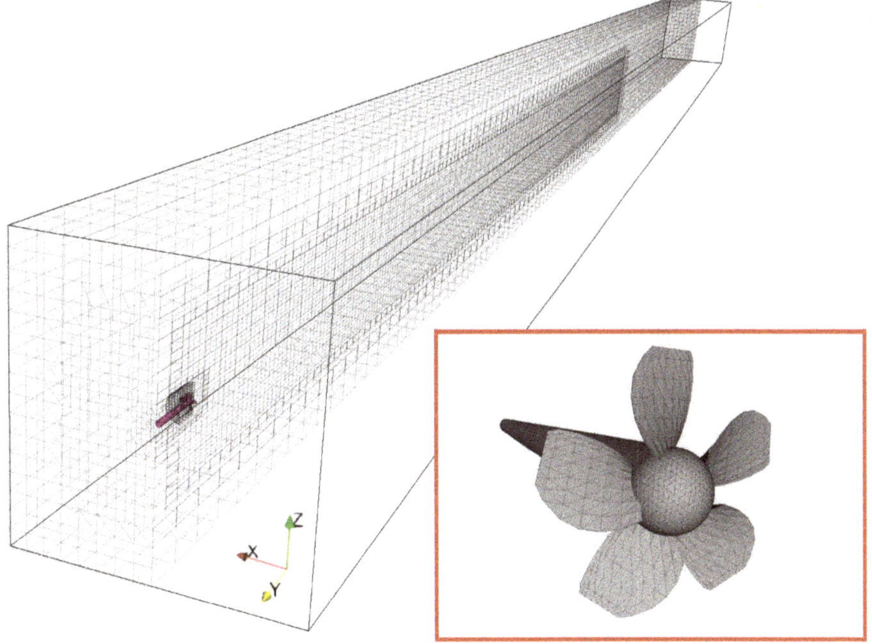

Figure 1. An illustration of the propeller system (coloured pink) within the three-dimensional domain. A cross-sectional slice (in the y–z plane) through the computational mesh is included to highlight the cylindrical and cuboidal regions of enhanced resolution. The vertices within the cylindrical region dynamically rotate throughout the simulation to model the motion of the propeller. A computer-aided design (CAD) rendering of the full propeller system, including the shaft and hub, is inset.

The computational grid included a series of mesh refinements performed using the OpenFOAM meshing algorithms. A coarse discretisation was first performed by introducing a uniform structured grid of $16 \times 128 \times 16$ (~32,000) solution points throughout the entire domain. A cylindrical region of enhanced resolution which encapsulated the propeller was then introduced. A cuboidal region of high resolution was also placed downstream to resolve the mixing action of the propeller. Cells were then removed using OpenFOAM's snappyHexMesh utility to create a propeller-shaped void in the mesh. The cells in the vicinity of the propeller were further refined, with the first layer of cells typically having a non-dimensional thickness (y^+ value) of ~0.8 (approximately 10^{-3} m). The resulting mesh comprised a total of approximately 346,000 solution points.

2.4. Propeller

The propeller featured five KCD-32 blades, appropriately scaled-down to fit the laboratory-scale domain. These blades, defined by [32], were meshed and written to a file in stereolithography (STL) format. Each blade had an approximate length of 0.0022 m. A shaft and hub (0.02 m in length, 0.002 m in diameter, inset in Figure 1) were designed using FreeCAD (https://www.freecadweb.org/) [33]. These were used to mount each of the blades to form the complete propeller system. The total diameter D of the propeller was therefore approximately 0.0064 m, similar to that of the oscillating grid (D = 0.00635 m) used in the experiments of Merritt [4].

The propeller was rotated at a rate of 120 revolutions per minute (rpm). The motion of the propeller was accomplished using the DyM and AMI functionality within OpenFOAM; this functionality rotated the vertices of the cylindrical region (encapsulating the propeller) embedded in the computational mesh, and also enabled the finite volume fluxes to be interpolated from the former mesh topography onto the modified (rotated) topography [23–26].

2.5. Initial Conditions

The initial velocity field assumed a uniform downstream flow throughout the entire domain such that $u(x, t = 0) = [0, -0.01, 0]$ ms^{-1}. The pressure field p was initially set to zero, but throughout the simulation the solver computed this field based on the hydrostatic pressure and any pressure fluctuations encountered (such as those at the propeller blades).

The temperature of the fluid at the bottom of the wave tank T_b was set to 293.15 K. The temperature increased linearly towards the surface of the wave tank, with the surface temperature T_s = 294.45 K. This yielded a stable stratification. Internal waves oscillate in a stratified environment with a maximum frequency known as the Brunt–Väisälä frequency [1], defined as

$$N = \frac{1}{2\pi}\sqrt{-\frac{|g_z|}{\rho_0}\frac{\partial \rho}{\partial z}}. \tag{4}$$

The Brunt–Väisälä frequency of the stratification considered here was N = 0.018 s^{-1} (i.e., a buoyancy period of ~55 s, as per one of the experiments by Merritt [4]).

The Froude number expresses the ratio between inertia and buoyancy forces. When based on the Brunt–Väisälä frequency N and the diameter of the propeller D, it is defined as

$$Fr = \frac{U}{ND}, \tag{5}$$

where U is the free-stream flow speed of 0.01 ms^{-1}. For this particular setup, Fr = 87.6 which is within the range of Froude numbers considered by Merritt [4] and Lin & Pao [5].

2.6. Boundary Conditions

The inlet flow speed was set to a constant 0.01 ms^{-1}, while a zero pressure boundary condition at the outlet allowed fluid to flow freely out of the domain. Free-slip conditions were assumed at the walls of the domain, representing the smooth walls of the wave tank. The temperature field at the top and bottom of the tank was set to 294.45 K and 293.15 K, respectively, while a zero-gradient condition was applied to all other walls of the domain. A zero `movingWallVelocity` boundary condition was enforced at the propeller. A `cyclicAMI` boundary condition was applied to enable the rotation of the propeller.

2.7. Discretisation

OpenFOAM uses a finite volume method to spatially discretise the domain. Upwind conditions were applied at the faces between each computational cell [34].

The forward Euler method was chosen to temporally discretise the governing equations and advance the simulation forwards in time until $t = 200$ s. This time-frame was sufficient to allow the mixed patch and its various stages of evolution to become established throughout the length of the domain. The timestep Δt was automatically adapted throughout the simulation, subject to a maximum Courant number constraint of 2.0.

2.8. Solution Algorithms

The PIMPLE algorithm was used to iteratively solve the incompressible Navier–Stokes equations. Two iterations of the PIMPLE algorithm were performed per timestep. Note that PIMPLE is a combination of the PISO (Pressure Implicit with Splitting of Operator) and SIMPLE (Semi-Implicit Method for Pressure Linked Equations) algorithms [34,35].

The PIMPLE algorithm requires the solution to several systems of equations using numerical linear algebra methods. The stabilised biconjugate gradient method, preconditioned with incomplete lower–upper (LU) factorisation, was used to solve the velocity field. The pressure field was solved using the conjugate gradient method, preconditioned with incomplete Cholesky factorisation [36].

2.9. Hardware

The numerical model was executed in parallel over 18 cores, embedded in a single Intel® Core™ i9-9980XE processor, using the Message Passing Interface (MPI) and 64 GB of random-access memory (RAM). The model required approximately 5 days to complete a simulation.

3. Results

Between $t = 0$ and $t = 200$ s the motion of the propeller successfully mixed the stratified fluid to create a mixed patch. The rotating blades rapidly transported warmer fluid situated above the propeller into the lower, cooler part of the domain, and vice-versa. This is visualised by the warped contours in the centre of Figure 2. Similar temperature profiles have been observed in propeller studies involving actuator line models [11,12]. The temperature of the patch was not perfectly uniform as a result of this continuous entrainment from the undisturbed regions of the stratification.

Experimental studies have shown that a mixed patch undergoes several stages of evolution [4,5]. Immediately downstream of the mixing source (i.e., the propeller), the developing mixed patch expands (both vertically and horizontally) in a similar manner to that of a mixed patch developing in a non-stratified environment. This growth is driven by the conversion of the turbulent kinetic energy provided by the propeller into potential energy. The results from the numerical simulation agree well with this observation. The flow speed perturbation field $\|u - u_0\|$ is used to illustrate this behaviour in Figures 3 and 4, where $u_0 \equiv u(t = 0)$ and $\|\cdot\|$ denote the Euclidean norm.

Further downstream the kinetic energy provided by the propeller is no longer sufficient to sustain the patch's growth. Buoyancy forces begin to dominate and the patch subsequently collapses vertically in order to restore the original stratification. This in turn induces lateral motions such that the patch continues to widen at a much faster rate. Once again, the numerical results in Figures 3 and 4 reinforce the experimental observations. However, unlike the experiments which considered a patch generated by a moving grid [4,5], the propeller-generated patch was characterised by a significant amount of swirl/vorticity which caused it to become slightly asymmetric in shape. Nevertheless, the stratification continued to recover as expected, with internal waves persisting within the wave tank.

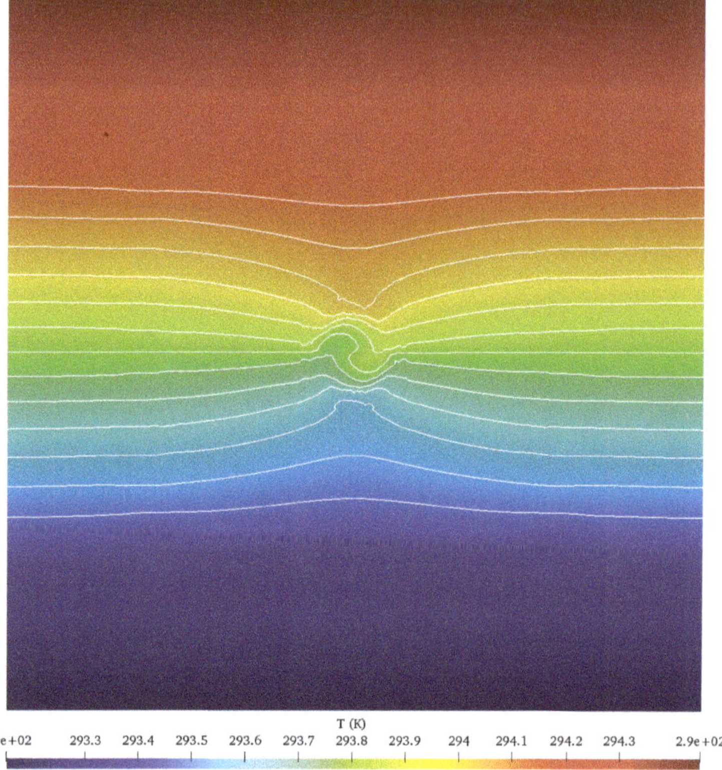

Figure 2. An x–z cross-section of the temperature field T immediately downstream of the propeller at $y = -0.01$ m, $t = 200$ s. This illustrates the mixing action of the propeller and the entrainment of surrounding fluid into the mixed patch.

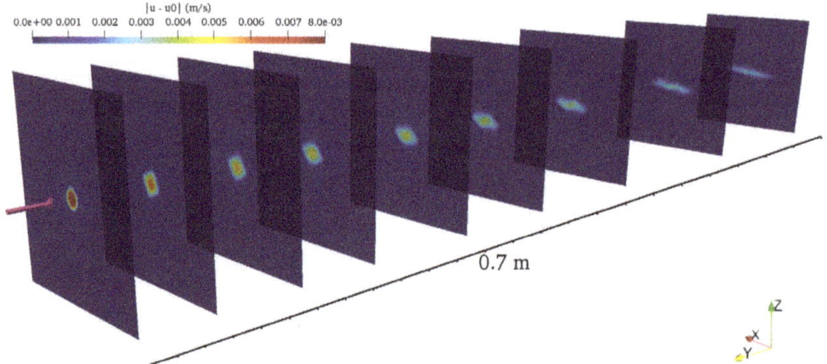

Figure 3. Visualisations of flow speed perturbation at $t = 200$ s. This illustrates the generation, growth, and collapse stages of the mixed patch as the wake propagates downstream. All cross-sections were taken in the x–z plane up to 0.7 m downstream of the propeller's hub (i.e., up to $y = -0.7$ m); the full length of the domain is not shown. The propeller system is coloured pink.

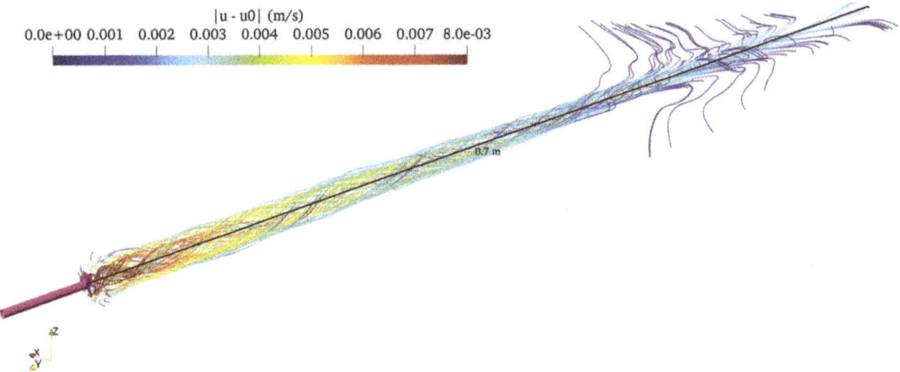

Figure 4. Streamlines of the flow speed perturbation downstream of the propeller at $t = 200$ s. One hundred streamlines are shown. This illustrates the swirl generated by the propeller and the spreading of the streamlines further downstream where the mixed patch is collapsing. The streamlines extend up to 0.7 m downstream of the propeller's hub (i.e., up to $y = -0.7$ m); the full length of the domain is not shown. The streamlines were terminated if $\|u - u_0\|$ became lower than 8×10^{-5} ms^{-1}. The propeller system is coloured pink.

In general, the numerical model was able to capture the key stages of mixed patch evolution and yielded a qualitative agreement with the experimental observations. In order to provide a quantitative assessment of the model's validity, the dimensions of the mixed patch were compared with data from the analogous experiment by Merritt [4]. The mixed patch's dimensions were measured consistently by introducing several sets of probe points at various locations downstream of the propeller. The x–z plane was populated with two intersecting lines of probe points along $x = 0$ m and $z = -0.05$ m, yielding a 'plus' shape. As a first approximation, the vertical and horizontal extent of the mixed patch were assumed to be the distance between the outermost points (along the corresponding line of probes) at which a threshold value of 10% of $\max(\|u - u_0\|)$ was attained.

The measurements plotted in Figure 5 indicate that the mixed patch grows to approximately 2.5 times the propeller's diameter at a distance of $20D$ m downstream, which is close to the expansion factor of ~ 3 measured in the experiments. As the mixed patch collapses further downstream the vertical extent eventually reaches a plateau at $\sim 80D$ m as the isopycnals are restored to their original positions in the stratification. The numerical data closely agrees with the experimental data for vertical extent. However, the horizontal extent in the simulation is typically up to one propeller diameter (D) smaller than the analogous experiment. This discrepancy may have been due to the asymmetric nature of the collapse; the swirl/vorticity present in the flow may have hindered the lateral spreading along the straight line of probe points. Overall, however, this comparison with experimental data represents a successful step towards the validation of the numerical model.

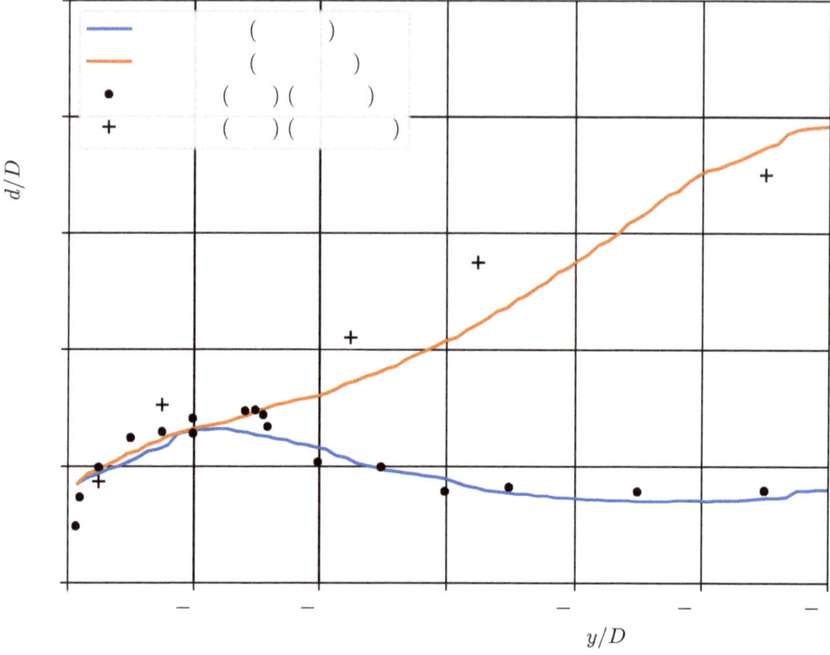

Figure 5. The evolution of the height and width of the mixed patch downstream of the propeller. All quantities are normalised by the patch's initial diameter. For the numerical results this is assumed to be equal to the diameter of the propeller, D. Data from an analogous experiment by Merritt [4] are included for comparison.

Scaling Laws

Another useful validation exercise involves comparing the mixed patch's dimensions against empirical scaling laws for height and width. It is helpful to define these scaling laws in terms of a non-dimensional buoyancy time Nt_e, where t_e represents the elapsed time since the mixed patch was generated (i.e., the elapsed time following the passage of the propeller). Removing the dependence on non-dimensional downstream distance (y/D) allows the point of mixed patch collapse to be determined without knowledge of the mixed patch's initial diameter D. The spatial-temporal relation

$$Nt_e = \frac{1}{\mathrm{Fr}}\frac{|y|}{D},$$

(6)

is applied in order to accomplish this. Note that the absolute value of y is considered here since downstream locations are represented by a negative value in the computational domain.

For example, a scaling law for the initial mixed patch growth stage (from $Nt_e = 0$ to $Nt_e \approx 0.2$), derived by Lin & Pao [5] by fitting to their experimental data, is given by

$$\frac{d}{D} = 0.8\mathrm{Fr}^{0.25}(Nt_e)^{0.25},$$

(7)

where d is the dimension of the mixed patch in either the vertical or horizontal direction. Similar scaling laws are given by Merritt [4] for each stage of the mixed patch dynamics. A plot of the numerical results in Figure 6 illustrates a generally good agreement with these scaling laws (labelled (a) to (f)).

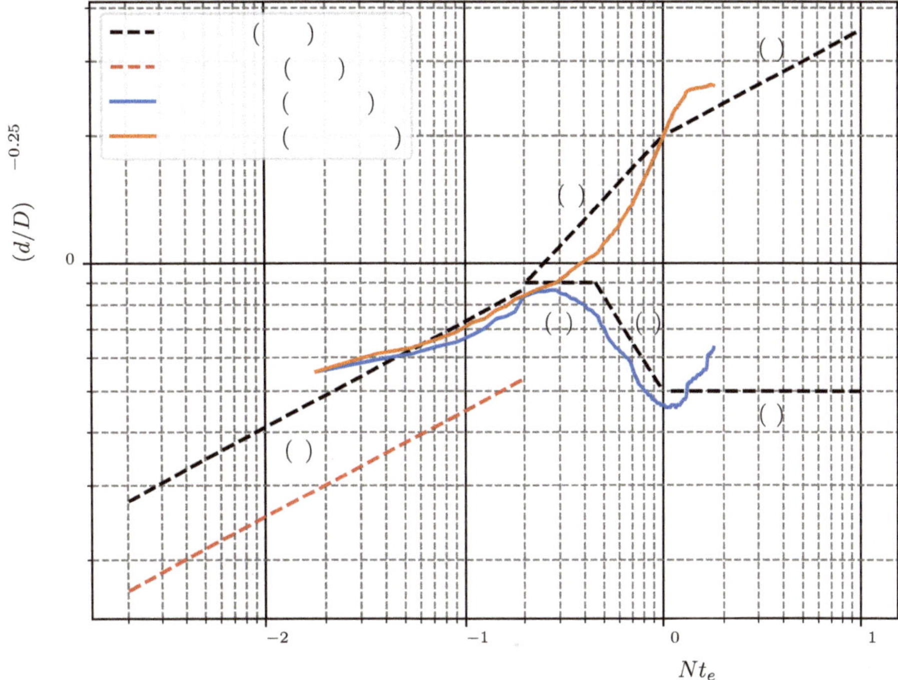

Figure 6. The evolution of the height and width of the mixed patch. Note that t_e represents the elapsed time since the mixed patch was generated (i.e., the elapsed time following the passage of the propeller). The scaling laws of Merritt [4] and Lin & Pao [5], labelled (a) to (f), are included for comparison.

Between $Nt_e = 0$ and $Nt_e = 0.2$ the expanding mixed patch in a stratification is expected to grow at a rate proportional to $(Nt_e)^{0.25}$ (scaling laws in region (a)); the same rate observed for a non-stratified environment. The numerical results between $Nt_e = 0.08$–0.2 successfully exhibit this rate of expansion. The slower rate observed immediately downstream of the propeller (between $Nt_e = 0$ to 0.08) may be attributed to a lower rate of turbulent mixing by the propeller blades compared to the oscillating grid apparatus used in the experiments.

Eventually the potential energy of the enlarged mixed patch becomes equal to the turbulent kinetic energy introduced by the propeller. The mixed patch reaches a (maximum) stationary point and vertical growth ceases (b). In the numerical model this maximum occurred at $Nt_e = 0.2$–0.3, which concurs with the value of $Nt_e = 0.23$ from the empirical fit of Lin & Pao [5]. At this point, buoyancy forces start to dominate inertia, subsequently resulting in the rapid vertical collapse (c) and lateral spreading (d) of the mixed patch. The numerical results generally agree with the rate of vertical collapse. However, the rate of lateral spreading varied significantly; initial slower spreading followed by a faster spreading than the scaling law suggests. This may have been due to residual vorticity effects from the propeller's motion.

The scaling laws at later times ($Nt_e \geq 1$, scaling laws (e) and (f)) suggest that the mixed patch's thickness tends towards a vertical asymptote as the isopycnals continue to spread out post-collapse (albeit at a slower rate than the collapse stage) in order to recover the original stratification. However, it is likely that ambient internal waves generated by the collapse process cause a small amount of variation in the measured thickness as the disturbed region of fluid oscillates about its equilibrium point. This may explain the slight increase in vertical extent in the numerical results for $Nt_e \geq 1$.

4. Conclusions

This work successfully modelled a scaled-down KCD-32 propeller system rotating in a stratified fluid environment using the OpenFOAM modelling framework. This was accomplished by modifying the buoyantPimpleFoam solver using the DyM and AMI functionality readily available within OpenFOAM. The numerical results presented in this paper represent a successful step towards validating the numerical model. The dynamics throughout the simulation generally agree well with published experimental data and scaling laws. Any minor differences/deviations may have been caused by the significant vorticity introduced by the propeller (compared to the grid-based approaches of Merritt [4] and Lin & Pao [5]). It is worth noting that these scaling laws can potentially be validated and extended by large-scale applications or additional experiments covering a wider range of mixed patch parameters.

It is recommended that future work considers the effect of a propeller's angle of attack on the dynamics. A non-linear stratification could also be introduced to determine the effect a strong pycnocline has on the dynamics. Furthermore, the mixed patch characteristics are known to depend on the Reynolds number of the flow [5]. The Reynolds number for the simulation presented in this paper, based on the flow speed and the diameter of the propeller, was 64. Considering a range of higher Reynolds numbers to produce strong eddies, and modelling or directly resolving the turbulence, would provide a more detailed understanding of turbulence levels in the mixed patch.

Funding: Dstl is part of the Ministry of Defence. The Ministry of Defence funded this research and the Article Processing Charge (APC) through contract MOD000298.

Acknowledgments: The author wishes to thank Neil Stapleton and 3 anonymous reviewers for their constructive feedback regarding this work. This paper (DSTL/JA125757) is based on Dstl Technical Report DSTL/TR120739 [37] and a conference paper (DSTL/CP121862) presented at the 15th OpenFOAM Workshop [38].

Conflicts of Interest: The author declares no conflict of interest.

References

1. Thorpe, S.A. *An Introduction to Ocean Turbulence*; Cambridge University Press: New York, NY, USA, 2007.
2. Sutherland, B.R.; Chow, A.N.F.; Pittman, T.P. The collapse of a mixed patch in stratified fluid. *Phys. Fluids* **2007**, *19*, 116602. [CrossRef]
3. Holdsworth, A.M.; Sutherland, B.R. The axisymmetric collapse of a mixed patch and internal wave generation in uniformly stratified rotating fluid. *Phys. Fluids* **2015**, *27*, 056602. [CrossRef]
4. Merritt, G.E. *Wake Laboratory Experiment*; Technical Report SC-5047-A-2; Cornell Aeronautical Laboratory, Inc.: Buffalo, NY, USA, 1972.
5. Lin, J.T.; Pao, Y.H. Wakes in Stratified Fluids. *Annu. Rev. Fluid Mech.* **1979**, *11*, 317–338. [CrossRef]
6. Swean, T.F., Jr.; Schetz, J.A. Flow about a Propeller-Driven Body in Temperature-Stratified Fluid. *AIAA J.* **1979**, *17*, 863–869. [CrossRef]
7. Meunier, P.; Spedding, G.R. Stratified propelled wakes. *J. Fluid Mech.* **2006**, *552*, 229–256. [CrossRef]
8. Rottman, J.W.; Dommermuth, D.G.; Innis, G.E.; O'Shea, T.T.; Novikov, E. Numerical Simulation of Wakes in a Weakly Stratified Fluid. In Proceedings of the Twenty-Fourth Symposium on Naval Hydrodynamics, Fukuoka, Japan, 8–13 July 2002.
9. Sanderse, B.; van der Pijl, S.; Koren, B. Review of computational fluid dynamics for wind turbine wake aerodynamics. *Wind. Energy* **2011**, *14*, 799–819. [CrossRef]
10. Tzimas, M.; Prospathopoulos, J. Wind turbine rotor simulation using the actuator disk and actuator line methods. *J. Physics: Conf. Ser.* **2016**, *753*, 032056. [CrossRef]
11. Jones, M.; Paterson, E.G. Evolution of the Propeller Near-Wake and Potential Energy in a Thermally-Stratified Environment. In Proceedings of the OCEANS 16, Monterey, CA, USA, 19–23 September 2016.
12. Jones, M.C.; Paterson, E.G. Influence of Propulsion Type on the Stratified Near Wake of an Axisymmetric Self-Propelled Body. *J. Mar. Sci. Eng.* **2018**, *6*, 46. [CrossRef]
13. Weick, F.E. *Propeller Design: Practical Application of the Blade Element Theory − I*; Technical Report 235; National Advisory Committee for Aeronautics: Washington, DC, USA, 1926.

14. Glauert, H. Airplane Propellers. In *Aerodynamic Theory: A General Review of Progress under a Grant of the Guggenheim Fund for the Promotion of Aeronautics*; Springer: Berlin/Heidelberg, Germany, 1935; pp. 169–360. [CrossRef]

15. Benini, E. Significance of blade element theory in performance prediction of marine propellers. *Ocean. Eng.* **2004**, *31*, 957–974. [CrossRef]

16. Carroll, J.; Marcum, D. Comparison of a Blade Element Momentum Model to 3D CFD Simulations for Small Scale Propellers. *SAE Int. J. Aerosp.* **2013**, *6*, 721–726. [CrossRef]

17. Colley, E. *Analysis of Flow around a Ship Propeller Using OpenFOAM*; Technical Report; Curtin University: Perth, Australia, 2012.

18. Mehdipour, R. Simulating Propeller and Propeller-Hull Interaction in OpenFOAM. Master's Thesis, Centre for Naval Architecture, Royal Institute of Technology, Stockholm, Sweden, 2013.

19. Esmaeilpour, M.; Ezequiel Martin, J.; Carrica, P.M. Near-field flow of submarines and ships advancing in a stable stratified fluid. *Ocean. Eng.* **2016**, *123*, 75–95. [CrossRef]

20. Weller, H.G.; Tabor, G.; Jasak, H.; Fureby, C. A tensorial approach to computational continuum mechanics using object-oriented techniques. *Comput. Phys.* **1998**, *12*, 620–631. [CrossRef]

21. Jasak, H. OpenFOAM: Open source CFD in research and industry. *Int. J. Nav. Archit. Ocean. Eng.* **2009**, *1*, 89–94. [CrossRef]

22. The OpenFOAM Foundation Ltd. *OpenFOAM User Guide, Version 7*; The OpenFOAM Foundation Ltd.: London, UK, 2019.

23. Beaudoin, M.; Jasak, H. Development of a Generalized Grid Interface for Turbomachinery simulations with OpenFOAM. In Proceedings of the Open Source CFD International Conference 2008, Berlin, Germany, 4–5 December 2008.

24. Jasak, H. Dynamic Mesh Handling in OpenFOAM. In Proceedings of the 47th AIAA Aerospace Sciences Meeting Including The New Horizons Forum and Aerospace Exposition, Orlando, FL, USA, 5–8 January 2009. [CrossRef]

25. Farrell, P.E.; Maddison, J.R. Conservative interpolation between volume meshes by local Galerkin projection. *Comput. Methods Appl. Mech. Eng.* **2011**, *200*, 89–100. [CrossRef]

26. Chandar, D.D.; Gopalan, H. Comparative Analysis of the Arbitrary Mesh Interface (AMI) and Overset Methods for Dynamic Body Motions in OpenFOAM. In Proceedings of the 46th AIAA Fluid Dynamics Conference, Washington, DC, USA, 13–17 June 2016. [CrossRef]

27. Brunetti, A.; Armenio, V.; Roman, F. Large eddy simulation of a marine turbine in a stable stratified flow condition. *J. Ocean. Eng. Mar. Energy* **2019**, *5*, 1–19. [CrossRef]

28. Tritton, D.J. *Physical Fluid Dynamics*; Oxford Science Publications, Oxford University Press Inc.: New York, NY, USA, 1988.

29. Turner, J.S. *Buoyancy Effects in Fluids*; Cambridge University Press: New York, NY, USA, 1979.

30. Jones, D.A.; Chapuis, M.; Liefvendahl, M.; Norrison, D.; Widjaja, R. *RANS Simulations Using OpenFOAM Software*; Technical Report DTRC-90/016; Defence Science and Technology Group: Victoria, Australia, 2016.

31. Dijkstra, H.A. *Dynamical Oceanography*; Springer: Berlin/Heidelberg, Germany, 2008.

32. Emerson, A.; Sinclair, L. Propeller Cavitation: Systematic Series Tests on 5- and 6-Bladed Model Propellers. *Trans. Soc. Naval Arch. Mar. Eng.* **1967**, *75*, 224–267.

33. Van Havre, Y.; The FreeCAD Community. *FreeCAD: A Manual*. 2019. Available online: https://github.com/yorikvanhavre/FreeCAD-manual (accessed on 21 November 2020).

34. Patankar, S.V. *Numerical Heat Transfer and Fluid Flow*; Hemisphere Publishing Corporation: San Francisco, CA, USA, 1980.

35. Van Doormaal, J.P.; Raithby, G.D. Enhancements of the SIMPLE Method for Predicting Incompressible Fluid Flows. *Numer. Heat Transf.* **1984**, *7*, 147–163. [CrossRef]

36. Golub, G.H.; Van Loan, C.F. *Matrix Computations*; The Johns Hopkins University Press: Baltimore, MD, USA, 1989.

37. Jacobs, C.T. *Modelling a Moving Propeller System in a Stratified Fluid Using OpenFOAM*; Technical Report DSTL/TR120739; Dstl: Salisbury, UK, 2020.
38. Jacobs, C.T. Modelling a moving propeller system in a stratified fluid using OpenFOAM. In Proceedings of the 15th OpenFOAM Workshop, Arlington, VA, USA, 22–25 June 2020.

Article

Mitigating Thermal NOx by Changing the Secondary Air Injection Channel: A Case Study in the Cement Industry

Domenico Lahaye [1],*, Mohamed el Abbassi [1], Kees Vuik [1], Marco Talice [2] and Franjo Juretić [3]

[1]   Delft Institute of Applied Mathematics, Faculty of Electrical Engineering, Mathematics and Computer Science, Technical University of Delft, 2628 CD Delft, The Netherlands; m.elabbassi@tudelft.nl (M.e.A.); c.vuik@tudelft.nl (K.V.)

[2]   PM2 Engineering, 09127 Cagliari, Italy; m.talice@pm2engineering.com

[3]   Creative Fields Ltd., 10000 Zagreb, Croatia; franjo-juretic@c-fields.com

*   Correspondence: d.j.p.lahaye@tudelft.nl; Tel.: +31-15-27-87-257

Received: 11 October 2020; Accepted: 17 November 2020; Published: 25 November 2020

Abstract: This work studies how non-premixed turbulent combustion in a rotary kiln depends on the geometry of the secondary air inlet channel. We target a kiln in which temperatures can reach values above 1800 degrees Kelvin. Monitoring and possible mitigation of the thermal nitric-oxide (NOx) formation is of utmost importance. The performed reactive flow simulations result in detailed maps of the spatial distribution of the flow, thermodynamics and chemical conditions of the kiln. These maps provide valuable information to the operator of the kiln. The simulations show the difference between the existing and the newly proposed geometry of the secondary air inlet. In the existing configuration, the secondary air inlet is rectangular and located above the base of the burner pipe. The secondary air flows into the furnace from the top of the flame. The heat release by combustion is unevenly distributed throughout the flame. In the new geometry, the secondary air inlet is an annular ring placed around the burner pipe. The secondary air flows circumferentially around the burner pipe. The new secondary air inlet geometry is shown to result in a more homogeneous spatial distribution of the heat release throughout the flame. The peak temperatures of the flame and the production of thermal NOx are significantly reduced. Further research is required to resolve limitations of various choices in our modeling approach.

Keywords: non-premixed combustion; turbulence; radiative heat transfer; conjugate heat transfer; thermal thermal nitric-oxide(NOx); rotary kiln

---

1. Introduction

Rotary kilns are long cylindrical furnaces [1]. They are widely employed by the material processing industry to manufacture, e.g., cement or zinc-oxide, to treat waste or to recycle materials such as bitumen.

We target a direct-fired counter-current flow rotary kiln in which the temperature reaches levels above 1800 degrees Kelvin. The kiln is placed horizontally with a slight inclination, as shown schematically in Figure 1. The raw material is fed from top end. It is forced to travel through the kiln by the rotary motion around its main axis. A burner producing a flame is installed at the low end the kiln. The material bed is mixed, heated and subjected to sintering reactions. The material leaves the furnace at the low end to be further processed. Sufficiently high temperatures are required to heat the material to reach the sintering reactions. This makes the process prone to the formation of thermal nitric-oxides (NOx).

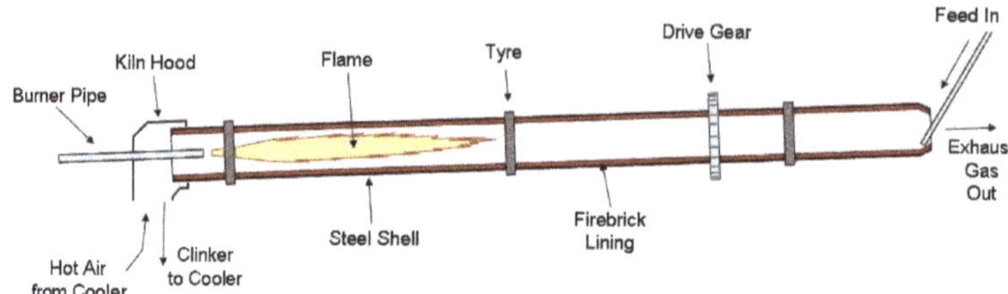

Figure 1. Schematic representation of a direct-fired counter-fed rotary kiln.

The numerical simulation of rotary kilns is important to obtain insight into the heat generation by the flame and transfer to the material bed. This understanding is vital to reduce off-spec production, fuel consumption and pollutant emission. In-situ experimental campaigns are hampered by high temperatures and harsh operating conditions. The combustion and material processing in rotary kilns has therefore been been studied extensively using both numerical simulations and lab-scale experiments. A recent survey is given in the PhD thesis by Pedersen [2].

The combustion in the cement clinker kiln that we consider is fed by a non-premixed turbulent flow of methane and air. The combustion air is separated into a primary and secondary air flow. The primary air and methane enter the kiln via the burner pipe. At the end of the burner pipe, a burner injects the methane and the primary air into the freeboard of the kiln. The burner consists of small nozzles that are mounted with either an axial or outwardly inclined orientation. These nozzles inject methane jets with high momentum. The nozzles are surrounded by a circular slot through which the primary air passes with lower momentum. During operation, the mixture of methane and air ignites and gives raise to a short and intense flame. The burner pipe is prevented from overheating by the flow of the primary air. Secondary air is blown into the kiln in co-flow with the fuel jets and primary air stream. This secondary air is entrained by the jets leaving the burner. This entrainment spurs the growth of the flame. A long and low burning (so-called *lazy*) diffusion flame with thin reaction fronts is formed. Representative images of the burner, burner pipe and flame are shown, e.g., in the recent PhD thesis byPedersen [2] and on the website of rotary kiln manufacturers such as FLSmith and Feeco. We refer to the mixed jets of the fuel and the primary combustion air as the primary jet. In the following three paragraphs, we review literature on non-reactive flow, turbulent combustion and thermal NOx formation in rotary kilns, respectively.

Early publications recognize that combustion in rotary kilns is mainly driven by the turbulent mixing of the primary jet and the secondary air [2]. The importance of mixing has given rise to a body of literature on the aerodynamics in rotary kilns. In this early literature, the effect of combustion is taken into account by considering the decrease of density of the primary jet due to heat release. The jet mixing, the entrainment of the secondary jet into the primary, the formation of recirculation zones and the length of the flame can be expressed in terms of the Craya–Curtet similarity parameter. This parameter is defined as the square root of the ratio of the momentum of the primary to the secondary jet. Representative values for the parameter for cement kilns are listed in [1]. The simplifying assumptions that allow the definition of a Craya–Curtet parameter, however, do not apply to the kiln configurations studied in this paper. More detailed CFD models, as developed in [3], are therefore required. This paper shows that the co-flow of jets at different velocities result in shear layers in which vortices are created. These vortices shed further downstream at a characteristic frequency. These vortices are studied experimentally for planar jets in [4]. The frequency of the vortex shedding has been determined for the non-reactive flow in a down-scaled rotary kiln both numerically and experimentally ([3] and references therein). The same authors discussed to what extent the non-stationary flow in the kiln can be approximated by stationary flow models.

The combustion in rotary kilns has been studied in various pilot-scale laboratory set-ups. The influence of the ratio of amount of primary and secondary air on the flame shape was investigated experimentally, e.g., by Mohanan et al. [5]. The analysis shows that a small amount of primary air ignites the mixture. The heat released by the combustion increases the momentum of the primary jet. The length of the flame increases with the amount of secondary air. The radiative heat transfer in kilns has been studied experimentally in [6].

The cement production process requires temperatures of the freeboard gasses well above 1800 K. These elevated temperatures render the process very sensitive to the formation of thermal NOx. Various counter measures have therefore been developed. These include the design of low NOx swirl burners, catalytic reduction techniques and systems that enrich the oxidizer with oxygen. A recent overview for iron-ore rotary kilns is given in [7]. The numerical simulation is deemed to be vital to render these counter measures even more effective. In this paper, we model the formation of thermal NOx using the Zel'dovich mechanism in the post-processing stage of reactive flow simulations. More information on measures to mitigate NOx formation can by found elsewhere (e.g., [8,9]).

In this work, we study the non-premixed turbulent combustion in the full-scale cement rotary kiln that we considered earlier in [10]. In this earlier work, simulations were performed using a commercial CFD package. Our goal here is to perform more comprehensive and detailed simulations using OpenFoam [11]. Our aim is to study how the thermal NOx formation is influenced by the shape of the secondary air inlet channel. We wish to compare the rectangular secondary air inlet with an annular one placed concentrically with the burner pipe. The annular configuration is expected to distribute the air evenly over the flame, result in evenly distributed reaction front and lower peak temperatures and therefore the formation of thermal NOx.

This paper is structured in nine sections. In Section 2, we describe the mathematical model for the turbulent non-premixed combustion in the rotary kiln that we solve. In Section 3, we describe tools available in OpenFoam to model this combustion process. In Section 4, we describe how these tools are used to model the combustion in the rotary kiln. In Section 5, we describe the mesh generation process using cfMesh. In Sections 6–8, we present simulation results for the non-reactive flow, the reactive flow and the thermal NOx formation in the kiln. In Section 9, we draw conclusions.

2. Mathematical Model

In this section, we describe the mathematical model for combustion in the kiln. The model consists of the following four components: a model for the combustion of methane in the freeboard of the kiln including radiative heat transfer, a model for the conductive heat transfer in the lining, a model for the interface between the freeboard and the lining and a model for the thermal NOx formation in the freeboard. In the following four subsections, these components are described individually. Details on the fluid flow, combustion, radiative heat transfer and conjugate heat transfer components of the model can be found elsewhere (see, e.g., [12–20], respectively).

2.1. Model for Combustion of Methane in the Freeboard

The mixture of gaseous fuel, primary and secondary combustion air and combustion products is modeled as a mixture of ideal gasses. Properties such as density ρ, heat capacity c_p, molecular viscosity μ_0 and thermal conductivity λ of this mixture are expressed in terms of weighted averages. The turbulent flow of this gas in the kiln is described using the Favre-averaged (or density-weighed Reynolds-Averaged) Navier–Stokes equations with the realizable k-epsilon model for the turbulent kinetic energy k and turbulent dissipation rate ϵ for the turbulence closure. The flow through the burner nozzles occurs at a high speed. The Mach number reaches the value of approximately 0.8–0.9 in proximity to the burner. The incompressible flow formulation is therefore no longer valid. A compressible formulation in which energy is formulated in terms of the enthalpy h and in which the thermodynamical properties are based on the density is adopted.

The chemistry of the combustion involves a number of chemical species indexed by ℓ and a number of reaction steps indexed by s. The mass fraction Y_ℓ of the ℓth species is governed by a convection-diffusion-reaction equation with source term $\dot{\omega}_\ell$. We assume that a single diffusion coefficient denoted by D can be used for all species and that the Lewis number can be set equal to one. The expression for reaction mechanism is specified by the combustion model. Here, the eddy-dissipation combustion model for methane with a single step reaction mechanism

$$CH_4 + 2\,O_2 + N_2 \rightarrow CO_2 + 2\,H_2O + N_2 \tag{1}$$

is chosen. The choice for fast chemistry is motivated by the high Damkohler number that governs the combustion. The Damkohler number Da is defined as $Da = \tau_{flow}/\tau_{chem}$, where $\tau_{flow} = k/\epsilon$ and $\tau_{chem} = \rho\, Y_\ell/\dot{\omega}_\ell$ are the time scales for the flow and the chemistry, respectively.

Let $\mathbf{v}$, p and T denote the velocity vector of the gas, the pressure and the temperature in the kiln, respectively. The enthalpy is the sum of the sensible enthalpy and the formation enthalpy of the mixture, i.e.,

$$h = \int_{T_0}^{T} c_p\, dT + \sum_\ell Y_\ell\, \Delta h_0^\ell \tag{2}$$

where T_0 is a reference temperature and Δh_0^ℓ is the enthalpy of formation of the ℓth species. To avoid the computational expense of transient simulations, we resort to stationary approximations of the flow conditions despite the transient nature of the flow. The conservation of mass, momentum and energy of the mixture and the conservation of the species $\ell \in \{CH_4, O_2, CO_2, H_2O\}$ can then be expressed as

$$\nabla \cdot (\rho\, \mathbf{v}) \;=\; 0 \tag{3}$$

$$\nabla \cdot (\rho\, \mathbf{v}\, \mathbf{v}) \;=\; \nabla \cdot (\mu\, \nabla\, \mathbf{v}) + \nabla \cdot \left(\mu\, (\nabla\, \mathbf{v})^T\right) - \frac{2}{3}\nabla(\mu\nabla \cdot \mathbf{v}) - \nabla p \tag{4}$$

$$\nabla \cdot (\rho\, \mathbf{v}\, h) \;=\; \nabla \cdot \left(\frac{\lambda}{c_p}\, \nabla\, h\right) + S_{rad} \tag{5}$$

$$\nabla \cdot (\rho\, \mathbf{v}\, Y_\ell) \;=\; \nabla \cdot (D\, \nabla\, Y_\ell) + \dot{\omega}_\ell\, , \tag{6}$$

where S_{rad} in the right-hand side of (5) denotes the source term due to radiative transport of heat. This term is discussed in more detail in the next paragraph. The system of conservation equations is closed by ideal gas law. The viscosity μ is the sum of the laminar viscosity μ_0 and turbulent viscosity μ_t. The latter is expressed in term of the turbulent quantities k and ϵ as

$$\mu = \mu_0 + \mu_t = \mu_0 + \rho\, C_\mu \frac{k^2}{\epsilon}\, , \tag{7}$$

where the model constant C_μ depends on the mean strain and rotation rate in the realizable k-epsilon turbulence model adopted. Inflow conditions are specified in terms of mass flow rates. The pressure is prescribed on the outlet. Standard wall functions for the turbulent kinetic energy k, turbulent dissipation ϵ and turbulent viscosity μ_t are applied. The boundary condition for the temperature is given by an interface condition of the interface between the freeboard and the lining.

The freeboard gasses are assumed to be gray (or frequency-independent) for the radiative transport of heat. This transport is modeled by a discrete ordinate method. The choice is motivated by the ability to use various numbers of rays to estimate the error made in modeling the radiative transport. The space is discretized in various angular directions $\vec{s}_\lambda$. In each direction, a transport equation for the radiative intensity $I(\vec{r}, \vec{s}_\lambda)$ is solved. This intensity depends on the position vector $\vec{r}$ and the direction $\vec{s}_\lambda$. The black body intensity $I_b(\vec{r})$ acts as a source term in this equation. We neglect the effect of scattering by soot in the radiative heat transfer given the high air–fuel ratio. Only emission and absorption of intensity therefore play a role. The effect of soot was omitted due to the high air–fuel ratio in the flame. The absorption coefficient κ is modeled using a weighted sum of gray gasses (WSGG) model. The radiative transport equation can then be written as

$$\frac{dI(\vec{r}, \vec{s}_\lambda)}{ds} = \kappa I_b(\vec{r}) - \kappa I(\vec{r}, \vec{s}_\lambda)\, . \tag{8}$$

The wall emissivity ϵ_w is taken into account in the boundary condition for $I(\vec{r}_w, \vec{s}_\lambda)$ as

$$I(\vec{r}_w, \vec{s}_\lambda) = \epsilon_w \, I_b(\vec{r}_w) + \frac{1 - \epsilon_w}{\pi} \sum_k w_k \, I_-(\vec{r}_w, \vec{s}_k) \, |\vec{n} \cdot \vec{s}_k| \,, \tag{9}$$

where $I_-(\vec{r}_w, \vec{s}_k)$ is the incident intensity and w_k are weights for the angular integration. Having solved for the individual rays, the radiative heat source S_{rad} in the energy Equation (5) can be expressed as

$$S_{rad}(\vec{r}) = 4\,\pi\,\kappa\,I_b(\vec{r}) - \kappa \sum_k w_k \, I_k(\vec{r}, \vec{s}_k) \,. \tag{10}$$

2.2. Model for Heat Conduction in the Lining

The combustion model of the freeboard gasses is coupled with a model that describes heat transfer by conduction in the refractory lining surrounding the freeboard. Heat is allowed to transfer to the exterior surrounding by prescribing a boundary condition for the radiative heat on the interface between the lining and the exterior.

2.3. Conditions on the Freeboard-Lining Interface

This coupling is realized by imposing the continuity of the temperature and the heat flux on the interface separating the freeboard and the lining. In the freeboard domain, the heat flux has a conductive, convective and radiative component, while, in the lining domain, it has a conductive component only.

The refractory lining absorbs the heat produced by the flame. This absorption causes the peak temperature in the flame to change. We modeled this effect in [21]. The change of peak temperature is important to consider in the modeling of thermal NOx formation. In practice, the presence of the material bed is expected to further redistribute the heat in the kiln. In this paper, however, the presence of the material bed is not taken into consideration. In future work, the presence of the material bed can be modeled by, for instance, changing the thermal properties of the lining.

2.4. Model for the Thermal NOx Formation in the Freeboard

The mathematical model that we thus obtain expresses the conservation equation for mass, momentum, enthalpy and chemical species. Having solved the model, we compute the thermal NOx concentration using the extended Zel'dovich mechanism with three reactions and seven species (H, N, N_2, NO, O, O_2 and OH) in the post-processing stage (see, e.g., Section 12.28 in [12]). This work employs the implementation of this model in OpenFoam realized in the master thesis by Kadar [22]. This implementation employs equilibrium chemistry to compute the mass fraction of the species O and OH [23].

We perform mesh generation using cfMesh. We verify our non-reactive flow computations using IB-Raptor. IB-Raptor is the immersed boundary finite volume solver for the compressible Navier–Stokes developed by PMSQUARED Engineering [24]. It uses an approximate Riemann method to compute the fluxes and a symmetric Gauss-Seidel LU method to advance the solution in time. In the current state of development, IB-Raptor does not allow the simulation of reactive flow. To validate the reactive flow computations, we therefore compare with our previous results in [10].

3. Combustion Processes in OpenFOAM

In this section, we describe recent progress in the modeling of turbulent non-premixed combustion using OpenFoam. We restrict our attention to the modeling of diffusion flames arising for burning methane in industrial furnaces. In this context, the term 'combustion processes' encompasses the turbulent reactive flow of fuel and oxidizer in the freeboard, the heat transfer by radiation and the conjugate heat transfer in the refractory lining. This section serves as a basis for further elaboration in the remainder of the paper.

The modeling of non-premixed combustion, radiative heat transfer and conjugate heat transfer in OpenFOAM has recently attracted considerable attention. Examples of the development of OpenFoam-based solvers based on the flamelet generated manifold concept are found in [25,26] (and references cited therein). Solvers based on the alternative eddy-dissipation concept are developed in [27,28]. The implementation of detailed models for radiative heat transfer is considered in [29,30]. The combination of combustion in the fluid domain and conjugate heat transfer in the solid domain was pursued, e.g., by the authors of [31–33].

4. Rotary Kiln in OpenFOAM

In this section, we describe tools available in OpenFoam to implement the non-reactive and reactive flow model in the kiln as well as the thermal NOx post-processor.

The non-reactive turbulent compressible flow in the kiln is simulated in stationary regime using rhoSimpleFoam. This solver adds thermodynamics to the aerodynamics implemented in the incompressible solver simpleFoam. The system expressing conservation of mass and momentum is augmented with an equation for energy. The augmented system is closed by an equation of state. Here, the ideal gas law is used. A formulation in terms of the density ρ is adopted. In this formulation, the pressure is an explicit function of the density. In the alternative formulation based on the compressibility ψ, the pressure depends implicitly on the density via the compressibility. The equations for pressure, velocity, turbulent quantities and energy are solved segregated using the SIMPLE algorithm. Details of a compressible solver similar to rhoSimpleFoam are given in Chapter 16 of [13]. The vortex shedding in the shear layer between the fuel jets and the coflow of air renders the flow non-stationary. We approximate this non-stationary behavior by running a stationary solver for a large number of iterations. The use of a steady state RANS approximation to study flows with vortex shedding was discussed by Iaccarino et al. [34]. The convergence of rhoSimpleFoam was initially hampered by the complexity of the flow pattern. We therefore extended the computational domain by attaching a large cube to the outlet path of the cylindrical domain. The side of the cube extends several kiln diameters. In this way, the outlet patch is moved further downstream and outflow boundary conditions no longer obstruct the flow. The flow field continues to change on a macroscopic level even after a small residual norm has been reached. Further research is required to describe conditions in which the additional cube on the outflow of the kiln is indispensable to obtain convergence of the flow solver.

The turbulent compressible reactive flow in the kiln is simulated in a stationary regime using multiRegionReactingFoam [35] with basic thermodynamical properties based on compressibility.

The thermal NOx post-processor is taken from the master thesis by Kadar [22]. Throughout this paper, OpenFoam-v1906 is used.

5. Mesh Generation

We perform mesh generation using cfMesh. cfMesh is a collection of mesh generation tools distributed as a library [36]. This library is designed to provide customizable meshing workflows for automatic generation of meshes with various cell types in complex geometries of industrial interest. cfMesh generates hex-dominant meshes, tetrahedral meshes, meshes consisting of arbitrary polyhedra and 2D quad-dominant meshes. The library uses both shared-memory with OpenMP and distributed-memory parallelization with MPI. Parallelization is encapsulated inside the meshing algorithms. Hence, it allows customization of the meshing process while preserving the benefits of the code executing in parallel.

The meshing process is controlled by a geometry file given as a surface triangulation, most often as an STL file, and a dictionary containing the parameters provided by the user. Once the geometry and the settings are given, the mesher generates the mesh automatically without any user intervention. The dictionary allows the user to specify the global cell size in the domain and local refinement zones. The latter can be specified via subsets, surface meshes and edges mesh, as well as via objects such as cylinders, boxes, spheres and lines. The dictionary also allows the control of the number of boundary layers at each boundary of the domain, their number and their thickness ratio. Dictionary settings allow mesh-sensitivity studies by changing a single parameter. The algorithms used by cfMesh are described in [36]. Figure 2 shows the mesh generated for two kiln geometries. In future work, we intend to further investigate to what extent these meshes satisfy the yplus criterium near the wall.

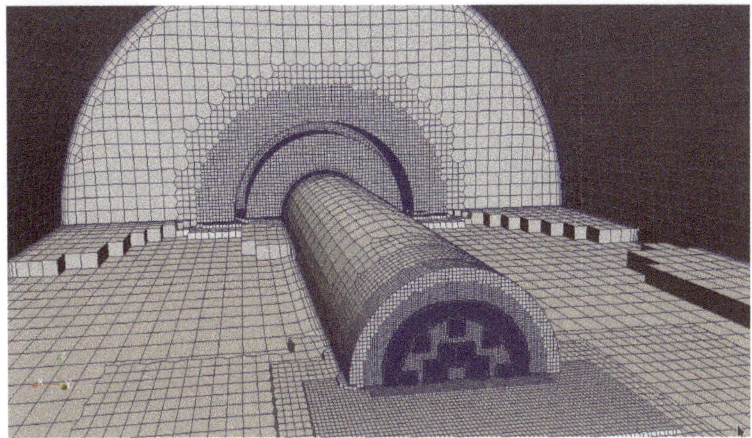

Figure 2. Mesh inside of the kiln with a modified (annular) secondary air inlet. The figures show the multi-nozzle burner and cooling slot in the foreground, the burner pipe and the rectangular secondary air inlet channel in the background. For increased visibility, a coarse mesh is shown here.

6. Non-Reactive Flow Results

In this section, we discuss simulation results for the non-reactive turbulent flow in the freeboard of the kiln. We consider the case of the rectangular secondary air inlet only. Non-isothermal effects are included by imposing the secondary combustion air temperature to be equal to 773 degrees Kelvin. The fuel and primary combustion air are kept at room temperature at their respective inlets. The flow consists of two features. The first feature is the mixing of the fuel jets and the primary air leaving the burner. The second feature is the backward facing step-like flow caused by the stream of secondary air. Both features interact and give rise to complex non-stationary flow patterns. We ran twenty thousand iterations of rhoSimpleFoam on a mesh consisting of approximately 2.8 million cells. Our goal in this section is to verify the answer provided by rhoSimpleFoam by comparing with IB-Raptor and results provided in the literature. This endeavor is justified by the fact that the combustion is mainly driven by the turbulent mixing of the fuel jets and air streams.

The comparison of the results generated by rhoSimpleFoam and IB-Raptor is hampered by three facts. The first is that the codes use entirely different numerics. The second is that the 2D snapshots of the flow that we show fail to capture three-dimensional flow phenomena. Care is required to give a physical interpretation to these 2D visualizations. The third is that vortices appear in the shear layers of the interaction of the fuel jets, the primary and secondary air stream. These vortices are transported by the flow and shed further downstream. The flow is thus non-stationary. The solution obtained by both codes continue to vary after convergence has been reached. We only present snapshots of the flow. It is however impossible to identify to which of the iteration indices the two codes should be compared. In future work, a more detailed comparison could by carried out by comparing, e.g., the frequency of the vortex shedding computed by both codes, as suggested, e.g., in [3,37]. These papers also discuss how the eddy viscosity ratio is overestimated due to an overestimation of the turbulent kinetic energy. In the remainder of this section, we therefore show results for the axial velocity component v_y and the eddy viscosity μ_t computed using rhoSimpleFoam and IB-Raptor.

The axial velocity v_y computed by IB-Raptor and rhoSimpleFoam is shown at the top and bottom of Figure 3, respectively. For illustration purposes, the velocity component was clipped to the range from $-2\,\text{m/s}$ to $2\,\text{m/s}$. Both codes capture the jet leaving the burner and the recirculation vortex underneath the burner pipe caused by the stream of secondary air. Apparent differences can be observed in the length of the primary jet, the entrainment of the secondary air by the primary jet, the length of the recirculation vortex underneath the burner pipe and the magnitude of the velocity at the center of the kiln. These differences are attributed to the aforementioned difficulties in comparing both codes.

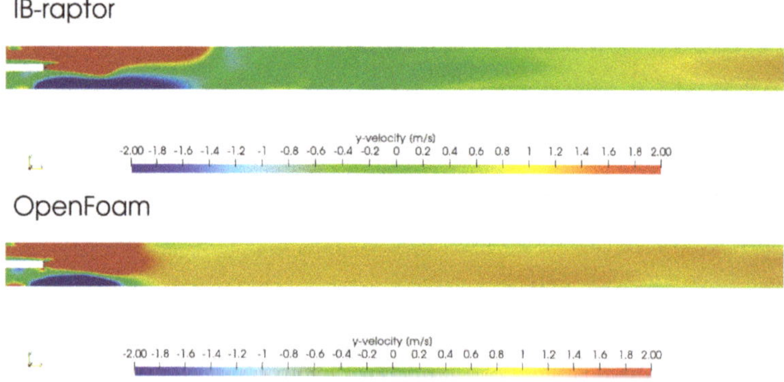

Figure 3. Clipped axial velocity in m/s computed by: IB-Raptor (**top**); and rhoSimpleFoam (**bottom**).

The eddy viscosity μ_t computed by IB-Raptor and rhoSimpleFoam is shown at the top and the bottom of Figure 4, respectively. Both codes capture large values with comparable magnitude near the burner and decreasing values towards the outlet. For rhoSimpleFoam, this decrease is monotonic, whereas it is not for IB-Raptor. More interestingly, both codes predict an eddy viscosity that remains large for a large distance downstream from the burner. This compares favorably with results in [3]. Larsson et al. [3] discussed how the two-equations turbulence model used in these simulations systematically overpredict the turbulent kinetic energy in regions of large strain. This entails an overprediction of the eddy viscosity and therefore a spurious damping of the oscillating motion of the flow. Compared to more advanced turbulence models, the primary jet and secondary stream travel further downstream before merging completely.

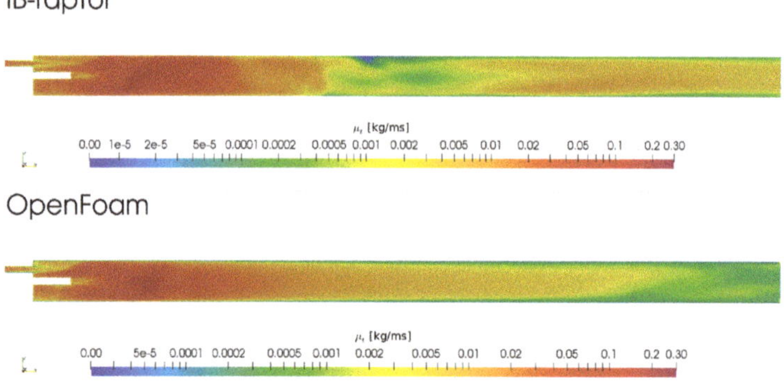

Figure 4. Clipped eddy viscosity ratio in kg/(m s) computed by: IB-Raptor (**top**); and rhoSimpleFoam (**bottom**).

7. Reactive Flow Results

In this section, we present the results of the reactive flow simulation in the kiln. We allow the fuel and air to mix and produce a flame. We use a volumetric air–fuel ratio equal to nine throughout this and the next section. Compared to the non-reactive simulation considered in the previous section, the primary jet is hotter, therefore lighter, and has thus more momentum at equal mass inflow rates. The change in balance of momentum contained in the primary and secondary jets significantly alters the parameters and thus the flow characteristics in the kiln. We consider the flame-wall interaction by modeling heat transfer via radiation, convection and conduction. We run 50k iterations of multiRegionReactingFoam [35] on a mesh consisting of approximately 4.2 million and 4.4 million cells for the configuration with rectangular and annular air inlet, respectively. Our goal is to validate the results of multiRegionReactingFoam for the velocity and heat distribution in the kiln. We compare the use of the rectangular and annular secondary air inlet configuration. In the remainder of this section we first compare the velocity of the primary jet in the non-reactive and reactive simulation. We subsequently compare results for the streamlines and the temperature with results in the literature. We in particular compare with earlier results obtained using commercial software published in [10].

The burner design determines the shape of the flame. This design is influenced by factors such as the number, location and orientation of the nozzles. The importance of the burner design motivates a closer look into the aerodynamics close to the burner. Figure 5 therefore shows the axial component (y-component) of the velocity of the primary jet on an axial section close to and downstream of the burner. The figure compares the non-reactive flow at the top and reactive flow at the bottom. The two subfigures on the left and the right show the jet leaving the axial and outwardly inclined fuel nozzles, respectively. The fuel jet from the axial nozzle moves towards the centerline while the jet leaving the inclined nozzles travels in the axial direction. The large increase in velocity due to the heat release can clearly be seen. The combustion of fuel jets and air similar to the configuration studied here has been extensively studied in the literature (e.g., [38]).

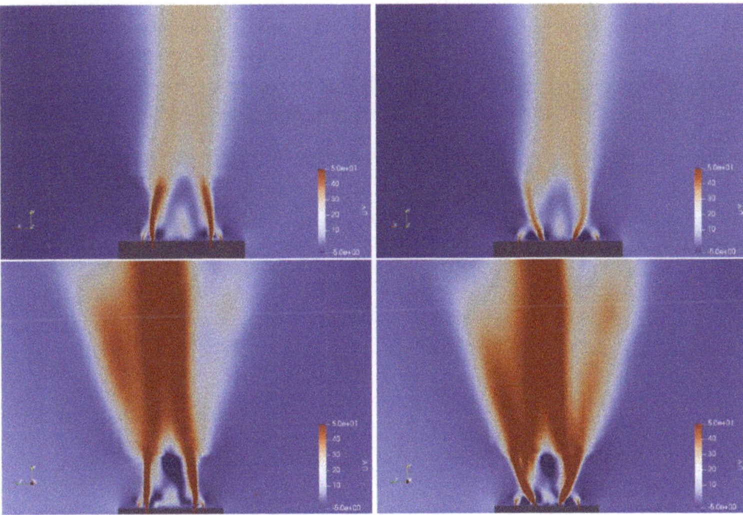

Figure 5. Clipped axial velocity in m/s of the fuel jets exiting from the axial (**left**) and inclined nozzles (**right**) and the primary air stream in the non-reactive (**top**) and reactive (**bottom**) simulation.

Figure 6 shows the computed stream for the original rectangular (top) and modified annular (bottom) secondary air inlet. Different recirculation patterns can clearly be seen. In the case of the rectangular air inlet, the recirculation vortex underneath the burner pipe supplies the flame with a hot mixture of products and oxidants. This mechanism ensures the stabilization of the flame. In the case of the annular air inlet, the recirculation vortex encircles the burner pipe.

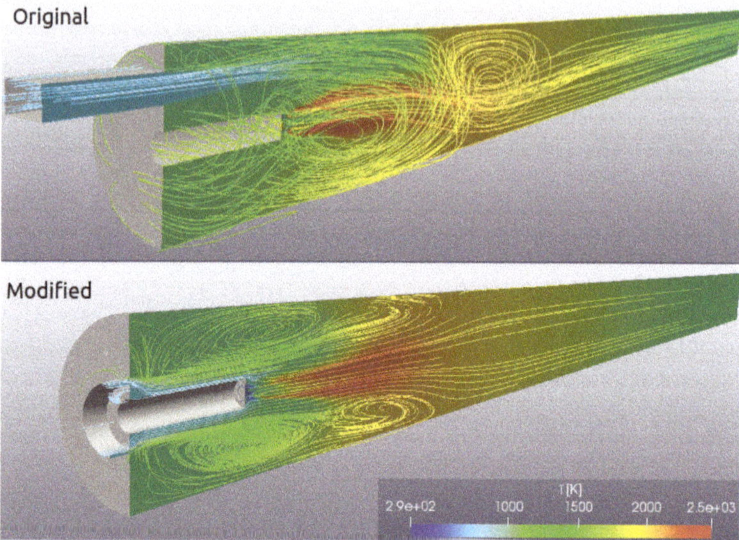

Figure 6. Computed streamlines colored with temperature values in K for the original rectangular (**top**) and modified annular (**bottom**) secondary air inlet. For both cases, a hot zone due to the flame and the recirculation due to the secondary air can be seen.

Figure 7 shows the vertical axial cross-section of the computed temperature near the flame region for the rectangular (top) and annular (bottom) secondary air inlet. A zone of high temperature is shown which approximates the location and shape of the flame. This zone corresponds to a thin reaction zone where the fuel and oxidizer meet. In the case of the rectangular inlet, the reaction zone can be seen to bend downwards due to the one-sided inflow on secondary from the top of the flame. In the case of the annular inlet, the reaction zone remains in the center of the enclosing walls.

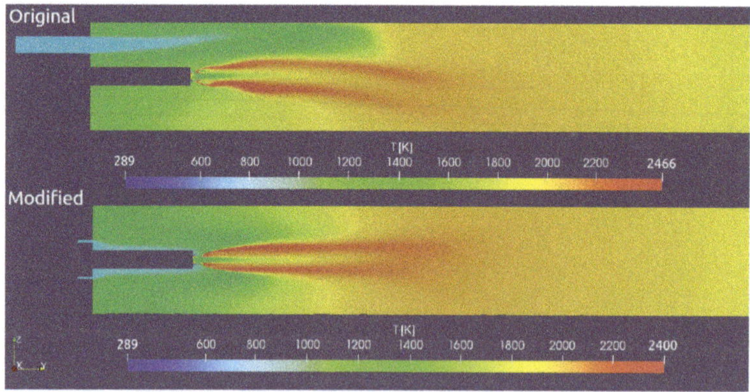

Figure 7. Computed temperature in K on the central axial slice for the original rectangular (**top**) and modified annular (**bottom**) secondary air inlet. A volumetric air–fuel ratio equal to nine is used. Different scales are chosen to highlight the maximum temperature.

Figure 8 shows the temperature on the interface between the freeboard gasses and the refractory lining. The solid and dashed dark lines correspond to the bottom and top wall in the rectangular configuration, respectively. The grey lines correspond to the annular inlet. All four graphs show a peak near the end of the flame and a linear decrease beyond the peak value. In the case of the rectangular inlet, the peak values of the bottom and top wall differ more than in case of the annular inlet. The peak values in the case of the annular inlet are lower than in the case of the rectangular inlet. At the top, the annular inlet yields the same peak temperature as the rectangular air inlet. Further downstream, the temperature of the annular inlet is slightly lower. These results can be explained by the fact that in the case of the annular inlet, the flame temperature is lower and the heat is mainly transported via radiation. The results for the rectangular air inlet are in good agreement with Figure 7 in [10].

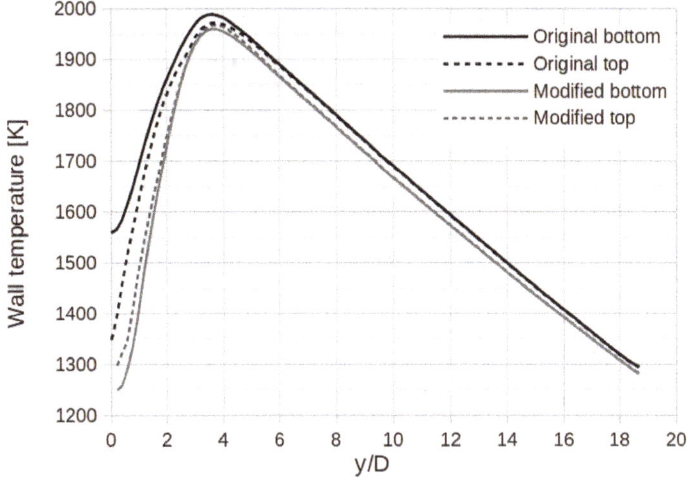

Figure 8. Computed top and bottom wall temperature in K along the axis of the kiln with the original rectangular and modified annular secondary air inlet with volumetric air–fuel ratio set equal to nine.

8. Thermal NOx Evaluation

In this section, we present the results of the computation on thermal NOx in the kiln. Figure 9 shows the computed thermal NOx mass fraction on an axial plane close to the burner for the rectangular (top) and annular (bottom) secondary air inlet. The figure shows that, in the case of the rectangular inlet, the thermal NOx concentration has a peak at the lower part of the flame close to the burner. This peak corresponds to a peak in the temperature at the same location shown at the top of Figure 7. This temperature peak is caused by the the recirculation of the secondary air prior to reaching the lower part on the flame. This recirculation is shown at the top of Figure 6. The recirculation causes the air to pass through a hot reaction zone and to be additionally heated before reaching the lower part of the flame. In the case of the co-axial air inlet, the combustion air is supplied more homogeneously around the burner. Hotspots are mitigated and a homogeneous temperature field is formed. The co-axial configuration results in a significant reduction of thermal NOx formation.

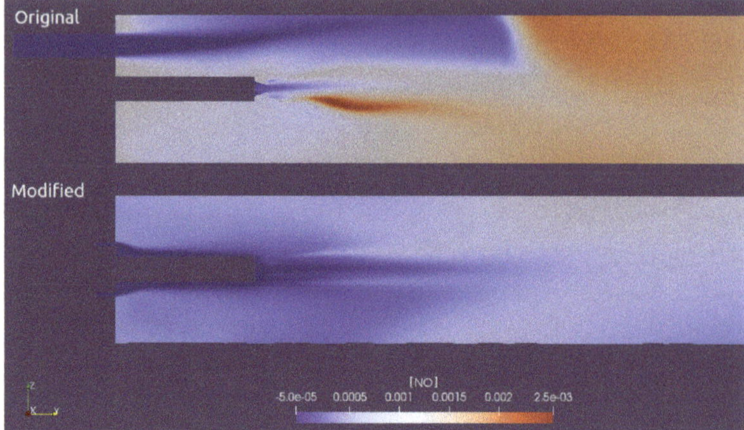

Figure 9. Contour plot of the computed NOx mass fraction (dimensionless) near the burner on the central axial slice for the original rectangular (**top**) and modified (**bottom**) secondary air inlet.

9. Conclusions

In this paper, we discuss the non-reactive flow, the reactive flow and the thermal NOx concentration in a rotary kiln in which the temperature reaches values above 1800 degrees Kelvin. The results of the non-reactive flow computations using the rhoSimpleFoam solver show reasonable agreement with results obtained using the IB-Raptor solver and results on scale models published in the literature. The results of the reactive flow computations show good agreements with the commercial code used before and results on the flame structure and temperature distribution published in the literature. The study of the aerodynamics in the kiln improved our understanding of the thermal NOx formation. We compared the effect two of the secondary air inlet geometries. The rectangular secondary air inlet geometry causes a one-sided inflow of oxidizer, spatially inhomogeneous heat release, high peak temperature and significant thermal NOx formation. The co-axial secondary air inlet instead results in an even distribution supply of oxidizer, heat release spatially distributed in space, lower peak temperatures and significantly lower thermal NOx formation. Overall, the paper shows that realistic computations on the kiln considered can be performed using cfMesh and OpenFoam. Simplified flow models in 1D or 2D fail to convey the same information. Models for the flow, combustion and radiative heat transfer can be replaced by more complex models to arrive at even more realistic predictions of heat transfer and pollutant formation in the kiln.

Author Contributions: Conceptualization, D.L., M.e.A.; Methodology, D.L., M.e.A.; Software, M.e.A., M.T., F.J.; Validation, D.L., M.e.A., M.T., F.J.; Formal Analysis, M.e.A., D.L.; Investigation, M.e.A., D.L.; Resources, D.L.; Data Curation, M.e.A., M.T., F.J.; Writing—original draft preparation, D.L., M.e.A.; Writing—review & Editing, D.L., M.e.A.; Visualization, M.e.A.; Supervision, D.L., K.V.; Project Administration, D.L. All authors have read and agreed to the published version of the manuscript.

Funding: This research received no external funding.

Conflicts of Interest: The authors declare no conflict of interest.

References

1. Boateng, A. *Rotary Kilns: Transport Phenomena and Transport Processes*, 2nd ed.; Elsevier Science: Amsterdam, The Netherlands, 2015.
2. Pedersen, M.N. Co-firing of Alternative Fuels in Cement Kiln Burners. Ph.D. Thesis, Technical University of Denmark, Kgs. Lyngby, Denmark, 2018.
3. Larsson, I.S.; Lundström, T.S.; Marjavaara, B.D. Calculation of kiln aerodynamics with two RANS turbulence models and by DDES. *Flow Turbul. Combust.* **2015**, *94*, 859–878. [CrossRef]

4. Dutta, A. Experimental Analysis of Coherent Structures in a Plane Confined Jet. Master's Thesis, University of Waterloo, Waterloo, ON, Canada, 2019.

5. Mohanan, G.; Tran, H.; Bussmann, M.; Manning, R. Effect of ring formation on burner flame stability in lime kilns. *TAPPI J.* **2018**, *17*, 285–293. [CrossRef]

6. Gunnarsson, A.; Bäckström, D.; Johansson, R.; Fredriksson, C.; Andersson, K. Radiative heat transfer conditions in a rotary kiln test furnace using coal, biomass, and cofiring burners. *Energy Fuels* **2017**, *31*, 7482–7492. [CrossRef]

7. Edland, R.; Smith, N.; Allgurén, T.; Fredriksson, C.; Normann, F.; Haycock, D.; Johnson, C.; Frandsen, J.; Fletcher, T.H.; Andersson, K. Evaluation of NOx-Reduction Measures for Iron-Ore Rotary Kilns. *Energy Fuels* **2020**, *34*, 4934–4948. [CrossRef]

8. Flagan, R.C.; Seinfeld, J. *Fundamentals of Air Pollution Engineering*; Dover Publications, New York, NY, USA, 2012.

9. Baukal, C. *Industrial Combustion Pollution and Control*; Environmental Science & Pollution; CRC Press: Boca Raton, FL, USA, 2003.

10. Pisaroni, M.; Sadi, R.; Lahaye, D. Counteracting ring formation in rotary kilns. *J. Math. Ind.* **2012**, *2*, 1–19. [CrossRef]

11. Weller, H.G.; Tabor, G.; Jasak, H.; Fureby, C. A Tensorial Approach to Computational Continuum Mechanics Using Object-oriented Techniques. *Comput. Phys.* **1998**, *12*, 620–631. [CrossRef]

12. Versteeg, H.; Malalasekera, W. *An Introduction to Computational Fluid Dynamics: The Finite Volume Method*, 2nd ed.; Pearson Education Limited: London, UK, 2007.

13. Moukalled, F.; Mangani, L.; Darwish, M. *The Finite Volume Method in Computational Fluid Dynamics*; Springer: Berlin/Heidelberg, Germany, 2016; Volume 113.

14. Peters, N. *Turbulent Combustion, Cambridge Monographs on Mechanics*; Cambridge University Press: Cambridge, UK, 2014.

15. Poinsot, T.; Veynante, D. *Theoretical and Numerical Combustion*, 2nd ed.; R.T. Edwards, Inc.: Philadelphia, USA, 2005.

16. Warnatz, J.; Maas, U.; Dibble, R. *Combustion: Physical and Chemical Fundamentals, Modeling and Simulation, Experiments, Pollutant Formation*; Springer: Berlin/Heidelberg, Germany, 2013.

17. Siegel, R.; Howell, J. *Thermal Radiation Heat Transfer*, 3rd ed.; Hemisphere Publishing: Washington, DC, USA, 1992.

18. Modest, M.F. *Radiative Heat Transfer*; Academic Press: Cambridge, MA, USA, 2003.

19. Modest, M.F.; Haworth, D.C. *Radiative Heat Transfer in Turbulent Combustion Systems: Theory and Applications*; Springer: Berlin/Heidelberg, Germany, 2016.

20. Dorfman, A. *Conjugate Problems in Convective Heat Transfer*; CRC Press: Boca Raton, FL, USA, 2009.

21. el Abbassi, M.; Fikri, D.; Lahaye, D.; Vuik, C. *Non-Premixed Combustion in Rotary Kilns Using Open-FOAM: The Effect of Conjugate Heat Transfer and External Radiative Heat Loss on the Reacting Flow and the Wall*; ICHMT Digital Library Online, Begel House Inc.: Danbury, CT, USA, 2018.

22. Kadar, A.H. Modelling Turbulent Non-Premixed Combustion in Industrial Furnaces. Master's Thesis, Technical University of Delft, Delft, The Netherlands, 2015.

23. ANSYS-Fluent. ANSYS Fluent Users Guide. Available online: https://www.afs.enea.it/project/neptunius/docs/fluent/html/ug//main_pre.htm (accessed on 1 November 2020)

24. PMSQUARED Engineering. Available online: www.pm2engineering.com (accessed on 1 November 2020)

25. Ma, L.; Roekaerts, D. Modeling of spray jet flame under MILD condition with non-adiabatic FGM and a new conditional droplet injection model. *Combust. Flame* **2016**, *165*, 402–423. [CrossRef]

26. Ottino, G.; Fancello, A.; Falcone, M.; Bastiaans, R.; de Goey, L. Combustion modeling including heat loss using flamelet generated manifolds: A validation study in OpenFoam. *Flow Turbul. Combust.* **2016**, *96*, 773–800. [CrossRef]

27. Li, Z.; Cuoci, A.; Sadiki, A.; Parente, A. Comprehensive numerical study of the Adelaide Jet in Hot-Coflow burner by means of RANS and detailed chemistry. *Energy* **2017**, *139*, 555–570. [CrossRef]

28. Li, Z.; Ferrarotti, M.; Cuoci, A.; Parente, A. Finite-rate chemistry modelling of non-conventional combustion regimes using a Partially-Stirred Reactor closure: Combustion model formulation and implementation details. *Appl. Energy* **2018**, *225*, 637–655. [CrossRef]

29. Gupta, A. Large-eddy simulation of turbulent flames with radiation heat transfer. Ph.D. Thesis, Pennsylvania State University, State College, PA, USA, 2011.
30. Ge, W.; Marquez, R.; Modest, M.F.; Roy, S.P. Implementation of High-Order Spherical Harmonics Methods for Radiative Heat Transfer on OpenFoam. *J. Heat Transf.* **2015**, *137*, 052701. [CrossRef]
31. Gariani, G.; Maggi, F.; Galfetti, L. Numerical simulation of HTPB combustion in a 2D hybrid slab combustor. *Acta Astronaut.* **2011**, *69*, 289–296. [CrossRef]
32. Dröske, N.C.; Förster, F.J.; Weigand, B.; von Wolfersdorf, J. Thermal investigation of an internally cooled strut injector for scramjet application at moderate and hot gas conditions. *Acta Astronaut.* **2017**, *132*, 177–191. [CrossRef]
33. Decan, G.; Broekaert, S.; Lucchini, T.; D'Errico, G.; Vierendeels, J.; Verhelst, S. Evaluation of wall heat flux calculation methods for CFD simulations of an internal combustion engine under both motored and HCCI operation. *Appl. Energy* **2018**, *232*, 451–461. [CrossRef]
34. Iaccarino, G.; Ooi, A.; Durbin, P.; Behnia, M. Reynolds averaged simulation of unsteady separated flow. *Int. J. Heat Fluid Flow* **2003**, *24*, 147–156. [CrossRef]
35. Source Code of multiRegionReactingFoam. Available online: github.com/TonkomoLLC (accessed on 1 November 2020)
36. Juretic, F. cfMesh version 1.1 Users Guide. Available online: http://cfmesh.com/wp-content/uploads/2015/09/User_Guide-cfMesh_v1.1.pdf (accessed on 1 November 2020)
37. Del Taglia, C.; Blum, L.; Gass, J.R.; Ventikos, Y.; Poulikakos, D. Numerical and experimental investigation of an annular jet flow with large blockage. *J. Fluids Eng.* **2004**, *126*, 375–384. [CrossRef]
38. Subramanian, V.; Domingo, P.; Vervisch, L. Large eddy simulation of forced ignition of an annular bluff-body burner. *Combust. Flame* **2010**, *157*, 579–601. [CrossRef]

Publisher's Note: MDPI stays neutral with regard to jurisdictional claims in published maps and institutional affiliations.

 *fluids*

Article

Numerical Simulation of an Air-Core Vortex and Its Suppression at an Intake Using OpenFOAM

Martin Kyereh Domfeh [1,*], **Samuel Gyamfi** [1], **Mark Amo-Boateng** [1], **Robert Andoh** [2], **Eric Antwi Ofosu** [1] **and Gavin Tabor** [3]

[1] School of Engineering, University of Energy and Natural Resources, BS-0061-2164 Sunyani, Ghana; samuel.gyamfi@uenr.edu.gh (S.G.); m.amoboateng@gmail.com (M.A.-B.); ericofosuantwi@gmail.com (E.A.O.)

[2] AWD Consult Inc., South Portland, ME 04106, USA; bobandoh@me.com

[3] College of Engineering, Mathematics and Physical Sciences, University of Exeter, Exeter EX4 4QF, UK; G.R.Tabor@exeter.ac.uk

* Correspondence: mardomfeh@gmail.com

Received: 14 October 2020; Accepted: 14 November 2020; Published: 26 November 2020

Abstract: A common challenge faced by engineers in the hydraulic industry is the formation of free surface vortices at pump and power intakes. This undesirable phenomenon which sometimes entrains air could result in several operational problems: noise, vibration, cavitation, surging, structural damage to turbines and pumps, energy losses, efficiency losses, etc. This paper investigates the numerical simulation of an experimentally observed air-core vortex at an intake using the LTSInterFoam solver in OpenFOAM. The solver uses local time-stepping integration. In simulating the air-core vortex, the standard $k - \varepsilon$, realizable $k - \varepsilon$, renormalization group (RNG) $k - \varepsilon$ and the shear stress transport (SST) $k - \omega$ models were used. The free surface was modelled using the volume of fluid (VOF) model. The simulation was validated using a set of analytical models and experimental data. The SST $k - \omega$ model provided the best results compared to the other turbulence models. The study was extended to simulate the effect of installing an anti-vortex device on the formation of a free surface vortex. The LTSInterFoam solver proved to be a reliable solver for the steady state simulation of a free surface vortex in OpenFOAM.

Keywords: intake; air-core vortex; LTSInterFoam; OpenFOAM; volume of fluid (VOF)

---

1. Introduction

A common challenge faced by engineers in the hydraulic industry is the formation of free surface vortices at pump and power intakes, a phenomenon which may entrain air [1–3]. This undesirable phenomenon often results in several operational problems: noise, vibration, cavitation, energy losses, efficiency losses, etc. [4,5]. Free surface vortices have also seen beneficial applications in water vortex hydropower plant systems [6–8], water treatments [9], vortex drop shafts in sewer systems [10], vortex settling basins [11] and vortex confined chambers [12].

Free surface vortices involving incipient air-entrainment occur at low submergence depth, a depth often referred to as the critical submergence [2]. This implies that ensuring adequate submergence presents an efficient way to avert the occurrence of free surface vortices. Apart from ensuring adequate submergence during the design of intakes, the use of anti-vortex devices also provides another cost-effective means of suppressing the formation of free surface vortices [5,13].

Refs. [14–17] proposed analytical models to describe the key vortex characteristics. Several validation test cases on these analytical studies also exist in the literature [18–20]. Some researchers [21–23] have proposed mathematical models and design plots which predict the critical submergence of intakes.

Time and cost constraints often render physical experimental modelling prohibitive and this has necessitated the use of numerical tools, which, in most instances, offer a reasonable complement. Numerical studies on free surface vortices have gained popularity due to the increasing trend of computational power [24]. According to [25], computational fluid dynamics (CFD) is capable of solving challenges associated with scale effects in model testing of free surface vortex problems through the simulation of the actual prototype with appropriate boundary conditions. In this regard, many studies have utilized CFD tools to simulate the occurrence of free surface vortices.

Ref. [7] numerically simulated a concentrated, full air-core vortex using the ANSYS-CFX tool and reported that the Reynolds stress model was the most suitable turbulence model. The velocity distribution in a vortex settling basin was also numerically assessed by [11] using FLOW-3D CFD software. The study reported a good agreement between the experimental and numerical results. Ref. [24] performed a numerical simulation of a free surface vortex formed in an intake channel of a small scale hydropower plant using the large eddy simulation (LES) model, where the free surface depression was successfully predicted. In a comparison study between the standard $k - \varepsilon$ and the RNG $k - \varepsilon$ turbulence models in simulating a vertical vortex, Ref. [25] found that the RNG $k - \varepsilon$ turbulence model provided a better prediction of the rapidly strained and curving streamline flows at the hydraulic intake. Free surface vortices in two geometric variations of a simplified intake were numerically simulated by [26] using the commercial ANSYS-CFX software along with the $k - \varepsilon$ turbulence model. Similarly, Ref. [27] numerically predicted a free surface vortex formation at the intake of a tidal power plant using the commercial ANSYS-CFX tool and with the shear stress transport curvature correction (SST-CC) turbulence model. Ref. [28] numerically quantified air entrainment rates as a result of intake vortices in different hydraulic conditions using the large eddy simulation (LES) model in the FLOW-3D CFD tool. The study observed a good agreement between the experimental and numerical results. Ref. [29] investigated key characteristics of a vertical vortex by utilizing the tracer technique and provided improved mathematical relations for the vortex by using the method of separation of variables.

A number of numerical studies have also been devoted to the simulation of free surface vortices in the presence of anti-vortex devices. For instance, Ref. [13] performed a numerical simulation of a Prosser disc and funnel-type anti-vortex devices using the large eddy simulation (LES) model and observed good agreement with experimental results. Ref. [30] looked into the efficacy of a rectangular anti-vortex plate at vertical pipe intakes, whereas [31] looked into the use of horizontal perforated and solid plates as anti-vortex devices. A detailed experimental assessment of 13 variant anti-vortex devices was conducted by [32] based on the physical model of the Siah Bisheh Pumped Storage Dam. Their study concluded that the horizontal plate anti-vortex device had the best performance. Some authors also looked into other types of anti-vortex devices—submerged water jets [33–35], rectangular plates [36] and funnel-shaped devices [37].

Despite the numerous research studies in this field, very little is known about the use of open-source CFD tools, such as OpenFOAM, for numerical simulation of free surface vortices. This study, therefore, presents a means by which free surface vortices could be numerically simulated using the OpenFOAM CFD tool. In this study, the steady state LTSInterFoam solver was used for the simulation of an experimentally observed air-core vortex. The solver uses local time-stepping integration and has, until now, been associated with studies involving the maneuverability of ships and boats. The study also investigated the performance of different turbulence models in predicting the occurrence of free surface vortices as well as the numerical simulation of the performance of an anti-vortex device installed at an intake.

Vortex Models

Vortex-related studies are often validated with either experimental data or analytical models. In this regard, this section focuses on the review of some selected mathematical vortex models.

Ref. [14] describes a vortex as being a solid rotating body comprising of an inner core and an outer free vortex. The equations for the tangential velocity of the inner core and outer core vortices are given by Equations (1) and (2), respectively.

$$r < r_m \quad V_\theta = \omega r = \frac{\Gamma}{2\pi} \frac{r}{r_m^2} \tag{1}$$

$$r > r_m \quad V_\theta = \frac{\Gamma}{2\pi r} = \omega \frac{r_m^2}{r} \tag{2}$$

Refs. [17,20,38] also proposed other vortex models and these are given in Equations (3)–(5), respectively.

$$V_\theta = \frac{\Gamma}{2\pi r_m} \frac{2R}{1 + 2R^2} \tag{3}$$

$$V_\theta = \frac{\Gamma}{2\pi r_m} \frac{R}{\sqrt{(1 + R^4)}} \tag{4}$$

$$V_\theta = \frac{\Gamma}{2\pi r_m} \frac{0.73R}{1 - R + R^2} \tag{5}$$

where V_θ represents the tangential velocity; ω refers to the angular velocity of the vortex centre; r refers to the radius; r_m is the radius at maximum tangential velocity; Γ refers to the constant circulation of the outer zone; and the normalised radius, $R = r/r_m$.

2. Materials and Methods

2.1. Experimental Set-Up

The first set of numerical investigations involving an intake without an anti-vortex device was based on the experimental work by [39], as illustrated in Figure 1. The experimental work entails the formation of an air-core vortex at an intake. To be able to study the free surface vortex occurrence in the set-up shown in Figure 1, water was made to circulate in a closed-loop set-up, such that a constant water level was maintained at the elevation sump. The intake had a diameter (D) of 16 cm and the water level in the sump was adjusted by using the pump to vary the volume of water in circulation. Visualisation of flow within the prototype sump, intake geometry and the tunnel was made possible by the use of clear Perspex in these areas of interest. The set-up was initially made to run for several hours and readings were only taken when steady conditions had been attained at a constant water level. A stable air-core vortex was observed when the relative critical submergence, S/D, and intake Froude number, F_r, were set at 1.5 and 0.6, respectively.

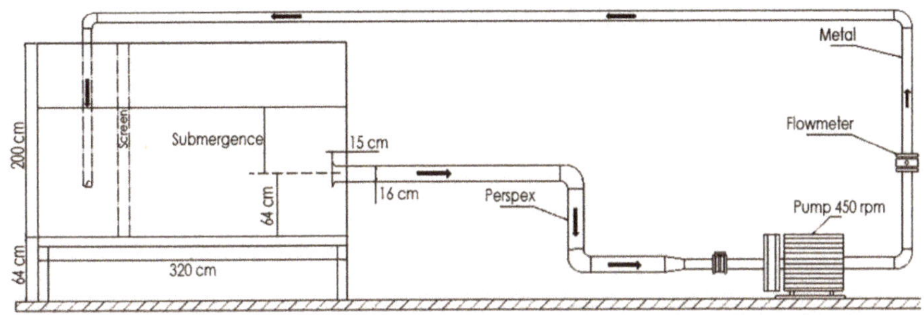

Figure 1. Experimental set-up by [39].

The second set of numerical investigations, which sought to assess the impact of installing an anti-vortex device, was based on the experimental work by [31]. Both studies implemented similar experimental set-ups but with an anti-vortex plate which has a 66% perforated opening installed at the intake of the experimental set-up by [31]. The plate had a thickness of 2 mm and measured 1.5D by 1D in length and width, respectively. The discharge flow and relative submergence for this experimental case were set at 0.015 m^3/s and 1.5, respectively.

2.2. Numerical Approach

2.2.1. Governing Equations

The governing Reynolds-Averaged Navier–Stokes (RANS) equations that underline the LTSInterFoam solver are the continuity, transport of phase-fraction and the momentum equations which are given by Equations (6)–(8), respectively. The flow parameters are decomposed using the Reynolds decomposition approach such that $\mathbf{u} = \bar{\mathbf{u}} + \mathbf{u}'$ and $p = \bar{p} + p'$.

$$\nabla \cdot \bar{\mathbf{u}} = 0 \tag{6}$$

$$\frac{\partial \alpha}{\partial t} + \nabla \cdot (\alpha \bar{\mathbf{u}}) = 0 \tag{7}$$

$$\frac{\partial (\rho \bar{\mathbf{u}})}{\partial t} + \nabla \cdot (\rho \bar{\mathbf{u}}\,\bar{\mathbf{u}}) = -\nabla \bar{p} + \rho g + \nabla \cdot \left[\mu_{eff} \left(\nabla \bar{\mathbf{u}} + (\nabla \bar{\mathbf{u}})^T \right) \right] \tag{8}$$

where $\mathbf{u}$ refers to the velocity; $\bar{\mathbf{u}}$ is the time-averaged velocity; $\mathbf{u}'$ refers to the velocity fluctuation; p is the pressure; $\bar{p}$ represents the time-averaged pressure; p' is the pressure fluctuation; α represents the volume fraction which ranges from 0 to 1 (with cells filled with gas given a value of 0 whilst cells filled with liquid are assigned a value of 1); $\mathbf{g}$ is the acceleration due to gravity; and μ_{eff} is the effective viscosity.

Equations (9) and (10) are used to compute the density ρ and effective viscosity μ_{eff} of the fluid, respectively. The subscripts l and g in the equations represent liquid and gas respectively.

$$\rho = \alpha \rho_l + (1 - \alpha) \rho_g \tag{9}$$

$$\mu_{eff} = (\mu_{eff})_l \alpha + \left(\mu_{eff} \right)_g (1 - \alpha) \tag{10}$$

The turbulence models selected for the study are the Standard $k - \varepsilon$ [40], RNG $k - \varepsilon$ [41,42], realizable $k - \varepsilon$ [43], $k - \omega$ [44,45] and the SST $k - \omega$ [46,47] models. Additional equations used for computing the various parameters in the different turbulence models are available in the literature.

2.2.2. OpenFOAM Application and Procedures

The governing differential equations are discretized using the finite volume method (FVM) along with the local time stepping (LTS) scheme for temporal discretization in order to ensure a rapid steady-state solution. The pressure–velocity equations are resolved using the PIMPLE algorithm. The linearUpwind grad (U) was selected for the convection of U. The divergence related to k and ω was computed using the linearUpwind limitedGrad scheme, whilst for that of epsilon, the upwind scheme was selected. For the divergence of alpha, div (phi,alpha), and the compression of the interface, div(phirb,alpha), the van Leer and the linear schemes were applied, respectively. The Gauss linear corrected scheme was used for the Laplacian schemes.

The building and meshing of the geometries were done in CAD and cfMesh, respectively. The cells were mainly hexahedra, comprising of coarse (average cell size 0.08 m), medium (average cell size 0.02 m) and fine (average cell size 0.01 m) cells. The meshed geometry with its associated boundaries is shown in Figure 2.

In CFD analysis, boundary conditions are used to specify the spatial or temporal variable values or behavior required to produce a unique solution. The boundary conditions used for the simulation are provided in Table 1 where U, p_rgh, *alpha.water*, *nut*, k and *omega* refer to the velocity, reduced pressure, phase fraction, turbulent viscosity, turbulence kinetic energy and specific rate of dissipation of turbulence kinetic energy, respectively. The fixedValue represents a *Dirichlet* Boundary Condition which is specified by the user. The pressureInletOutletVelocity boundary condition applies a zero-gradient for outflow, whilst the inflow velocity is the patch-face normal component of the internal-cell value. The fixedFluxPressure provides an adjustment to the pressure gradient, such that the flux on the boundary is the one specified by the velocity boundary condition. The totalPressure boundary condition is computed as static pressure reference plus the dynamic component due to velocity. The inletOutlet boundary condition provides a zero-gradient outflow condition for a fixed value inflow. The kqRWallFunction is the wall function for the turbulence kinetic energy while the nutkRoughWallFunction is the rough wall function for kinetic eddy viscosity and omegaWallFunction represents the wall function for frequency. The groovyBC boundary condition (contained in the swak4Foam library) was used to impose a constant water level.

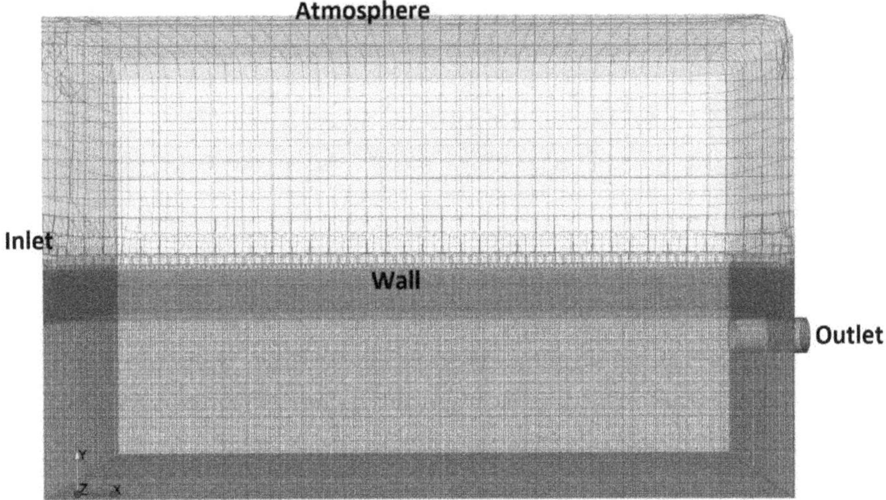

Figure 2. Meshed geometry with its associated boundaries.

Table 1. Boundary conditions.

Variable	Inlet	Outlet	Walls	Atmosphere
U	flowRateInletVelocity	flowRateInletVelocity	fixedValue	pressureInlet-OutletVelocity
p rgh	fixedFluxPressure	zeroGradient	zeroGradient	totalPressure
alpha.water	groovyBC	zeroGradient	zeroGradient	inletOutlet
nut	calculated	calculated	nutkRoughWall-Function	zeroGradient
k	fixedValue	inletOutlet	kqRWallFunction	inletOutlet
omega	fixedValue	inletOutlet	omegaWallFunction	inletOutlet

The simulations were run on 28 cores (2 × 14 core Intel Xeon E5-2660 2.00 GHz) of a desktop workstation which has 256 GB RAM. The simulation convergence was assessed using a normalized residual value of order 1×10^{-5}. Visualisation of the numerical results was performed in ParaView.

3. Results and Discussions

3.1. Visualisation of the Air-Core Vortex

Using an isosurface of $\alpha = 0.96$, the stable air-core vortex was depicted as a conical drop extending from the free surface towards the intake, as illustrated in Figure 3. In fluid flow, the presence of a vortex is often depicted by a roughly circular or spiral pattern of streamlines [48,49]. The streamlines, which were drawn with surface line integral convolution (Figure 4), exhibit a spiral pattern, thus confirming the presence of a vortex. The velocity vector plot (Figure 5) illustrates how the fluid flows spirally into the intake. From the observations in Figures 4 and 5, the fluid flow can be described as moving spirally from the water surface towards the intake and also rotating around the vortex axis. The plot depicted in Figure 6 shows the contour plot of the normalized velocity field expressed as a percentage, $(V_{XZ}/V)\%$, where V_{XZ} refers to the velocity in x-z plane whilst V is the intake flow velocity. The plot is made at a horizontal depth of D (16 cm) above the axis of the intake. The contours of the velocity field indicate the boundaries of the conical flow towards the intake [39].

Figure 3. Vortex formed at the intake (side view).

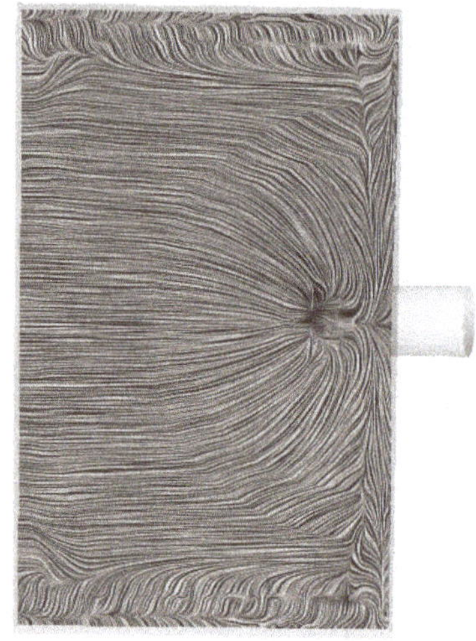

Figure 4. Streamlines illustrated with surface line integral convolution.

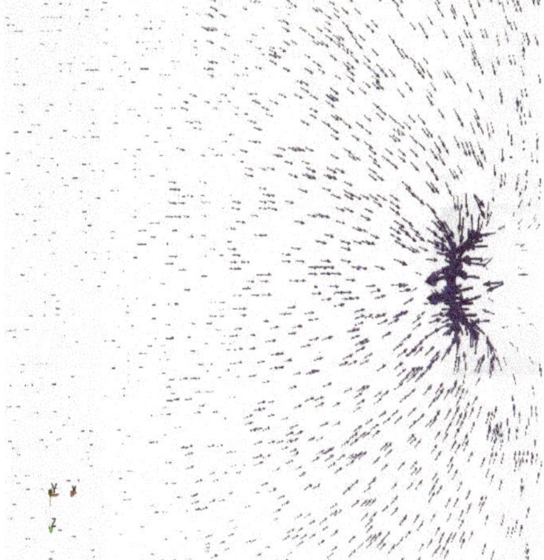

Figure 5. Velocity vector plot.

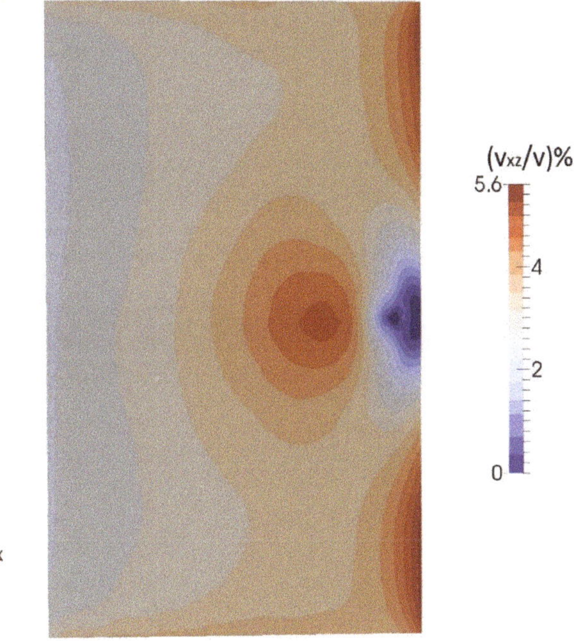

Figure 6. Velocity field plot.

3.2. Assessment of Different Turbulence Models

During the numerical simulation, the authors observed that the use of the standard $k - \varepsilon$ and RNG $k - \varepsilon$ models resulted in the development of flow instabilities at the free surface, especially at the upstream end. This phenomenon was particularly pronounced at the high-pressure gradient region, thus causing the numerical simulation to fail abruptly. This situation may be linked to the fact that adverse pressure gradient conditions often cause the standard $k - \varepsilon$ and RNG $k - \varepsilon$ models to exhibit poor performances [50].

On the other hand, the realizable $k - \varepsilon$, $k - \omega$ and the SST $k - \omega$ models successfully predicted the occurrence of the air-core vortex. The realizable $k - \varepsilon$ model outperformed the other family of $k - \varepsilon$ models (standard $k - \varepsilon$ and RNG $k - \varepsilon$) due to its ability to perform well with rotation and separation flow, a situation which is linked to the inclusion of a realizability constraint on the predicted stress tensor of the realizable $k - \varepsilon$ model [43].

However, regarding the realizable $k - \varepsilon$ model, we observed an amount of unusual free surface deformation along the walls, although the simulation progressed successfully. This observation was generally absent when the $k - \omega$ models were used. These observations may be attributed to the fact that the $k - \omega$ models often exhibit better numerical prediction at the walls compared to the $k - \varepsilon$ models [43,48].

A comparison of results in terms of the normalized tangential velocity distribution from the three (3) different turbulence models (realizable $k - \varepsilon$, $k - \omega$ and the SST $k - \omega$ models) has been illustrated in Figure 7a, where R refers to the normalized radius. From Figure 7b, the three (3) turbulence models were assessed by comparing the normalized tangential velocity distribution of the turbulence models with some selected analytical vortex models by [14,17,20,38] as well as other related experimental data by [12,20,51]. The plot showed a generally fair agreement with the realizable $k - \varepsilon$ model and the $k - \omega$ model. However, an amount of over-prediction of the tangential velocity occurred at the outer core of the vortex. Furthermore, at the point of maximum tangential velocity ($R = 1$), the realizable $k - \varepsilon$ model over-predicted the tangential velocity. The ability of the $k - \omega$ model to provide a better prediction at the boundary or near-wall layers could be the reason why the model outperformed the realizable $k - \varepsilon$ model [43,48]. The best agreement was, however, obtained from the SST $k - \omega$ model, which outperformed the $k - \omega$ model, and this could be attributed to the fact that even though the $k - \omega$ model is suitable for moderate adverse pressure gradients, the model could exhibit some amount of failure with pressure-induced separation [52], which includes the formation of an air-core vortex. Furthermore, the SST $k - \omega$ model has been found to be generally robust, with an enhanced performance in terms of flow at walls, adverse pressure gradients as well as flow separation [46,47,51–54]. From the illustration in Figure 7b, the ranking of the performance of the three turbulence models beginning from the most suitable model is, thus, SST $k - \omega$ model, $k - \omega$ model and realizable $k - \varepsilon$ model.

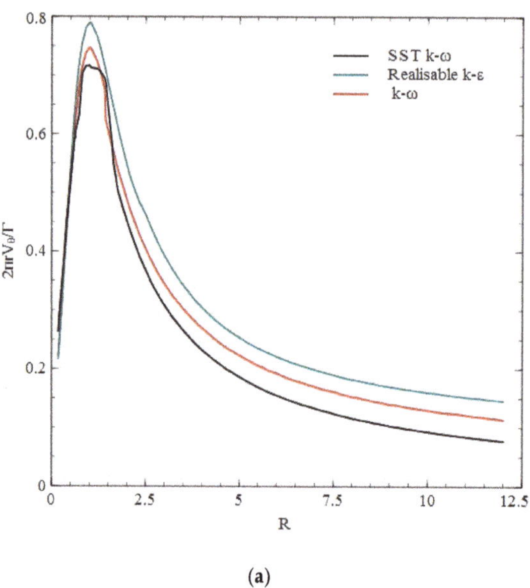

(**a**)

Figure 7. *Cont.*

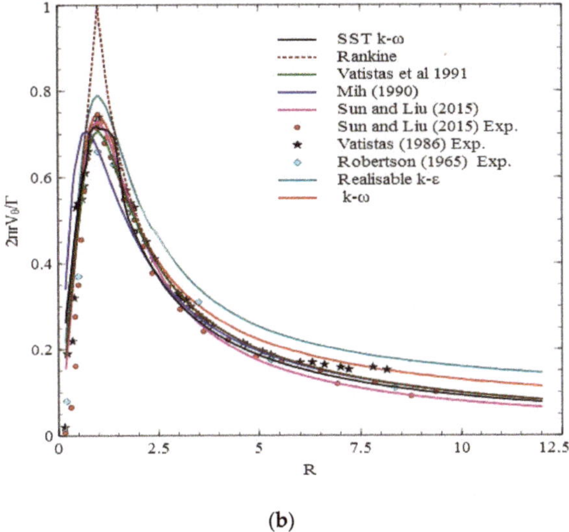

(b)

Figure 7. (a): Comparison of results from the three different turbulence models. **(b)**: Comparison of the results from the different turbulence models with other studies.

3.3. Suppression of Air Entrainment Using an Anti-Vortex Device

As reported by [31] and observed in Figure 8, the use of the anti-vortex plate resulted in the suppression of the air-core vortex. The region for the anticipated vortex only showed the formation of a slight depression. Ref. [31] observed a reduction of a Type 6 vortex (a stable air-core vortex) to a Type 1 vortex (coherent surface swirl) when a perforated plate serving as an anti-vortex device was installed. The anti-vortex device suppressed the formation of the air-core vortex by inducing turbulence and friction, which succeeded in cutting the path of the vortex core, thus ensuring a relatively smoother flow towards the intake [31]. The formation of the suppressed vortex is illustrated in Figure 8.

Figure 8. A slight surface depression formed in the presence of the anti-vortex plate.

4. Conclusions

Optimizing the intake of a hydropower plant often involves the suppression or prevention of the formation of free surface vortices whose occurrence could jeopardize the operations at the facility. In general, numerical simulations have been found to be relevant to such interventions by providing a key complement to experimental studies. In this regard, the study investigated the numerical simulation of an experimentally observed air-core vortex using different turbulence models as well as the suppression of the air-core vortex at an intake. Based on the results from the study, the following conclusions have been drawn:

1. The new approach of using the LTSInterFoam solver, as highlighted in this study, proved to be a reliable means for the steady-state simulation of a free surface vortex at an intake. The LTSInterFoam solver, until now, has been associated with hydrodynamic studies involving ships. However, being a steady state solver, it is unable to account for the transient process of evolution of a free surface vortex.
2. The SST $k - \omega$ model provided the best results compared to the other turbulence models.
3. The approach highlighted in this study can be used to investigate the effectiveness of an anti-vortex device in suppressing the formation of an air-core vortex.

The authors, thus, recommend that future studies explore the use of the interFoam solver in OpenFOAM to simulate the occurrence of free surface vortices in order to account for the transient process associated with the formation of free surface vortices.

Author Contributions: Conceptualization, R.A.; methodology, M.K.D. and G.T.; software, M.K.D. and G.T.; validation, M.K.D.; formal analysis, M.K.D.; investigation, M.K.D.; resources, G.T. and R.A.; data curation, M.K.D.; writing—original draft preparation, M.K.D.; writing—review and editing, M.K.D.; visualization, M.K.D.; supervision, G.T., S.G., R.A., M.A.-B. and E.A.O.; project administration, G.T., S.G. and R.A.; funding acquisition, M.K.D., G.T., S.G. and R.A. All authors have read and agreed to the published version of the manuscript.

Funding: This research was funded by the Commonwealth Scholarship Commission (CSC) funded by the UK Department for International Development, University of Exeter, UK and the AWD Consult Inc., South Portland, ME, USA.

Acknowledgments: The authors would like to acknowledge the support received from the Commonwealth Scholarship Commission (CSC) funded by the UK Department for International Development, University of Exeter, UK, AWD Consult Inc., South Portland, ME, USA. and the University of Energy and Natural Resources, Sunyani, Ghana. GRT was partly funded on this work by EPSRC Grant #EP/M022382/1 A CCP on Wave/Structure Interaction: CCP-WSI, and #EP/T026782/1 CCP-WSI+ Collaborative Computational Project on Wave Structure Interaction +.

Conflicts of Interest: The authors declare no conflict of interest. The funders had no role in the design of the study; in the collection, analyses, or interpretation of data; in the writing of the manuscript, or in the decision to publish the results.

References

1. Guyot, G.; Pittion-Rossillon, A.; Archer, A. A free surface vortex modelling with a 3D CFD comparison between an experimental case and a numerical one. In Proceedings of the 3rd IAHR Europe Congress, Porto, Portugal, 14–16 April 2014.
2. Yildirim, N. Critical submergence for a rectangular Intake. *J. Eng. Mech.* **2004**, *130*, 1195–1210. [CrossRef]
3. Farell, C.; Cuomo, A.R. Characteristics and modeling of intake vortices. *J. Eng. Mech.* **1984**, *110*, 723–742. [CrossRef]
4. Knauss, J. Swirling flow problems at intakes. In *IAHR Hydraulic Structures Design Manual 1*; Balkema: Leiden, The Netherlands, 1987; pp. 13–38.
5. Rindels, A.J.; Gulliver, J.S. *An Experimental Study of Critical Submergence to Avoid Free-Surface Vortices at Vertical Intakes*; The University of Minnesota: Minneapolis, MN, USA, 1983.
6. Mulligan, S.; Casserly, J.; Sherlock, R. Effects of geometry on strong free-surface vortices in subcritical approach flows. *J. Hydraul. Eng.* **2016**, *142*, 1–12. [CrossRef]
7. Mulligan, S.; Casserly, J.; Sherlock, R. Experimental and numerical modelling of free-surface turbulent flows in full air-core water vortices. In *Advances in Hydroinformatics*; Gourbesville, P., Cunge, J.A., Caignaert, G., Eds.; Springer WaterSpringer: Singapore, 2016; pp. 549–569, ISBN 9789812876140.
8. Shabara, H.M.; Yaakob, O.B.; Ahmed, Y.M.; Elbatran, A.H. CFD simulation of water gravitation vortex pool flow for mini hydropower plants. *J. Teknol.* **2015**, *74*, 77–81. [CrossRef]
9. Levi, E. A fluidic vortex device for water treatment processes. *J. Hydraul. Res.* **1983**, *21*, 17–31. [CrossRef]

10. Andoh, R.; Osei, K.; Fink, J.; Faram, M. Novel drop shaft system for conveying and controlling flows from high level sewers into deep tunnels. In Proceedings of the World Environmental and Water Resources Congress, Honolulu, HI, USA, 12–16 May 2008; pp. 1–9.

11. Huang, T.-H.; Jan, C.-D.; Hsu, Y.-C. Numerical simulations of water surface profiles and vortex structure in a vortex settling basin by using FLOW-3D. *J. Mar. Sci. Technol.* **2017**, *25*, 531–542.

12. Vatistas, G.H.; Lin, S.; Kwock, C.K. Theoretical and experimental studies on vortex chamber flows. *AIAA J.* **1986**, *24*, 635–642. [CrossRef]

13. Rabe, B.K.; Najafabadi, S.H.G.; Sarkardeh, H. Numerical simulation of anti-vortex devices at water intakes. *Proc. Inst. Civ. Eng.-Water Manag.* **2016**, *171*, 18–29. [CrossRef]

14. Rankine, W.J.M. *Manual of Applied Mechanics*; C. Griffen Co.: London, UK, 1858.

15. Scully, M. *Computation of Helicopter Rotor Wake Geometry and Its Influence on Rotor Harmonic Airloads*; Massachusetts Institute of Technology: Cambridge, MA, USA, 1975.

16. Burgers, J.M. A mathematical model illustrating the theory of turbulence. *Adv. Appl. Mech.* **1948**, *1*, 171–199.

17. Vatistas, G.H.; Kozel, V.; Mih, W.C. A simpler model for concentrated vortices. *Exp. Fluids* **1991**, *11*, 73–76. [CrossRef]

18. Hite, J.E.; Mih, W.C. Velocity of air-core vortices at hydraulic intakes. *J. Hydraul. Eng.* **1994**, *120*, 284–297. [CrossRef]

19. Wang, Y.; Jiang, C.; Liang, D. Comparison between empirical formulae of intake vortices. *J. Hydraul. Res.* **2011**, *49*, 113–116. [CrossRef]

20. Sun, H.; Liu, Y. Theoretical and experimental study on the vortex at hydraulic intakes. *J. Hydraul. Res.* **2015**, *53*, 787–796. [CrossRef]

21. Gordon, J.L. Vortices at Intake. *Water Power* **1970**, *22*, 137–138.

22. Reddy, Y.R.; Pickford, J.A. Vortices at Intakes in Conventional Sump. *Water Power* **1972**, *24*, 108–109.

23. Prosser, M.J. *The Hydraulic Design of Pump Sumps and Intake*; British Hydromechanics Research Association: Cranfield, UK, 1977.

24. Nakayama, A.; Hisasue, N. Large eddy simulation of vortex flow in intake channel of hydropower facility. *J. Hydraul. Res.* **2010**, *48*, 415–427. [CrossRef]

25. Chen, Y.; Wu, C.; Wang, B.; Du, M. Three-dimensional Numerical Simulation of Vertical Vortex at Hydraulic Intake. *Procedia Eng.* **2012**, *28*, 55–60. [CrossRef]

26. Suerich-Gulick, F.; Gaskin, S.; Villeneuve, M.; Holder, G.; Parkinson, E. Experimental and numerical analysis of free surface vortices at a hydropower intake. In Proceedings of the 7th International Conference on Hydroscience and Engineering, Philadelphia, PA, USA, 10–13 September 2006; pp. 1–11.

27. Ahn, S.; Xiao, Y.; Wang, Z.; Zhou, X.; Luo, Y. Numerical prediction on the effect of free surface vortex on intake flow characteristics for tidal power station. *Renew. Energy* **2017**, *101*, 617–628. [CrossRef]

28. Sarkardeh, H. Numerical calculation of air entrainment rates due to intake vortices. *Meccanica* **2017**, *52*, 3629–3643. [CrossRef]

29. Chen, Y.; Wu, C.; Ye, M.; Ju, X. Hydraulic characteristics of vertical vortex at hydraulic intakes. *J. Hydrodyn. Ser. B* **2007**, *19*, 143–149. [CrossRef]

30. Kabiri-Samani, A.R.; Borghei, S.M. Effects of anti-vortex plates on air entrainment by free vortex. *Sci. Iran.* **2013**, *20*, 251–258.

31. Amiri, S.M.; Roshan, R.; Zarrati, A.R.; Sarkardeh, H. Surface vortex prevention at power intakes by horizontal plates. *Proc. Inst. Civ. Eng. Manag.* **2011**, *164*, 193–200. [CrossRef]

32. Taghvaei, S.M.; Roshan, R.; Safavi, K.; Sarkardeh, H. Anti-vortex structures at hydropower dams. *Int. J. Phys. Sci.* **2012**, *7*, 5069–5077. [CrossRef]

33. Monshizadeh, M.; Tahershamsi, A.; Rahimzadeh, H. Vortex dissipation using a hydraulic-based anti-vortex device at intakes. *Int. J. Civ. Eng.* **2017**, *16*, 1137–1144. [CrossRef]

34. Monshizadeh, M.; Tahershamsi, A.; Rahimzadeh, H.; Sarkardeh, H. Comparison between hydraulic and structural based anti-vortex methods at intakes. *Eur. Phys. J. Plus* **2017**, *132*, 1–11. [CrossRef]

35. Tahershamsi, A.; Rahimzadeh, H.; Monshizadeh, M.; Sarkardeh, H. A new approach on anti-vortex devices at water intakes including a submerged water jet. *Eur. Phys. J. Plus* **2018**, *133*, 1–11. [CrossRef]

36. Borghei, S.M.; Kabiri-Samani, A.R. Effect of anti-vortex plates on critical submergence at a vertical intake. *Sci. Iran.* **2010**, *17*, 89–95.

37. Trivellato, F. Anti-vortex devices: Laser measurements of the flow and functioning. *Opt. Lasers Eng.* **2010**, *48*, 589–599. [CrossRef]

38. Mih, W.C. Discussion of Analysis of fine particle concentrations in a combined vortex. *J. Hydraul. Res.* **1990**, *28*, 392–395. [CrossRef]

39. Sarkardeh, H.; Zarrati, A.R.; Jabbari, E.; Tavakkol, S. Velocity field in a reservoir in the presence of an air-core vortex. *Proc. Inst. Civ. Eng. Manag.* **2014**, *167*, 356–364. [CrossRef]

40. Jones, W.P.; Launder, B.E. The prediction of laminarization with a two-equation model of turbulence. *Int. J. Heat Mass Transf.* **1972**, *15*, 301–314. [CrossRef]

41. Yakhot, V.; Orszag, S.A. Renormalization group analysis of turbulence. I. Basic theory. *J. Sci. Comput.* **1986**, *1*, 3–51. [CrossRef]

42. Yakhot, V.; Orszag, S.A.; Thangam, S.; Gatski, T.B.; Speziale, C.G. Development of turbulence models for shear flows by a double expansion technique. *Phys. Fluids* **1992**, *4*, 1510–1520. [CrossRef]

43. Andersson, B.; Andersson, R.; Håkansson, L.; Mortensen, M.; Sudiyo, R.; Van Wachem, B. *Computational Fluid Dynamics for Engineers*; Cambridge University Press: Cambridge, UK, 2011.

44. Wilcox, D. Reassessment of the scale-determining equation for advanced turbulence models. *AIAA J.* **1988**, *26*, 1299–1310. [CrossRef]

45. Wilcox, D.C. *Turbulence Modeling for CFD*, 2nd ed.; DCW Industries: La Canada, CA, USA, 1998.

46. Menter, F.R. Review of the shear-stress transport turbulence model experience from an industrial perspective. *Int. J. Comput. Fluid Dyn.* **2009**, *23*, 305–316. [CrossRef]

47. Menter, F.R.; Carregal, F.J.; Esch, T.; Konno, B. The SST turbulence model with improved wall treatment for heat transfer predictions in gas turbines. In Proceedings of the International Gas Turbine Congress, Tokyo, Janpan, 2–7 November 2003.

48. Robinson, S.K.; Kline, S.J.; Spalart, P.R. A review of quasi-coherent structures in a numerically simulated turbulent boundary layer. *NASA Tech. Rep.* **1989**, *TM-10*, 1–44.

49. Robinson, S.K. Coherent Motions in the Turbulent Boundary Layer. *Annu. Rev. Fluid Mech.* **1991**, *23*, 601–639. [CrossRef]

50. Moukalled, F.; Mangani, L.; Darwish, M. *The Finite Volume Method in Computational Fluid Dynamics. An Advanced Introduction with OpenFoam® and Matlab®*; Thess, A., Moreau, R., Eds.; Springer International Publishing: New York, NY, USA, 2016; ISBN 0926-5112.

51. Robertson, J.M. *Hydrodynamics in Theory and Application*; Prentice-Hall Inc.: Englewood Cliffs, NJ, USA, 1965.

52. Menter, F.R. Zonal two-equation k-ω turbulence model for aerodynamic flows. *AIAA Pap.* **1993**, 1993–2906.

53. Menter, F.R.; Kuntz, M.; Langtry, R. Ten years of industrial experience with the SST turbulence model. *Turbul Heat Mass Transf.* **2003**, *4*, 625–632.

54. Menter, F. Two-equation eddy-viscosity turbulence models for engineering applications. *AIAA J.* **1994**, *32*, 1598–1605. [CrossRef]

Publisher's Note: MDPI stays neutral with regard to jurisdictional claims in published maps and institutional affiliations.

 *fluids*

Article

Multi-Physics Modeling of Electrochemical Deposition

Justin Kauffman [1,*], John Gilbert [2], Eric Paterson [2,3]

[1] The Hume Center for National Security and Technology, Virginia Tech, Arlington, VA 22203, USA
[2] The Hume Center for National Security and Technology, Virginia Tech, Blacksburg, VA 24061, USA; johng12@vt.edu (J.G.); egp@vt.edu (E.P.)
[3] Department of Aerospace and Ocean Engineering, Virginia Tech, Blacksburg, VA 24061, USA
* Correspondence: jakauff@vt.edu

Received: 6 November 2020; Accepted: 8 December 2020; Published: 11 December 2020

Abstract: Electrochemical deposition (ECD) is a common method used in the field of microelectronics to grow metallic coatings on an electrode. The deposition process occurs in an electrolyte bath where dissolved ions of the depositing material are suspended in an acid while an electric current is applied to the electrodes. The proposed computational model uses the finite volume method and the finite area method to predict copper growth on the plating surface without the use of a level set method or deforming mesh because the amount of copper layer growth is not expected to impact the fluid motion. The finite area method enables the solver to track the growth of the copper layer and uses the current density as a forcing function for an electric potential field on the plating surface. The current density at the electrolyte-plating surface interface is converged within each PISO (Pressure Implicit with Splitting Operator) loop iteration and incorporates the variance of the electrical resistance that occurs via the growth of the copper layer. This paper demonstrates the application of the finite area method for an ECD problem and additionally incorporates coupling between fluid mechanics, ionic diffusion, and electrochemistry.

Keywords: multi-physics; electrochemistry; fluid mechanics; OpenFOAM; finite volume method; finite area method

---

1. Introduction

Electrochemical deposition (ECD) is an integral process in the fabrication of semiconductor devices. In addition, the applications spread through a variety of other fields such as the deposition of coatings for corrosion resistance [1] and water treatments [2]. ECD is not the only deposition technique used for growing copper layers in semiconductor devices, but it is a popular and inexpensive method, since it does not require a vacuum or heating of the system [3]. Additionally, this method is able to deposit high-aspect ratio features with a high degree of precision.

The process of ECD is shown in a simplified schematic in Figure 1.

The process applies a current I that flows from the anode to the cathode. Positive ions are shed from the anode and added to the electrolyte. Diffusion and migration of ions occurs in the electrolyte. The positive ions are deposited on the cathode surface and depletes the ionic concentration near the cathode-electrolyte interface. As time evolves, the copper layer continues to grow, and the rate of this growth is directly proportional to the applied current density. The copper layer is deposited edge high, center low. So it is imperative to control how the copper layer is deposited and one way to control the uniformity of the copper layer is to induce mixing of the electrolyte. In this paper the anode and cathode are copper and the electrolyte is a copper sulfate-acid solution.

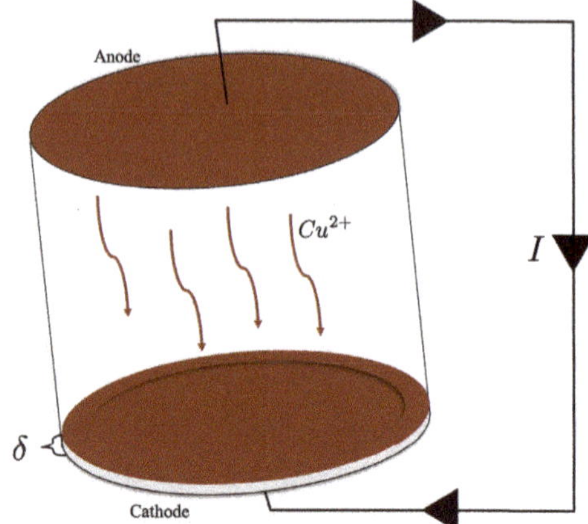

Figure 1. A schematic of a simplified electrolytic cell being used for electrochemcial deposition.

Mathematical models of electrochemical phenomena in the literature showcase a spectrum of weak to strong coupling between electric potential, ionic concentration, and fluid motion. The work by Karcz [4], which investigates a tubular fuel cell, chose to account for the current density with a zero-dimensional or one-dimensional model, while maintaining a three-dimensional model for the fluid motion and ionic transport. Karcz concluded that the characteristics of the current density were similar in both models and the one-dimensional model compared favorably to experimental data. Introducing relationships such as the Butler-Volmer equation, which computes a current density based on a nonlinear dependence of the voltage difference between the anode and cathode and ionic concentration of the species that participates in chemical kinetics [5], strengthens the coupling of the electrochemical model and can be observed in work performed by Ritter et al. [3], Drese [6], Hughes, Baily and McManus [7], and Hughes et al. [8], where four different current distributions, at the cathode-electrolyte interface, are presented. These four regimes are tertiary, secondary, primary, and diffusion limited current distributions. This work uses the tertiary current distribution. The strongest coupling presented in the literature shows dependence of ionic concentration and electric potential in the equations that govern ionic transport and electrochemistry, see [2,9–11].

The overall objective of this work is to determine how turbulent mixing impacts copper layer growth. With that being said, this paper is the first building block to achieve that higher goal, where the focus will be on the electrochemistry and ionic transport. There have been others that have looked into mixing via a reciprocating paddle [10]. Motion of a paddle within the fluid or a deforming boundary due to depletion or plating of copper require a numerical approach that will account for these changes. One such approach is to use the Arbitrary Lagrangian-Eulerian (ALE) [10] or a dynamic mesh method [2] that moves the mesh based on the motion of a boundary. Other approaches include a level set method [7–9], an explicit interface tracking method (EITM) [12,13], and the finite area method (FAM) [1]. Here the FAM is used because the growth of the copper layer is small in comparison to the domain size, and, therefore, would have negligible effects on the motion of the fluid.

The paper is organized as follows: Section 2 provides a description of the mathematical models and key assumptions used in this work. Section 3 summarizes the numerical methods employed to solve the physics model. Section 4 presents the results of a validation study performed to assess the accuracy of the proposed approach. Section 5 extends the validation results by considering concentration dependence. Finally, Section 6 provides a summary of the research and some possible future extensions.

2. Problem Formulation

The physics that governs ECD couples fluid mechanics, ionic diffusion, and electrochemistry. To computationally model a complex system such as ECD, several mathematical, physical, and geometric assumptions must be made that dictate the level of coupling. It is necessary to have a strong enough level of coupling between the different physical models to demonstrate and predict the required physical behavior observed in experiments and practice.

2.1. Governing Equations

The equations that govern the fluid mechanics are conservation of mass and conservation of momentum. The conservation of mass for an incompressible fluid is

$$\nabla \cdot \mathbf{u} = 0, \tag{1}$$

and the conservation of momentum, the Navier-Stokes equations, are

$$\frac{\partial \mathbf{u}}{\partial t} + \nabla \cdot (\mathbf{u} \otimes \mathbf{u}) - \nabla \cdot (\nu \nabla \mathbf{u}) = -\frac{1}{\rho} \nabla p + \mathbf{F}, \tag{2}$$

where $\mathbf{u}$ is the fluid velocity, ν is the kinematic viscosity, ρ is the fluid density, p is the fluid pressure, and $\mathbf{F}$ is any body force acting on the fluid.

To mathematically model the transport of copper ions within the electrolyte, the concentration of copper ions is tracked via diffusion, migration, and convection (if fluid motion is present). The governing equation for the transport of Cu^{2+} ions is

$$\frac{\partial C}{\partial t} = -\nabla \cdot \left[-D\nabla C - \left(\frac{FnD}{RT} C \right) \nabla \varphi - C\mathbf{u} \right], \tag{3}$$

where C is the mass fraction of copper ion concentration, D is the diffusivity of copper ions, F is Faraday's constant, n is the valence electrons for copper ions, R is the ideal gas constant, T is chamber temperature, and φ is the electric potential of the electrolyte. This equation is known as the Nernst-Planck equation. The first term on the right-hand side of Equation (3) represents diffusion, the middle term is the migration of the ionic concentration (this is a coupling term between the electric potential and ionic concentration), and the last term is convection due to fluid motion.

The electrochemistry is governed by ensuring that the electric flux in the system is divergence-free, where the flux is defined as

$$\mathbf{N} = -Fn \left[D\nabla C - \left(\frac{FnD}{RT} C \right) \nabla \varphi + C\mathbf{u} \right]. \tag{4}$$

Therefore, the governing equation is $\nabla \cdot \mathbf{N} = 0$ or

$$\nabla^2 (FnDC) + \nabla \cdot \left[\left(\frac{F^2 n^2 D}{RT} C \right) \nabla \varphi \right] = 0, \tag{5}$$

since the fluid motion is quiescent and Equation (1) still holds $\nabla \cdot (C\mathbf{u})$ vanishes. If fluid motion was present the term $\nabla C \cdot \mathbf{u}$ would still be included.

The equations that have been presented to this point only account for the electrolyte and do not consider any electrodynamics that occur from the growth of the copper layer. To account for the changes in electric potential on the plating surface the following equation must be solved:

$$\nabla \cdot \left(\sigma_{eff} \nabla \varphi_s \right) = -i, \tag{6}$$

where σ_{eff} is the effective surface conductivity at the interface between the seed layer and electrolyte and i is the current density on the plating surface. The effective conductivity is calculated by considering the initial seed layer in addition to the growing copper layer; it is defined as:

$$\sigma_{eff} = \sigma\delta + \sigma_0\delta_0, \tag{7}$$

where σ is the conductivity of solid copper, σ_0 is the conductivity of the initial seed layer material, and δ_0 is the initial seed layer thickness. The current density at the electrolyte-cathode interface is computed using the following:

$$i = \frac{FDc_b}{M_{Cu}}\left(\nabla C \cdot \mathbf{n} - \frac{nF}{RT}C\nabla\varphi \cdot \mathbf{n}\right), \tag{8}$$

where c_b is the bulk concentration of copper ions in the electrolyte and M_{Cu} is the molar mass of copper.

The Butler-Volmer equation Equation (9) is another expression for the current density, but here it used to calculate the overpotential η, which requires using a Newton-Raphson [14] method to solve the nonlinear equation. The Bulter-Volmer equation is

$$i = i_0\left[\exp\frac{\alpha_A nF\eta}{RT} - C\exp\frac{-\alpha_C nF\eta}{RT}\right], \tag{9}$$

where i_0 is the exchange current density (cathodic current at equilibrium [15]), and α_A and α_C are the charge transfer coefficient for the anode and cathode, respectively. η is then used to update the interface potential,

$$\varphi_I = \varphi_s - \eta, \tag{10}$$

which is used as a boundary condition for the current algorithm. Equation (9) also enforces convergence of the current density inside the PISO loop, where the full algorithm will be discussed in more detail in Section 3.2.

Finally, the equation that governs the deposited copper is

$$\frac{\partial\delta}{\partial t} = \frac{iM_{Cu}}{\rho_{Cu}nF}, \tag{11}$$

where δ is the deposited copper thickness and ρ_{Cu} is the density of copper. All of the parameters that need to be set in the simulations are defined in Table 1.

Table 1. Material and Fluid properties that will be calculated through the course of each simulation. Symbols, descriptions, and dimensions are provided here.

Constants	Description	Dimensions (SI)
ρ	Fluid density	$[\text{kg}/\text{m}^3]$
ν	Kinematic viscosity of the fluid	$[\text{m}^2/\text{s}]$
D	Cu ion diffusivity	$[\text{m}^2/\text{s}]$
F	Faraday's constant	$[\text{A s}/\text{mol}]$
n	Ion valence	$[-]$
R	Universal gas constant	$[\text{kg m}^2/(\text{K mol s}^2)]$
T	Operating temperature	$[\text{K}]$
c_b	Bulk concentration of copper ions in the electrolyte	$[\text{kg}/\text{m}^3]$
M_{Cu}	Molar mass of copper	$[\text{kg}/\text{mol}]$
ρ_{Cu}	Density of copper	$[\text{kg}/\text{m}^3]$
δ_0	Initial seed layer thickness	$[\text{m}]$
σ_0	Initial seed layer conductivity	$[\text{s}^3\,\text{A}^2/(\text{kg m}^3)]$
σ	Electric conductivity of solid copper	$[\text{s}^3\,\text{A}^2/(\text{kg m}^3)]$
i_0	Exchange current density	$[\text{A}/\text{m}^2]$
$\alpha_{A;C}$	Charge transfer coefficients for the anode and cathode	$[-]$

2.2. Material and Fluid Properties

Tables 1 and 2 list the fluid and material properties and the fields being computed throughout the ECD simulations, respectively.

Table 2. Material and Fluid fields that will be calculated through the course of each simulation. Symbols, descriptions, and dimensions are provided here.

Field	Description	Dimensions (SI)
$\mathbf{u}$	Fluid velocity	$[\text{m/s}]$
p	Fluid pressure	$[\text{kg/}\left(\text{m s}^2\right)]$
$\mathbf{F}$	Body force acting on the fluid	$[\text{m/s}^2]$
C	Mass fraction of copper ion concentration	$[-]$
φ	Electric potential of the electrolyte	$[\text{kg m}^2/\left(\text{A s}^3\right)]$
φ_s	Electric potential copper layer	$[\text{kg m}^2/\left(\text{A s}^3\right)]$
φ_I	Interface potential	$[\text{kg m}^2/\left(\text{A s}^3\right)]$
$\mathbf{N}$	Electric flux	$[\text{A m/mol}]$
δ	Deposited copper layer thickness	$[\text{m}]$
i	Current density	$[\text{A/m}^2]$
σ_{eff}	Effective conductivity to account for δ_0 and δ	$[\text{s}^3\,\text{A}^2/\left(\text{kg m}^2\right)]$
η	Overpotential	$[\text{kg m}^2/\left(\text{A s}^3\right)]$

3. Numerical Methods

3.1. Discretization

In this work, a cell-centered finite volume method (FVM) is used to discretize the bulk fluid (Equations (1)–(3)) and electric potential (Equation (5)) equations and a face-centered finite area method (FAM) is used to discretize the electric potential equation along the plating surface (Equation (6)). Both the FVM and FAM approaches are based on integral forms of the governing equations. Unless otherwise noted, all time derivatives are discretized using the second-order accurate, implicit scheme referred to as the backward scheme [16]. The following sections will briefly describe the FVM and FAM spatial discretizations.

3.1.1. Finite Volume Method (FVM)

In a finite volume approach, the computational space is divided into non-overlapping control volumes (CV) that completely fill the space. Figure 2a shows an example of a hexahedral control volume, V_p, centered around a point, P. The face f with surface area S_f and unit normal $\mathbf{n}_f$ is shared by a neighboring CV with centroid N.

Integral forms of the governing equations are discretized in the FVM by transforming continuous surface integrals into discrete summations over CV faces. For example, the FVM form of the Laplacian term in Equation (16) is given by

$$\int_V \nabla \cdot \left(\frac{F^2 n^2 D}{RT} \nabla \varphi\right) dV = \int_S \frac{F^2 n^2 D}{RT} \nabla \varphi \cdot d\mathbf{S} = \sum_f \frac{F^2 n^2 D}{RT} \mathbf{S}_f \cdot (\nabla \varphi)_f. \tag{12}$$

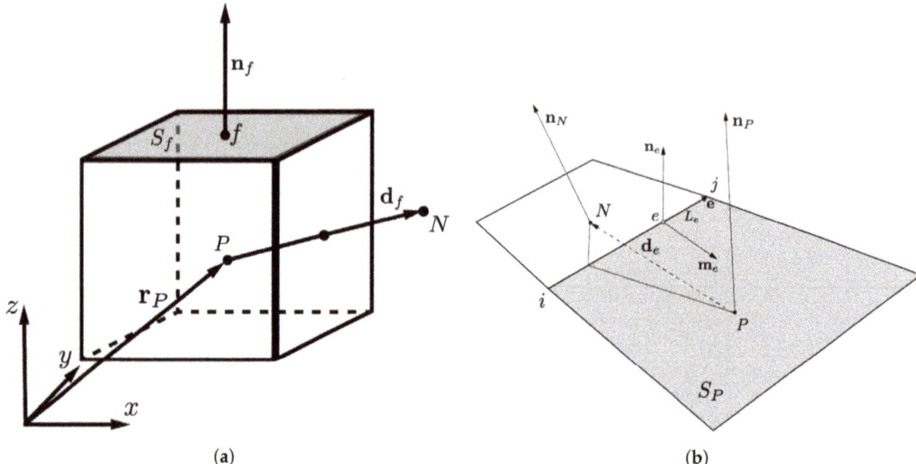

Figure 2. Illustrations of a (**a**) hexahedral control volume, and (**b**) quadrilateral control area based off of figures in [17].

The face normal gradient, $\mathbf{S}_f \cdot (\nabla \varphi)_f$, is evaluated according to

$$\mathbf{S}_f \cdot (\nabla \varphi)_f = |S_f| \frac{\varphi_N - \varphi_P}{|\mathbf{d}|}, \tag{13}$$

where $\mathbf{d}$ is the length vector between the center of cell P and neighboring cell N and φ_P and φ_N are the value of φ at the points P and N, respectively. In the case of non-orthogonal meshes, an additional correction term is introduced (see [16]). Similar discretizations can be introduced for the remaining terms in the governing equations. The interested reader is referred to [18] for additional information on the FVM discretizations used within the OpenFOAM® framework [19]. For more detailed information on the FVM the interested reader is referred to [20].

3.1.2. Finite Area Method (FAM)

The finite area method can be thought of as a two-dimensional version of the FVM over curved surfaces, and is well-suited to problems for which the thickness of the region of interest is much less than the lateral dimensions of the domain and through-thickness gradients can be neglected. For example, in [17], the FAM is employed to solve the transport of surfactants along the surface of a free-rising bubble. In this work, a face-centered FAM is used to solve for the electric potential field within the thin deposited copper layer along the plating surface, see Equation (6).

In the FAM, similar to the FVM, the surface is decomposed into non-overlapping polyhedral control areas. Figure 2b provides an illustration of two adjacent quadrilateral control areas with centroids P and N, areas S_P and S_N, and outward facing normals $\mathbf{n}_P$ and $\mathbf{n}_N$. Integral equations are then discretized as summations over the control area edges, analogous to the FVM procedure. For example, the FAM discretization of the Laplacian term in Equation (17) is expressed as

$$\int_{\partial S} \nabla \cdot (\sigma_{eff} \nabla \varphi_s) = \sum_e \sigma_{eff} L_e \mathbf{m}_e \cdot (\nabla \varphi_s)_e, \tag{14}$$

where the subscript e denotes edge-centered values, L_e is the length of the edge, and $\mathbf{m}_e$ is the outward pointing unit bi-normal of the edge. For more detailed information on the FAM the interested reader is referred to [21].

3.2. Solver Algorithm

Figure 3 describes the full solution algorithm that the ECD simulation follows. It should be noted that there is logic that allows the user to turn on or off the ionic transport of copper. The inputs into the algorithm are the initial concentration, seed layer thickness, and electric current. Here the overpotential is initially chosen to be zero and iterated to convergence via a Newton-Raphson [14] scheme.

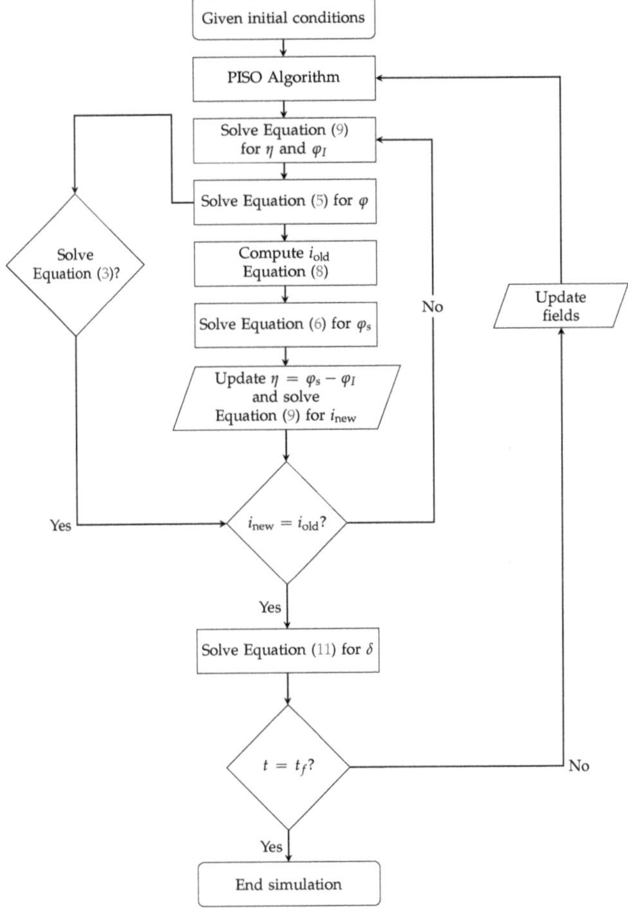

Figure 3. Solution algorithm for electrochemical deposition. This algorithm is executed each timestep. C_0, η_0, δ_0 are initial copper concentration, overpotential, and seed layer thickness, respectively. t_f represents the end time of the simulation.

3.3. Solver Parameters

The numerical solvers and schemes used in the ECD solver are noted in Tables 3 and 4. Numerical solvers are the linear algebraic solvers used to solve the discretized equations and numerical schemes are the methods used to approximate the differential operators in those equations. The relaxation factors for the equations that govern the fields φ and φ_s are set to 0.7. These relaxation factors were chosen to reduce oscillations that occur in the electric potential fields during the loop that ensures the current density converges, but a more formal investigation into the optimal relaxation factors is necessary and is part of future work.

Table 3. Numerical solvers used in the ECD simulations. LU = Lower-Upper; GAMG = Geomtric-Alegbraic Multi-Grid; FDIC = Faster Diagonal Incomplete-Cholesky; DICGaussSeidel = Diagonal Incomplete-Cholesky with Gauss Seidel; PBiCGStab = Stabilized Preconditioned Bi-Conjugate Gradient; DILU = Diagonal Incomplete-LU; DILUGaussSeidel = Diagonal Incomplete-LU with Gauss Seidel; PCG = Preconditioned Conjugate Gradient; DIC = Diagonal Imcomplete-Cholesky.

Finite Volume Method				
Field	**Equation**	**Solver**	**Preconditioner**	**Smoother**
p	(1), (2)	GAMG	FDIC	DICGaussSeidel
$\mathbf{u}$	(1), (2)	PBiCGStab	DILU	DILUGaussSeidel
C	(3)	PBiCGStab	DILU	DILUGaussSeidel
φ	(5)	PCG	DIC	-
Finite Area Method				
φ_s	(6)	PCG	DIC	-
δ	(11)	PBiCGStab	DILU	DILUGaussSeidel

Table 4. Numerical schemes to approximate the differential operators. For the divergence operators, $\nabla \cdot$ the choice is either Gauss upwind or Gauss linear, which depends on the specific form of the divergence.

Operator	Scheme	Description
$\partial/\partial t$	backward	Transient, 2nd order, potentially unbounded, implicit
$\nabla(\cdot)$	Gauss linear 1.0	Central differencing, bounded
$\nabla\cdot$	Gauss upwind/Gauss linear	1st order, bounded/ 2nd order, unbounded
$\nabla(\cdot)\cdot\mathbf{n}$	corrected	Explicit non-orthogonal correction at cell faces, 2nd order, conservative
∇^2	Gauss linear corrected	2nd order, unbounded, non-orthogonal correction, conservative

4. Validation Test Case

The geometric assumptions for this problem will limit the computational domain to two-dimensions (2D), as a simplification. Another simplification for the present state of the computational model is to have quiescent fluid motion. In other words there will be no mixing of the fluid, and therefore, it has no motion. The focus on this effort was to describe the coupling between ionic transport and electrochemistry. First, it is demonstrated that the approach agrees with previously published resultsin [1], but without ionic transport.

4.1. Description

The performance of the proposed model described in Sections 2 and 3 is validated against the published results of [1] for copper electroplating of an L-shaped surface at the bottom of an electrolyte bath. In this simple test case, a constant electric current is supplied along the top surface of the bath and copper is deposited on the bottom L-shaped surface, highlighted in orange as shown in Figure 4. Due to the configuration of the plating surface, the thickness of the Cu layer is expected to be non-uniformly distributed along the plating surface, with the highest deposition rate occurring at the exposed corner of the L-shaped surface. The sampling line for used in the figures below starts in the center of the cyan line (this is the zero distance in the line plots below) then extends up and around the corner of the orange L-shaped plating surface and ends at the left most point on the orange plating surface.

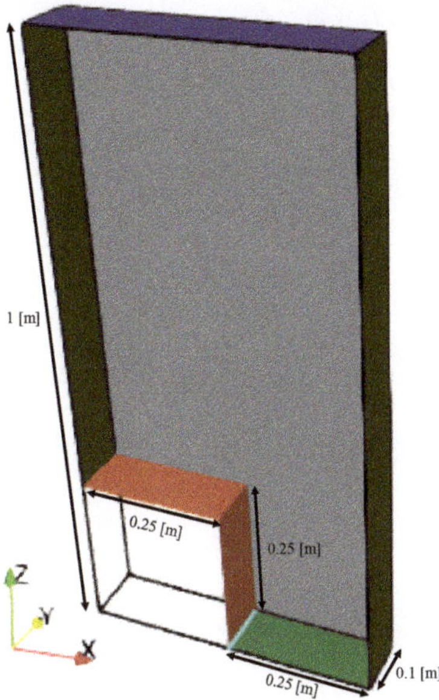

Figure 4. L-shaped plating surface at the bottom of an electrolyte bath. The orange surface is the plating surface (cathode), the blue surface is the anode, the green surfaces are insulated, and the cyan edge is the ground on the FA mesh.

4.2. Model Formulation and Assumptions

For this test case, the bulk fluid in the electrolyte bath is assumed to be quiescent ($\mathbf{u} = 0$), and the electrolyte solution is sufficiently mixed such that ionic transport within the bulk fluid can be safely neglected. Under these assumptions, the governing equations simplify to

$$\nabla p = -\rho \mathbf{g}, \tag{15}$$

$$\nabla \cdot \left(\frac{F^2 n^2 D}{RT} \nabla \varphi \right) = 0, \tag{16}$$

$$\nabla \cdot (\sigma_{eff} \nabla \varphi_s) = -i, \tag{17}$$

where $F^2 n^2 D / RT = \kappa$ when comparing to [1] which is representative of the conductivity of the electrolyte and the Butler-Volmer equation reduces to

$$i = i_o \left[\exp \left(\frac{\alpha_A n F \eta}{RT} \right) - \exp \left(\frac{-\alpha_C n F \eta}{RT} \right) \right]. \tag{18}$$

Equation (15) expresses the hydrostatic equilibrium existing within the bulk fluid with the gravitational force vector defined as $\mathbf{g} = (0, 0, g)$. Equations (16) and (17) describe the electric potential in the bulk fluid and deposited copper layer, respectively. The conductivity of the growing copper layer is given by

$$\sigma_{eff} = \sigma \delta + \sigma_{s0}, \tag{19}$$

where σ is the electric conductivity of the deposited metal, δ is the thickness of the deposited layer, and σ_{s0} is the electric conductivity of the initial deposited surface. This formulation allows for the initial seed layer and deposited metal to be of different materials with differing electric conductivities. Equation (18) is simply the Butler-Volmer Equation without the dependence on Cu^{2+} concentration.

The solution algorithm is still given by Figure 3 and the expressions for interface potential and deposition rate, given by Equations (10) and (11), remain unchanged.

4.3. Domain and Computational Mesh

Referring back to Figure 4, the overall domain is a 1.0 m $\times$ 0.5 m $\times$ 0.1 m box with a 0.25 m $\times$ 0.25 m $\times$ 0.1 m section removed. A fixed, uniform current density boundary condition ($\nabla \varphi \cdot \mathbf{n} = i/\kappa$) is applied along the top surface, colored in blue. Here, $\mathbf{n} = (n_x, n_y, n_z)$ is the boundary surface normal. The orange surface indicates the area over which electroplating occurs, and the bottom edge of the plated surface, colored in cyan, is grounded ($\varphi_s = 0$). All other surfaces are considered to be perfectly insulated ($\nabla \varphi \cdot \mathbf{n} = 0$). The computational domain is discretized with uniform hexahedral cells with a cell size of 1.25 cm. The total number of cells is approximately 22,400.

4.4. Parameters

The results presented in [1] included two test cases, *case-1* and *case-2*, to explore model parameter sensitivity. In this work, the proposed model is parameterized in accordance with values from *case-1* only, as the results from *case-2* do not exercise any additional physics and are qualitatively similar to the results of *case-1*. The relevant model parameters are given in Table 5. In order to account for the difference between the model presented here and the model presented in [1], namely that $F^2 n^2 D / RT = \kappa$, the inclusion of the diffusivity and bulk copper concentration of the electrolyte are included in the model parameters.

Table 5. Model parameters used in the L-shaped validation case [1]. Additional parameters are presented to account for the modifications to the model.

Parameter	Symbol	Value
Electric conductivity of electrolyte	κ	1.0 S/m
Electric conductivity of deposited metal	σ	5.95×10^7 S/m
Surface electric conductivity of initial deposited surface	σ_{s0}	59.5 S
Molar weight of deposited metal	m	63.546 g/mol
Valence of metallic ion in electrolyte	n	2
Mass density of deposited metal	ρ	8940 kg/m^3
Temperature	T	300 K
Total current	J	5 A
Exchange current density	i_0	150 A/m^2
Charge transfer coefficients	α_A and α_C	0.5, 0.5
Time step	Δt	10 s
Tolerance of potential relaxation subcycle	ϵ	1.0×10^{-6}
Copper diffusivity	D	0.67×10^{-9} m^2/s
Bulk copper concentration	c_b	6.3536 kg/m^3

4.5. Solution

Model predictions for the validation case described above are shown in Figures 5–7. The model is run for a total time of 1500 s. Qualitatively a comparison between the results of the electric potential in the bulk electrolyte can be made between the predictions from [1] and to the predictions from the proposed model, the results of the electric potential from the proposed model are shown in Figure 5, a time $t = 1500$ s. From this qualitative analysis the proposed model is in agreement with the results presented in [1].

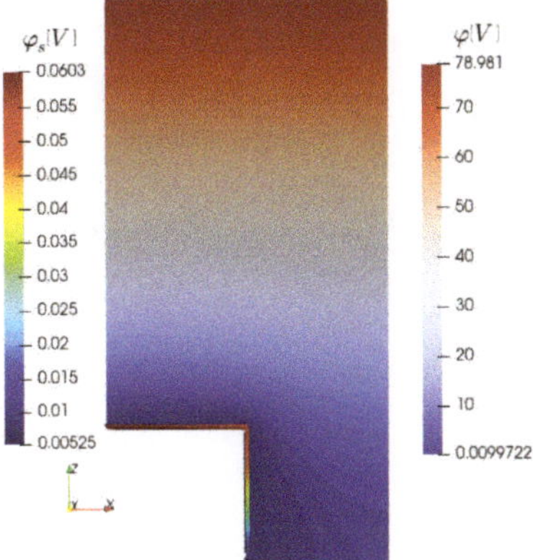

Figure 5. Contours of electric potential in the bulk electrolyte predicted by the proposed model at $t = 1500$ s. Comparisons can be made to Figure 3a in [1].

Quantitatively comparing the time distributions of the current density and the copper layer thickness from [1] the proposed model time-dependent behavior of the interface current and deposition thickness distributions are provided in Figure 6a,b. The time-dependent behaviors correspond well with the results presented in [1].

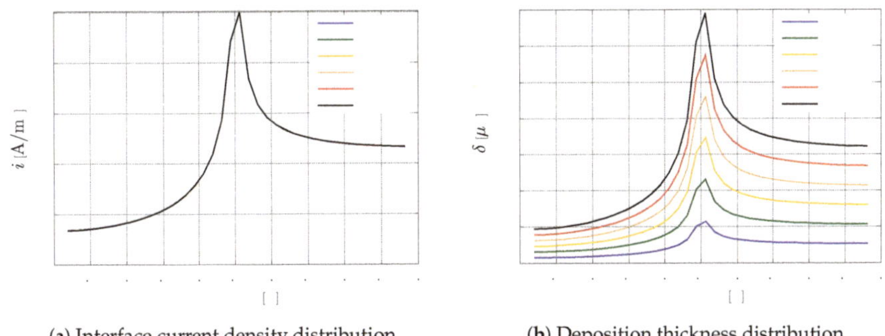

(**a**) Interface current density distribution.   (**b**) Deposition thickness distribution.

Figure 6. Time dependent behavior of the (**a**) interface current density distribution and (**b**) deposition thickness distribution along the plating surface predicted by the proposed model. Comparisons can be made to Figure 4a,b in [1], respectively.

Figure 7a,b show distributions of interface current density and deposition layer thickness on the deposition surface, also at time $t = 1500$ s. A qualitative comparison between the plots produced by this model and those presented in [1] again demonstrate good agreement.

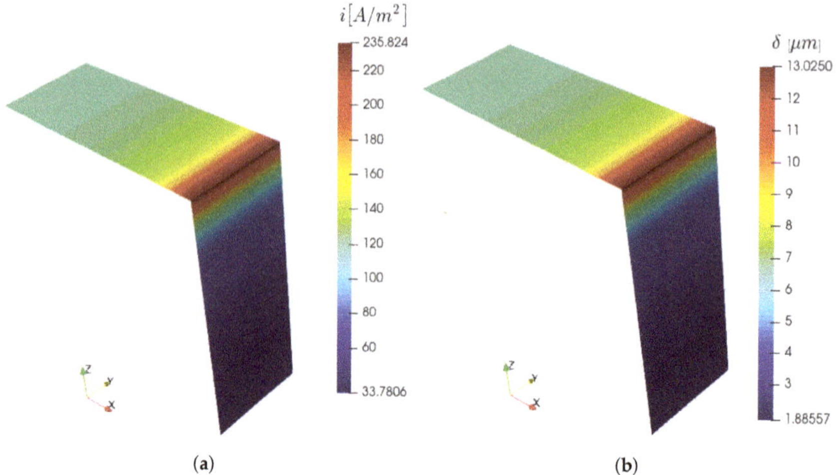

(a)   **(b)**

Figure 7. (**a**) Interface current density distribution and (**b**) Deposition thickness distribution along the plating surface predicted by the proposed model at $t = 1500$ s. Comparisons can be made to Figure 3b,c in [1], respectively.

4.6. Analysis and Discussion

Model predictions of electric potential in the bulk electrolyte, shown in Figure 5, agree qualitatively with the published results of [1]. Unfortunately, a more quantitative comparison of the electrolyte potential is not possible; however, based on the range of scales published in [1], maximum and minimum values between the two model predictions for electrolyte potential (interface potential) agree to within 0.8% (6%) and 5.5% (187%), respectively. It is important to note that the large percent difference between the minimum value of interface potential between the two models is relative to a value $<< 1$, and that small differences in meshing or solver parameters can lead to small numerical differences that will accumulate over time.

Figure 6a,b shows the interface current distribution and deposition thickness evolution over time, respectively, and provide a quantitative comparison between the two model predictions. Both models predict an interface current distribution that is constant over time, with the highest values occurring at the exposed corner of the plating surface, around a distance of 0.25 m. The proposed model predicts a peak current density of approximately 251 A/m^2, while the model from [1] predicts a value of 250 A/m^2.

Qualitative comparisons are shown in Figure 7a,b for the current density and Cu layer thickness distributions along the plating surface. Maximum and minimum values for both interface current density and Cu thickness agree with the results in [1] to within 0.2%.

Qualitatively similar results are present in the prediction of deposition thickness, as shown in Figure 6b. Both models predict increasing Cu layer thicknesses over time, with the thickest deposition occurring at the corner of the plating surface. These results are consistent with the deposition rate's linear dependence on interface current density (see Equation (11)).

The results of this code-to-code comparison provide confidence that the proposed approach can successfully model the electrochemical deposition process in cases where fluid motion is quiescent and ionic transport can be neglected. The performance of the proposed model which includes ionic transport will be discussed in the following section.

5. Ionic Transport Case

An extension case from Section 4 is presented to demonstrate the coupling of copper ionic transport to the electric potential, current density, and copper layer growth. This case uses the same problem parameters presented in Table 5. The difference here is that the branch shown in Figure 3 for ionic transport is turned on so the transport of copper ions, and depletion near the plating surface, impacts the current density and electric potential fields near the electrolyte-copper layer interface. This depletion of copper ions at the plating surface also indicates that the electroltye is no longer well-mixed.

Analysis and Discussion

Figure 8 shows the electric potential field contours at $t = 1500$ s, when copper ionic transport is included in the physical system. From this, it can be observed that there are slight changes in the surface electric potential at the corner of the plating surface, and the bulk electric potential is decreased.

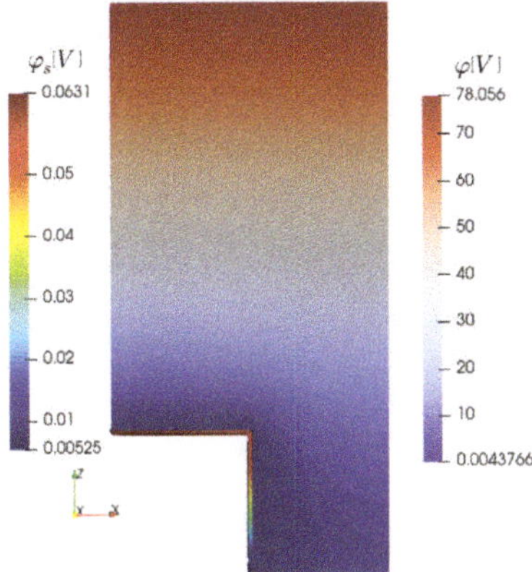

Figure 8. Contours of electric potential in the bulk electrolyte predicted by the proposed model, with copper ionic transport, at $t = 1500$ s.

This change in electric potential at the interface of the plating surface leads to a decrease of the current density over time, but overall the current density at this interface is larger (because of the copper ion concentration coupling), see Figure 9a. The main change in current density occurs where the majority of copper is deposited (see Figure 9b), which is at the corner of the plating surface.

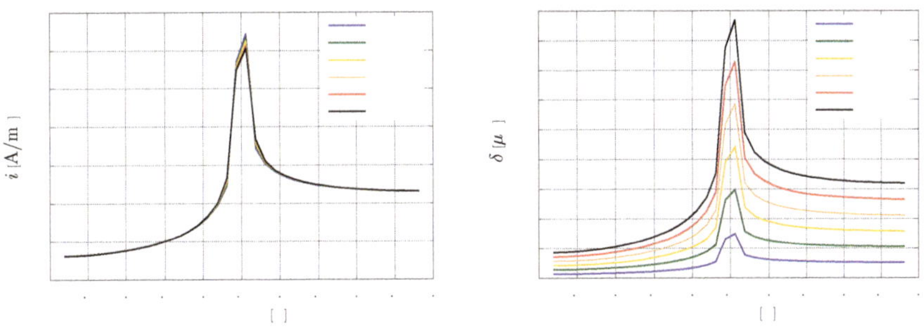

(**a**) Interface current density distribution.　　　　(**b**) Deposition thickness distribution.

Figure 9. Time dependent behavior of the (**a**) interface current density distribution and (**b**) deposition thickness distribution along the plating surface predicted by the proposed model, with copper ionic transport.

Figure 9a,b show the time dependent evolution of the current density and copper layer growth, respectively. It can be observed from these plots that the peak value of the current density, at the corner of the plating surface, decreases because of the increased resistivity due the growth of the copper layer. It is much less obvious, but the rate at which the copper layer is growing is starting to decrease, meaning that less copper is being deposited in the same amount of time. This occurs because there is less copper ions in the vicinity and because the current density has decreased. The overall current density is greater than in the case presented in Section 4, but it can be observed in Figure 10a that the current density has decreased as the simulation progressed. Figure 10b shows a similar trend to that presented in Section 4.5.

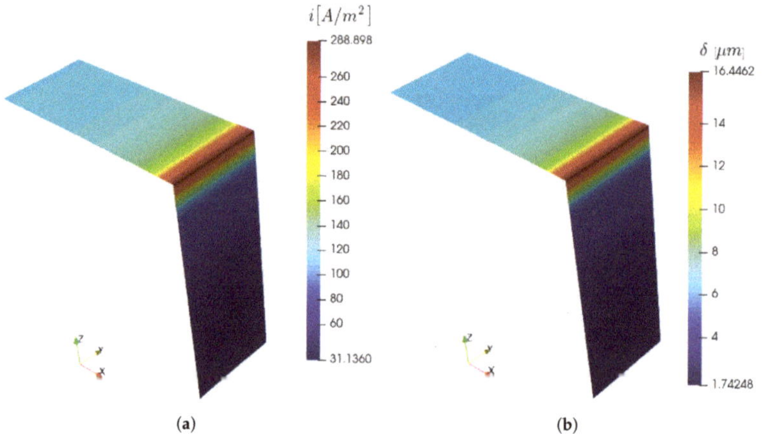

(a)   (b)

Figure 10. (**a**) Interface current density distribution and (**b**) Deposition thickness distribution along the plating surface predicted by the proposed model, with copper ionic transport, at $t = 1500$ s.

Additionally, Figure 11a,b shows the concentration surface contour along the plating surface at $t = 1500$ s and the copper ion deposition at the plating surface interface as time evolves. It can be observed in Figure 11b, unsurprisingly, that the regions of the plating surface that correspond to large deposition result in less copper concentration near the interface of the electrolyte-copper layer. It can also be observed that the growth on the corner of the plating surface results in a shadowing effect along the vertical portion of the plating surface which can be observed in the left side of Figures 9a,b, and 11b, based on the start of the sampling line.

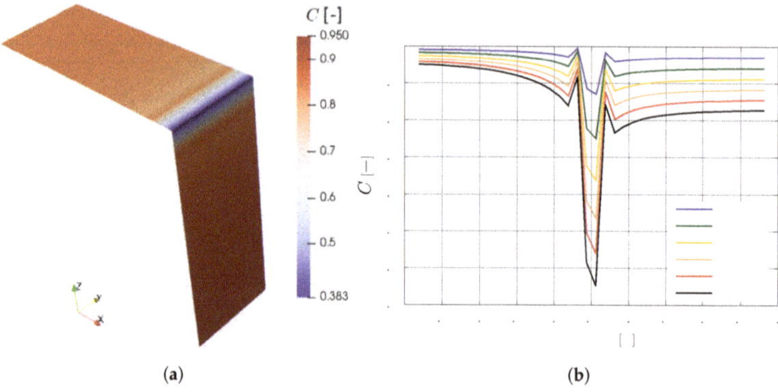

(a)   (b)

Figure 11. (**a**) Mass fraction of copper ion concentration distribution near the plating surface predicted by the proposed model, with copper ionic transport, at $t = 1500$ s. (**b**) Time dependent behavior of the mass fraction of copper ion concentration distribution near the plating surface predicted by the proposed model, with copper ionic transport.

6. Summary

This paper details the development, testing, and application of a multi-physics simulation code capable of predicting the spatio-temporal deposition of copper during an electrochemical deposition process. The details of the underlying mathematical models, several implementation details, and modeling considerations have been presented. The results of a validation study are also included, in which the proposed model is used to predict copper deposition along an L-shaped plating surface. The proposed model results presented for validation are in agreement with the results presented in [1], but the additional assumption that the electrolyte remained well mixed was too restrictive. To address this, the proposed model includes the addition of copper ion transport, which demonstrates how deposition changes as the electrolyte is depleted of copper ions near the plating surface. An extension of the validation problem, which incorporates ionic transport, has also been studied. It can be observed that the depletion of copper ions results in a larger resistivity on the plating surface (the deposited copper layer), which, in turn, decreases the current density on the plating surface, specifically at the corner (where the copper concentration is being depleted). This also slows the rate at which copper is deposited as time evolves.

As mentioned in Section 1 the overall objective is to study how turbulent mixing affects the deposition of copper, and is future work to build upon this initial effort. Mixing the electrolyte will ensure that copper ions are not depleted at the plating surface-electrolyte interface. Mixing is also believed to help with the uniformity of deposition. Incorporating either a forcing function within the fluid, or a moving mesh approach, such as the ALE method, introduces additional challenges in the numerical method.

Author Contributions: Conceptualization, E.P.; Formal analysis, J.K. and J.G.; Investigation, J.K. and J.G.; Methodology, E.P. and J.G. and J.K.; Software, J.K. and J.G.; Supervision, E.P.; Validation, J.K. and J.G.; Visualization, J.K.; Writing—original draft, J.K. and J.G. All authors have read and agreed to the published version of the manuscript.

Funding: This research was funded by SPTS Technologies Ltd.

Acknowledgments: The authors would like to thank Toby Jeffery, Ben Wharton, and Chien-Hung Lai for valuable discussions about the electrochemical deposition process and Danielle Kauffman for assisting with the creation of figures.

Conflicts of Interest: The authors declare no conflict of interest.

Abbreviations

The following abbreviations are used in this manuscript:

ECD	Electrochemical deposition
PISO	Pressure Implicit with Splitting Operator
ALE	Arbitrary Lagrangian-Eulerian
EITM	Explicit Interface Tracking Method
OpenFOAM	Open Field Operation And Manipulation
FVM	Finite Volume Method
FAM	Finite Area Method
CV	Control Volume

References

1. Takayama, T.; Yoneda, M. Implementation of a Model for Electroplating in OpenFOAM R. In Proceedings of the 8th International OpenFOAM Workshop, Jeju, Korea, 11–14 June 2013; p. 13.
2. Litrico, G.; Vieira, C.B.; Askari, E.; Proulx, P. Strongly coupled model for the prediction of the performances of an electrochemical reactor. *Chem. Eng. Sci.* **2017**, *170*, 767–776. [CrossRef]
3. Ritter, G.; McHugh, P.; Wilson, G.; Ritzdorf, T. Two-and three-dimensional numerical modeling of copper electroplating for advanced ULSI metallization. *Solid-State Electron.* **2000**, *44*, 797–807. [CrossRef]

4.   Karcz, M. From 0D to 1D modeling of tubular solid oxide fuel cell. *Energy Convers. Manag.* **2009**, *50*, 2307–2315. [CrossRef]

5.   Kendall, K.; Kendall, M. *High-Temperature Solid Oxide Fuel Cells for the 21st Century: Fundamentals, Design and Applications*; Elsevier: Amsterdam, The Netherlands, 2015.

6.   Drese, K.S. Design rules for electroforming in the LIGA process. *J. Electrochem. Soc.* **2004**, *151*, D39. [CrossRef]

7.   Hughes, M.; Bailey, C.; McManus, K. Multi physics modelling of the electrodeposition process. In Proceedings of the 2007 International Conference on Thermal, Mechanical and Multi-Physics Simulation Experiments in Microelectronics and Micro-Systems, EuroSime 2007, London, UK, 16–18 April 2007; pp. 1–8.

8.   Hughes, M.; Strussevitch, N.; Bailey, C.; McManus, K.; Kaufmann, J.; Flynn, D.; Desmulliez, M.P. Numerical algorithms for modelling electrodeposition: Tracking the deposition front under forced convection from megasonic agitation. *Int. J. Numer. Methods Fluids* **2010**, *64*, 237–268. [CrossRef]

9.   Zhu, Y.; Ma, S.; Sun, X.; Chen, J.; Miao, M.; Jin, Y. Numerical modeling and experimental verification of through silicon via (TSV) filling in presence of additives. *Microelectron. Eng.* **2014**, *117*, 8–12. [CrossRef]

10. Fukukawa, M.; Tong, L. Effect of Mass Flow Induced by a Reciprocating Paddle on Electroplating. In Proceedings of the 2017 COMSOL Conference, Boston, MA, USA, 4–6 October 2017; p. 5.

11. Oliaii, E.; Litrico, G.; Désilets, M.; Lantagne, G. Mass transport and energy consumption inside a lithium electrolysis cell. *Electrochim. Acta* **2018**, *290*, 390–403. [CrossRef]

12. Strusevich, N.; Bailey, C.; Costello, S.; Patel, M.; Desmulliez, M. Numerical modeling of the electroplating process for microvia fabrication. In Proceedings of the 2013 14th International Conference on Thermal, Mechanical and Multi-Physics Simulation and Experiments in Microelectronics and Microsystems (EuroSimE), Wroclaw, Poland, 15–17 April 2013; pp. 1–6.

13. Strusevich, N. Numerical Modelling of Electrodeposition Process for Printed Circuit Boards Manufacturing. Ph.D. Thesis, University of Greenwich, London, UK, 2013.

14. Quarteroni, A.; Sacco, R.; Saleri, F. *Numerical Mathematics*; Springer Science & Business Media: Berlin, Germany, 2010; Volume 37.

15. Rieger, P.H. *Electrochemistry*, 2nd ed.; Chapman and Hall Inc.: New York, NY, USA, 1994.

16. Jasak, H. Error Analysis and Estimation for the Finite Volume Method with Applications to Fluid Flows. Ph.D. Thesis, Imperial College of Science, Technology and Medicine, London, UK, 1996.

17. Tukovic, Z.; Jasak, H. Simulation of Free-Rising Bubble with Soluble Surfactant using Moving Mesh Finite Volume/Area Method. In Proceedings of the 6th International Conference on CFD in Oil & Gas, Trondheim, Norway, 10–12 June 2008; p. 11.

18. Greenshields, C.J. ProgrammersGuide.pdf. 2015. Available online: http://foam.sourceforge.net/docs/Guides-a4/ProgrammersGuide.pdf (accessed on 21 September 2020).

19. OpenCFD Ltd., United Kingdom. *OpenFOAM—The Open Source CFD Toolbox—User's Guide*, 2nd ed.; OpenCFD Ltd.: Bracknell, UK, 2018.

20. Ferziger, J.H.; Peric, M. *Computational Methods for Fluid Dynamics*, 3rd ed.; Springer Science & Business Media: Berlin, Germany, 2012.

21. Rauter, M.; Tuković, Ž. A finite area scheme for shallow granular flows on three-dimensional surfaces. *Comput. Fluids* **2018**, *166*, 184–199. [CrossRef]

Publisher's Note: MDPI stays neutral with regard to jurisdictional claims in published maps and institutional affiliations.

Article

Anisotropic RANS Turbulence Modeling for Wakes in an Active Ocean Environment [†]

Dylan Wall *,[‡,§] **and Eric Paterson** [‡,§]

Department of Aerospace and Ocean Engineering, Virginia Polytechnic Institute and State University, Blacksburg, VA 24061, USA; egp@vt.edu

* Correspondence: dylanjw@vt.edu; Tel.: +1-865-816-0632
† This paper is an extended version of our paper published in 15th OpenFOAM Workshop.
‡ Current address: Randolph Hall, RM 215, 460 Old Turner Street, Blacksburg, VA 24061, USA.
§ These authors contributed equally to this work.

Received: 14 November 2020; Accepted: 14 December 2020; Published: 18 December 2020

Abstract: The problem of simulating wakes in a stratified oceanic environment with active background turbulence is considered. Anisotropic RANS turbulence models are tested against laboratory and eddy-resolving models of the problem. An important aspect of our work is to acknowledge that the environment is not quiescent; therefore, additional sources are included in the models to provide a non-zero background turbulence. The RANS models are found to reproduce some key features from the eddy-resolving and laboratory descriptions of the problem. Tests using the freestream sources show the intuitive result that background turbulence causes more rapid wake growth and decay.

Keywords: stratified wakes; turbulence; RANS; stress-transport

---

1. Introduction

Oceanographic flows include a broad variety of turbulence-generating phenomena, and the associated unsteady motions are in general inhomogeneous, non-stationary, and anisotropic. The thermohaline stratification of the ocean introduces a conservative body force which must be considered when examining flows in such an environment. Numerous effects including buoyancy, shear, near-free-surface damping, bubbles, and Langmuir circulations complicate any attempt to describe turbulent motions. The variety of production mechanisms includes (but is not limited to) wind shear, wave breaking, internal gravity waves, double diffusion, and overturning due to the alternating heating and cooling of the ocean surface.

The numerical simulation of engineering-related problems in such an environment is a daunting prospect. For the case of wakes generated by ships and other man-made objects, the associated Reynolds number can be $\mathcal{O}(10^9)$ in the near field, while approaching very small Froude number in the far field. The broad separation of scales means that the use of scale-resolving methods such as large-eddy simulation (LES) or direct numerical simulation (DNS) at full scale is typically prohibitively computationally expensive for use in design and analysis problems. In previous work, we have tested a set of Reynolds stress models (RSMs) against laboratory representations of the oceanographic environment (see Wall and Paterson [1], publication pending). In this work, we then further develop the application of these models to the problem of wakes, testing against laboratory and scale-resolving model descriptions of stratified wakes. The models are then modified with source terms to produce a finite background turbulence, intended to model the environmental turbulence in the ocean.

As has been noted, one of the key complications associated with the ocean is the density stratification, which causes anisotropy in the stress tensor. In the late wake, the effects of the stratification inevitably dominate the character of turbulent motion. Models which account for buoyancy-induced stress tensor anisotropy in some way are therefore desirable for simulations of late-wake behavior, necessitating a second-moment closure. There is an extensive history of second-moment approaches in modeling geophysical problems (early examples are the hierarchy of Mellor and Yamada [2] or the summary of Rodi [3]). Similar approaches have also been adopted in dealing with stratified wakes. Algebraic closures paired with two-equation models have been used to good effect by, for example, Hassid [4], Voropaeva et al. [5], or Voropaeva et al. [6]. Such models reproduce key behaviors of stratified wakes, including the vertical collapse and production of internal waves. Full stress-transport models have also been applied to the problem, with the earliest example being Lewellen et al. [7]. More recently, Voropaeva [8] has even adopted algebraic and transport-equation based triple-moment closures.

Due to the expectation of strong stratification effects in the late wake, models which remain realizable in an approximately two-component turbulence state are desirable. The family of cubic tensorial expansion models developed at the University of Manchester were developed with the two-component limit (TCL) as an explicit constraint, and have been applied to a variety of flows with strong buoyant forcing (for example, see Craft et al. [9], Craft and Launder [10], Suga [11], or Craft et al. [12]) The so called TCL model has also been applied to doubly-stratified (simultaneous salinity and temperature stratifications) environs, as presented by Armitage [13].

In evaluating RANS closures, it is possible to employ a wealth of LES and DNS numerical experiments which have been conducted to complement previous experimental studies of stratified wakes. Temporal-model, or 3D+t simulations such as those conducted by Dommermuth et al. [14] or Brucker and Sarkar (Brucker [15], Brucker and Sarkar [16]), have helped to describe the distribution of energy within the wakes of both towed and self-propelled bodies. More recently, body-inclusive simulations such as those conducted by Chongsiripinyo and Sarkar [17] have done much to refine the understanding the scaling laws which can be applied a given stratified wake, and to qualify each stage encountered during its life. These stages were originally identified as the three-dimensional (3D), non-equilibrium (NEQ), quasi-two-dimensional (Q2D), and viscous three-dimensional (V3D) stages by Spedding [18]. It is an interesting digression to note that these stages roughly align with the stages of a full-scale ship wake as defined by Somero et al. [19] (the near wake, the far wake, and the persistent wake), though the rigorous definitions are different in each case. Some recent experiments and LES/DNS studies have also concerned themselves with the effect of non-trivial free-stream turbulence. Studies such as that of Amoura et al. [20] and Rind and Castro [21] have shown that environmental turbulent motions can have dramatic effects on wake behavior. In the case of stratified turbulence, the simulations of Radko and Lewis [22] include consideration of pre-existing double-diffusive fluctuations. The authors of that study also establish fairly simple wake-detection criteria based on the centerline deficits of dissipation rate ϵ and turbulent scalar variance ($\overline{\theta^2}$ or $\overline{s^2}$). It is fully understood that much of the physics described by these scale-resolving simulations will be lost in adopting a RANS approach; the results of these studies must then be carefully applied to refining RANS models.

Having addressed the scope of the problem, we now recapitulate a set of general criteria which must be satisfied by a turbulence model which might be applied to full-scale, far-field ship wakes in an oceanic environment:

- The model must be implementable as part of a general-use computational fluid dynamics package (in the case of this work, the finite-volume code OpenFOAM)
- The model must be able to accommodate the anisotropy that arises in stratified turbulent flows. Paramount is accounting for anisotropy in the energy-containing eddies, however, under many stratification conditions anisotropy may also arise throughout the turbulence wavenumber spectrum (see, for example, Khani and Waite [23]).

- The model must gracefully handle free-stream environmental turbulence, with differing levels of turbulent variance in both the temperature and salinity fields. Ideally, the impact of pre-existing turbulence on wake similarity would be accurately accounted for as well.
- The model must reproduce classic stratified wake experiments such those of Lin and Pao [24]
- The model must reproduce key behaviors observed using scale resolving methods, including decay rates of turbulence kinetic energy (TKE) and turbulence potential energy (TPE), and the growth (or lack thereof) of the wake in the vertical and horizontal directions.

The novelty in the work presented here is primarily in the application and testing of stress transport RANS models to the wake problem, and the addition of sustaining turbulence sources to begin addressing the issue of environmental turbulence. As such, the simulations and evaluation described for this study were conducted primarily to address the third and fifth bullet points. The simulation approach (including computational methods and initial/boundary conditions) and turbulence model closure methods employed are detailed in Section 2. The results of comparisons between the RANS model predictions and laboratory/LES models of stratified wakes are provided in Section 3, which also includes some commentary on these results. Section 4 provides some brief concluding remarks and discusses avenues for further work.

2. Simulation Methodology

The model system of equations was solved using extensions to the open-source finite-volume fluid dynamics package OpenFOAM. A "2D + t" approach was adopted, the same as that employed by, for example, Lewellen et al. [25] or Voropaeva [8]. Mean field transport and RANS model equations were solved on a two-dimensional grid, representative of a slice of fluid through which the wake progenitor has passed.

2.1. Mean Transport Equations

Momentum was transported according to the Reynolds-averaged incompressible Navier–Stokes equations under the Boussinesq approximation:

$$\frac{\partial U_i}{\partial t} + U_j \frac{\partial U_i}{\partial x_j} = -\frac{1}{\rho_0}\frac{\partial P}{\partial x_i} + \frac{\rho - \rho_0}{\rho_0} g_i + \frac{\partial}{\partial x_j}\left(v\frac{\partial U_i}{\partial x_j} - \overline{u_i u_j} \right) \tag{1}$$

$$\frac{\partial U_i}{\partial x_i} = 0 \tag{2}$$

where U_i is the mean-velocity vector, u_i is the fluctuating component of velocity, P is the mean kinematic pressure, g_i is the gravitational vector, v is the fluid viscosity, ρ is the fluid density, and $\overline{u_i u_j}$ is the Reynolds stress tensor. For the laboratory-scale simulations conducted in this work, a linear equation of state for the density ρ was deemed sufficient:

$$\frac{\rho - \rho_0}{\rho_0} = -\beta_S \left(S - S_0 \right) - \beta_\Theta \left(\Theta - \Theta_0 \right) \tag{3}$$

where the relevant scalar values are the mean temperature Θ and the mean salinity S, the 0 subscript denotes a reference value, and the expansion coefficients are defined by:

$$\beta_\Theta = -\frac{1}{\rho}\left.\frac{\partial \rho}{\partial \Theta}\right|_P, \qquad \beta_S = -\frac{1}{\rho}\left.\frac{\partial \rho}{\partial S}\right|_P \tag{4}$$

Note that all of the simulations presented in Section 3 are singly-stratified. In keeping with the methods employed for experimental study of stratified wakes, a salinity stratification was employed. The equation of state reduces to:

$$\frac{\rho - \rho_0}{\rho_0} = -\beta_S \left(S - S_0 \right) \tag{5}$$

For the remainder of the work, only the model for the transport of salinity will be provided. However, the same model can be applied as a temperature field under certain conditions. The scalar quantities are transported according to an advection–diffusion equation:

$$\frac{\partial \left(\delta S \right)}{\partial t} + U_i \frac{\partial \left(\delta S \right)}{\partial x_i} + U_i \frac{\partial S_B}{\partial x_i} = \frac{\partial}{\partial x_i} \left(\alpha_S \frac{\partial \left(\delta S \right)}{\partial x_i} - \overline{s u_i} \right) \tag{6}$$

where the total mean scalar field S is assumed to be the sum of a background S_B and a perturbation to that background δS, and $\overline{s u_i}$ is the turbulent flux of the scalar quantity. The quantities $\overline{u_i u_j}$ and $\overline{s u_i}$ are supplied by the turbulence model.

2.2. Stress/Flux Transport

In general for this work, the framework and nomenclature for second moment models laid out by Hanjalić and Launder [26] is adopted, where mean quantities are denoted by capital symbols (U, S, etc.), fluctuating quantities by lower case symbols (u, s), and averaging is denoted by an over-bar ($\overline{u_i u_j}$, $\overline{s u_i}$, $\overline{\theta^2}$). The stress tensor can be obtained by solving the associated transport equation:

$$\frac{\partial \overline{u_i u_j}}{\partial t} + U_k \frac{\partial \overline{u_i u_j}}{\partial x_k} = \mathcal{P}_{ij} + \mathcal{G}_{ij} + \Phi_{ij} - \epsilon_{ij} + \mathcal{D}_{ij} + \mathcal{P}_{ij_\infty} \tag{7}$$

where the terms on the right-hand side are the dissipation tensor ϵ_{ij}, the shear production:

$$\mathcal{P}_{ij} = - \left(\overline{u_i u_k} \frac{\partial U_j}{\partial x_k} + \overline{u_j u_k} \frac{\partial U_i}{\partial x_k} \right) \tag{8}$$

the production due to the buoyancy body force in a Boussinesq fluid:

$$\mathcal{G}_{ij} = - \left(\mathcal{F}_j g_i + \mathcal{F}_i g_j \right) \tag{9}$$

$$\mathcal{F}_i = \beta_S \overline{s u_i} + \beta_\Theta \overline{\theta u_i} \tag{10}$$

the re-distributive effects due to pressure interactions:

$$\Phi_{ij} = \frac{\overline{p}}{\rho} \overline{\left(\frac{\partial u_i}{\partial x_j} + \frac{\partial u_j}{\partial x_i} \right)} \tag{11}$$

and diffusive effects:

$$\mathcal{D}_{ij} = \frac{\partial}{\partial x_k} \left[\nu \frac{\partial \overline{u_i u_j}}{\partial x_k} - \overline{u_i u_j u_k} - \frac{\overline{p}}{\rho} \left(u_i \delta_{jk} + u_j \delta_{ik} \right) \right] \tag{12}$$

A free-stream sustaining source $\mathcal{P}_{ij_\infty}$ is also included, intended to maintain the TKE at some finite value (preferably associated with some background condition representative of the active ocean environment). The forms of the free-stream sources employed are given in Section 2.8. Similarly, the transport of the turbulent flux of a scalar quantity, such as the temperature or salinity in ocean water, was described according to the equation:

$$\frac{\partial \overline{su_i}}{\partial t} + U_k \frac{\partial \overline{su_i}}{\partial x_k} = \mathcal{P}_{si}^S + \mathcal{P}_{si}^U + \mathcal{G}_{si} + \Phi_{si} + \mathcal{D}_{si} \tag{13}$$

where the physical interpretation of each term is the same as given for (7). The source terms are:

$$\mathcal{P}_{si}^S = -\overline{u_i u_j} \frac{\partial S}{\partial x_j} \tag{14}$$

$$\mathcal{P}_{si}^U = -\overline{su_j} \frac{\partial U_i}{\partial x_j} \tag{15}$$

$$\mathcal{G}_{si} = -\beta_S g_i \overline{s^2} \tag{16}$$

The terms ϵ_{ij}, Φ_{ij}, $\mathcal{D}_{ij}$, Φ_{sj}, $\mathcal{D}_{sj}$, and the quantity $\overline{s^2}$ must be modeled in order to close the second-moment RSM. The following sections describe the closure approaches employed; the different models so constructed are summarized in Table 1.

Table 1. Summary of the model variations employed, with equation numbers. The models and implementation are described in detail in Section 2.

Model	$\overline{u_i u_j}$	ϵ	ϕ_{ij}	$\phi_{\theta i}$	ϵ_{ij}	$\mathcal{P}_{ij\infty}$
EVM1	(45), (48)	(46), (47)	N/a	N/a	N/a	N/a
RSM1	(7)	(39), (40)	(33)–(35)	(36)–(38)	(41)	(51)
RSM1a	(7)	(39), (40)	(33)–(35)	(36)–(38)	(42)	(51)
RSM1b	(7)	(39), (40)	(33)–(35)	(36)–(38)	(41)	(54)
RSM2	(7)	(39), (40)	(27)–(29)	(30)–(32)	(41)	(51)

2.3. Diffusive Process Closure

For all of the RSMs used in this work, the generalized gradient–diffusion hypothesis (GGDH) model originally proposed by Daly and Harlow [27] was employed to approximate the diffusive effects:

$$\mathcal{D}_{ij} = \frac{\partial}{\partial x_k} \left(\nu \frac{\partial \overline{u_i u_j}}{\partial x_k} - c_s \frac{k}{\epsilon} \overline{u_k u_l} \frac{\partial \overline{u_i u_j}}{\partial x_l} \right) \tag{17}$$

$$\mathcal{D}_{si} = \frac{\partial}{\partial x_k} \left(\gamma \frac{\partial \overline{\theta u_i}}{\partial x_k} - c_s \frac{k}{\epsilon} \overline{u_k u_l} \frac{\partial \overline{su_i}}{\partial x_l} \right) \tag{18}$$

Other, more complex closures for these terms have been applied stratified problems. The most pertinent example is the set of models employed by Voropaeva et al. [6], using both complex empirical algebraic expressions and even transport equations for a subset of the triple correlations. Craft and Launder [10] also recommend using transport equations which account for buoyancy effects on a subset of the triple correlations for strongly stratified flows. However, the effects described in that work were primarily associated with sharp pycnoclines, in which inhomogeneity in the triple correlations became significant. Such approaches would then likely be unnecessary for the linear-stratification environment in this work. Under certain environmental conditions, the diffusion closure may need further revision.

2.4. Pressure Strain/Scrambling Closure

In modeling the pressure strain and scrambling terms, it is common to adopt the approach of Chou [28], in which the pressure fluctuations are eliminated from (11) by taking the divergence of the transport equation for u_i and so obtaining a Poisson equation for p. The details of such a derivation are here elided. The resulting expressions for Φ_{ij} and Φ_{si} can be arranged into terms associated with different physical processes: the return to isotropy (Φ_1), isotropization of mean strain (Φ_2), isotropization of body forcing (Φ_3), and wall blocking effects (Φ_w). The wall blocking effects are neglected for this work, as the flow of interest is a free shear flow. The term (11) and analogous term from (13) can then be written as:

$$\Phi_{ij} = \Phi_{ij_1} + \Phi_{ij_2} + \Phi_{ij_3} \tag{19}$$

$$\Phi_{si} = \Phi_{si_1} + \Phi_{si_2} + \Phi_{si_3} \tag{20}$$

Typically, models make use of the stress anisotropy tensor a_{ij} and its invariants; the associated definitions are included for completeness:

$$a_{ij} = \frac{\overline{u_i u_j}}{k} - \frac{2}{3}\delta_{ij} \tag{21}$$

$$A_2 = a_{ij} a_{ji} \tag{22}$$

$$A_3 = a_{ij} a_{jk} a_{ki} \tag{23}$$

Lumley's flatness parameter is also employed by some models:

$$A = 1 - \frac{9}{8}\left(A_2 - A_3\right) \tag{24}$$

which takes the value of unity in isotropic turbulence, and the value of zero in two-component turbulence. Additionally, the symmetric (S) and an antisymmetric (W) portions of the velocity gradient tensor are employed by, for example, the cubic pressure–strain model employed by RSM1 and the Boussinesq eddy–viscosity model:

$$S_{ij} = \frac{1}{2}\left(\frac{\partial U_i}{\partial x_j} + \frac{\partial U_j}{\partial x_i}\right) \tag{25}$$

$$W_{ij} = \frac{1}{2}\left(\frac{\partial U_i}{\partial x_j} - \frac{\partial U_j}{\partial x_i}\right) \tag{26}$$

Two pressure–strain models were employed for the simulations presented in this work. The simpler was a linear model, and the other a cubic model based on the work of Craft et al. [9]. The linear model employed the return-to-isotropy model first proposed by Rotta [29], and the linear isotropization-of-production terms from Launder et al. [30]:

$$\Phi_{ij_1} = -c_1 \epsilon a_{ij}, \qquad c_1 = 1.8 \tag{27}$$

$$\Phi_{ij_2} = -c_2\left(\mathcal{P}_{ij} - \frac{1}{3}\mathcal{P}_{kk}\delta_{ij}\right), \qquad c_2 = 0.6 \tag{28}$$

$$\Phi_{ij_3} = -c_3\left(\mathcal{G}_{ij} - \frac{1}{3}\mathcal{G}_{kk}\delta_{ij}\right), \qquad c_3 = 0.6 \tag{29}$$

The accompanying model of the pressure-scrambling processes in the scalar flux equations is detailed by Gibson and Launder [31]:

$$\Phi_{si_1} = -c_{1s}\frac{\epsilon}{k}\overline{su_i}, \qquad c_{1s} = 3.5 \tag{30}$$

$$\Phi_{si_1} = - c_{2s}\mathcal{P}_{si}^{u}, \qquad c_{2s} = 0.5 \tag{31}$$

$$\Phi_{si_1} = - c_{3s}\mathcal{G}_{si}, \qquad c_{3s} = 0.5 \tag{32}$$

The cubic model was that developed at the University of Manchester, and is detailed in the book by Hanjalić and Launder [26]. The model is designed to be realizable approaching the so-called two-component limit (TCL), at which one of the normal stresses is approximately zero. The pressure–strain process models were originally described by Craft et al. [9], and have here been re-cast in terms of the symmetric and antisymmetric portions of the velocity gradient tensor:

$$\Phi_{ij_1} = - c_1 \left[a_{ij} + c_1' \left(a_{ik}a_{kj} - \frac{1}{3}A_2\delta_{ij} \right) \right] - \epsilon a_{ij}, \qquad c_1 = 3.1\,(A_2 A)^{1/2}, \; c_1' = 1.2 \tag{33}$$

$$\begin{aligned}
\Phi_{ij_2} =\,&c_2 k \left(a_{ik}S_{jk} + a_{jk}S_{ik} - \frac{2}{3}a_{kl}S_{kl}\delta_{ij} \right) + c_3 k \left(a_{ik}W_{jk} - a_{jk}W_{ik} \right) + c_4 k S_{ij} + c_5 k a_{ij}\mathcal{P}_{kk} \\
&+ c_6 k \left(a_{ik}a_{kl}S_{jl} + a_{jk}a_{kl}S_{il} - 2a_{kj}a_{li}S_{kl} - 3a_{ij}a_{kl}S_{kl} \right) + c_7 k \left(a_{ik}a_{kl}W_{jl} + a_{jk}a_{kl}W_{il} \right) \\
&+ c_8 k \left[a_{mn}^2 \left(a_{ik}W_{jk} + a_{jk}W_{ik} \right) + \frac{3}{2}a_{mi}a_{nj} \left(a_{mk}W_{nk} + a_{nk}W_{mk} \right) \right], \\
&c_2 = 0.6, \; c_3 = 0.866, \; c_4 = 0.8, \; c_5 = 0.3, \; c_6 = 0.2, \; c_7 = 0.2, \; c_8 = 1.2
\end{aligned} \tag{34}$$

$$\begin{aligned}
\Phi_{ij_3} = & - \left(\frac{3}{10} + \frac{3}{80}A_2 \right) \left(\mathcal{G}_{ij} - \frac{1}{3}\delta_{ij}\mathcal{G}_{kk} \right) + \frac{1}{6}a_{ij}\mathcal{G}_{kk} \\
& + \frac{2}{15}\mathcal{F}_m \left[g_i a_{mj} + g_j a_{mi} \right] - \frac{1}{3}g_k \left[a_{ik}\mathcal{F}_j + a_{jk}\mathcal{F}_i \right] \\
& + \frac{1}{10}\delta_{ij}g_k a_{mk}\mathcal{F}_m + \frac{1}{4}g_k a_{mk}\mathcal{F}_m a_{ij} \\
& + \frac{1}{8}g_k \left\{ \mathcal{F}_m \left(a_{ki}a_{mj} + a_{kj}a_{mi} \right) - a_{mk} \left(a_{mj}\mathcal{F}_i + a_{mi}\mathcal{F}_j \right) \right\} \\
& - \frac{3}{40} \left\{ a_{mk}\mathcal{F}_k \left(g_i a_{mj} + g_j a_{mi} \right) - \frac{2}{3}\delta_{ij}g_k a_{mk}a_{mn}\mathcal{F}_n \right\}
\end{aligned} \tag{35}$$

The pressure-scrambling process models were further developed by Craft and Launder [10], and for a doubly-stratified system (such as the temperature/salinity stratification in the ocean) by Armitage [13]:

$$\begin{aligned}
\Phi_{si_1} = & - 1.7 \left(1 + 1.2\sqrt{A_2 A} \right) r^{1/2}\frac{\epsilon}{k} \left(\overline{su_i}\,(1 + 0.6A_2) - 0.8 a_{ik}\overline{su_k} + 1.1 a_{ik}a_{kj}\overline{su_j} \right) \\
& - 0.2 A^{1/2} r k a_{ij}\frac{\partial S}{\partial x_j}
\end{aligned} \tag{36}$$

$$\begin{aligned}
\Phi_{si_2} = & \,0.8\overline{su_i}\frac{\partial U_i}{\partial x_k} - 0.2\frac{\partial U_k}{\partial x_i}\overline{su_k} + \frac{1}{6}\frac{\epsilon}{k}\overline{su_i}\frac{\mathcal{P}_{kk}}{\epsilon} \\
& - 0.4\overline{su_k}a_{il} \left(\frac{\partial U_m}{\partial x_l} + \frac{\partial U_l}{\partial x_m} \right) + 0.1\overline{su_k}a_{ik}a_{ml} \left(\frac{\partial U_k}{\partial x_l} + \frac{\partial U_l}{\partial x_k} \right) \\
& - 0.1\overline{su_k}\frac{1}{k} \left(a_{im}\mathcal{P}_{mk} + 2a_{mk}\mathcal{P}_{im} \right) + 0.15 a_{ml} \left(\frac{\partial U_k}{\partial x_l} + \frac{\partial U_l}{\partial x_k} \right) \left(a_{mk}\overline{su_i} - a_{mi}\overline{su_k} \right) \\
& - 0.05 a_{ml} \left[7a_{mk} \left(\overline{su_i}\frac{\partial U_k}{\partial x_l} + \overline{su_k}\frac{\partial U_i}{\partial x_l} \right) - \overline{su_k} \left(a_{ml}\frac{\partial U_i}{\partial x_k} + a_{mk}\frac{\partial U_i}{\partial x_l} \right) \right]
\end{aligned} \tag{37}$$

$$\Phi_{si_3} = -\frac{1}{3}\mathcal{G}_{si} - \beta_s \overline{s^2} a_{ik} g_k \tag{38}$$

Notably, the coefficients in (35), (37), and (38) are not empirical, and are determined only by the realizability constraints. The principle justification for adopting such a complex model is that, even for a wake with an initially high Re and Fr, the flow will eventually decay to the point at which the turbulent Froude number $Fr_T = \epsilon/kN$ is small. The turbulence will then be dominated by stratification effects, and so approach the two-component limit. Recent LES/DNS studies such as those by Chongsiripinyo and Sarkar ([17,32]) have lent some credence to this notion, showing that the vertical normal stress decreases much more quickly than the horizontal normal stresses.

2.5. Scale-Equation Closure

The scale-determining equation is typically the most empirical part of a given RANS turbulence model. In describing stratified flows, a number of different quantities have been used to describe the scale of turbulent motion; the various transport equations for kL, ϵ, or ω each have virtues, but are ultimately somewhat interchangeable (see Umlauf and Burchard [33]). For this work, the empirical model of the ϵ equation developed by Craft et al. [9] was adopted:

$$\frac{\partial \epsilon}{\partial t} + U_i \frac{\partial \epsilon}{\partial x_i} = \frac{\epsilon}{k}\left(\frac{1}{2}c_{\epsilon_1}\mathcal{P}_{kk} - c_{\epsilon_2}\epsilon + \frac{1}{2}c_{\epsilon_3}\mathcal{G}_{kk}\right) + \frac{\partial}{\partial x_i}\left[\left(\nu\delta_{ij} + c_\epsilon\frac{k}{\epsilon}\overline{u_iu_j}\right)\frac{\partial \epsilon}{\partial x_j}\right] + \mathcal{P}_{\epsilon_\infty} \tag{39}$$

As with the stress transport equation, a free-stream source $\mathcal{P}_{\epsilon_\infty}$ is included. For the stress transport models, the coefficients for the model ϵ transport Equation (39) were taken to be:

$$c_{\epsilon_1} = 1.0, \quad c_{\epsilon_2} = \frac{1.92}{1.0 + 0.7 A_2^{1/2}A}, \quad c_{\epsilon_3} = 1.0, \quad c_\epsilon = 0.15 \tag{40}$$

The coefficient values and parameterizations in (40) are taken as-is from the works of, for example Craft and Launder [10], the combination of coefficients has been tested against a variety of free-shear flows (see Hanjalić and Launder [26] for a thorough accounting). The model for ϵ given by (39) and (40) is designed for free-shear flows, and has been found to be ill-posed for homogeneous turbulence (see Speziale [34]. It is therefore likely to be less-suited to the far-wake than to near-wake regions. Furthermore, Pereira and Rocha [35] has noted a general deficiency in models like (39) in the case of strongly-stratified turbulence. The empirical model ϵ equation is therefore perhaps the best target for model improvement in future work. Two different models for the dissipation rate tensor were tested. The first assumes that ϵ_{ij} is isotropic:

$$\epsilon_{ij} = \frac{2}{3}\delta_{ij}\epsilon \tag{41}$$

The second is the model of Hallbäck et al. [36], and adopts a nonlinear dependence on the anisotropy of the stress tensor:

$$\epsilon_{ij} = \epsilon\left[\frac{2}{3}\delta_{ij}(1 - f_s) + \frac{\overline{u_iu_j}}{k}f_s\right] - \frac{3}{4}\epsilon\left(a_{ik}a_{jk} - \frac{1}{3}A_2\delta_{ij}\right), \quad f_s = 1 + \frac{3}{4}\left(\frac{1}{2}A_2 - \frac{2}{3}\right) \tag{42}$$

2.6. Scalar-Variance Closure

Closure of (13) also requires the variance of the scalar fluctuations, $\overline{s^2}$. Per Radko and Lewis [22], $\overline{s^2}$ is also a potentially useful quantity in its own right. A transport equation can be solved for $\overline{s^2}$:

$$\frac{\partial \overline{s^2}}{\partial t} + U_i\frac{\partial \overline{s^2}}{\partial x_i} = -2\overline{su_i}\frac{\partial U_k}{\partial x_i} - \epsilon_{ss} + \frac{\partial}{\partial x_i}\left[\left(\alpha_S\delta_{ij} + c_{ss_s}\frac{k}{\epsilon}\overline{u_iu_j}\right)\frac{\partial \overline{s^2}}{\partial x_j}\right] + \mathcal{P}_{ss_\infty} \tag{43}$$

which includes a free-stream source $\mathcal{P}_{ss_\infty}$. Dissipation of $\overline{s^2}$ was modeled using the algebraic expression of Craft et al. [9]:

$$\epsilon_{ss} = r\frac{\epsilon}{k}\overline{s^2}, \qquad r = 1.5\left(1 + \frac{\overline{su_i}\,\overline{su_i}}{k\overline{s^2}}\right) \tag{44}$$

Some preliminary simulations conducted using a transport equation for ϵ_{ss}, rather than (44), indicated that the algebraic expression was sufficient. However, if environmental conditions are expected to have significantly different scalar and mechanical time scales (fossilized turbulence, for example), this evaluation may need revision.

2.7. Eddy Viscosity Model

For the sake of comparison, the same set of wake conditions was also applied to an isotropic eddy–viscosity model. A standard k-ϵ model was employed, with the body-force effects on the scale equation modeled after the approach of Henkes et al. [37]:

$$\frac{\partial k}{\partial t} + U_i\frac{\partial k}{\partial x_i} = \mathcal{P}_k - \epsilon + \mathcal{G}_k + \frac{\partial}{\partial x_i}\left[\left(\nu + \frac{1}{\sigma_k}\nu_T\right)\frac{\partial k}{\partial x_j}\right] + \mathcal{P}_{k_\infty} \tag{45}$$

$$\frac{\partial \epsilon}{\partial t} + U_i\frac{\partial \epsilon}{\partial x_i} = \frac{\epsilon}{k}\left(c_{\epsilon_1}\mathcal{P}_k - c_{\epsilon_2}\epsilon + c_{\epsilon_3}\mathcal{G}_k\right) + \frac{\partial}{\partial x_i}\left[\left(\nu + \frac{1}{\sigma_\epsilon}\nu_T\right)\frac{\partial \epsilon}{\partial x_j}\right] + \mathcal{P}_{\epsilon_\infty} \tag{46}$$

$$c_{\epsilon_1} = 1.44, \quad c_{\epsilon_2} = 1.92, \quad c_{\epsilon_3} = \tanh\left(\frac{|U_3|}{\sqrt{U_1^2 + U_2^2}}\right)c_{\epsilon_1}, \quad \sigma_k = 1.4, \quad \sigma_\epsilon = 1.3 \tag{47}$$

As with (40), the coefficient values and parameterizations in (47) are taken and tested as given in the literature (Henkes et al. [37], in this case). The remaining modeled values are closed with the Boussinesq approximation:

$$\overline{u_i u_j} = \tfrac{2}{3}k\delta_{ij} - 2\nu_T S_{ij}, \qquad \overline{su_i} = -\nu_T\frac{\partial S}{\partial x_i} \tag{48}$$

$$\mathcal{P}_k = \overline{u_i u_j}\frac{\partial U_i}{\partial x_j}, \qquad \mathcal{G}_k = \beta_S g_i\overline{su_i} \tag{49}$$

$$\nu_T = C_\mu\frac{k^2}{\epsilon}, \qquad C_\mu = 0.09 \tag{50}$$

2.8. Environmental Turbulence Sources

Finally, in order to accommodate the existence of environmentally generated background turbulence, free-stream source terms were introduced to the turbulence quantity transport equations, as proposed by Spalart and Rumsey [38]. The terms maintain the background turbulence quantities at a specified value, and have the added benefit of improving numerical stability and convergence. For the $\overline{u_i u_j}$, the simplest approach is to introduce an isotropic source (the ∞ subscript denotes a free-stream value):

$$\mathcal{P}_{ij_\infty} = \frac{2}{3}\delta_{ij}\epsilon_\infty \tag{51}$$

The sources for the ϵ, $\overline{s^2}$ equations, respectively, are then:

$$\mathcal{P}_{\epsilon_\infty} = c_{\epsilon 2} \frac{\epsilon_\infty^2}{k_\infty} \tag{52}$$

$$\mathcal{P}_{ss_\infty} = 1.5 \frac{\epsilon_\infty}{k_\infty} \overline{s^2}_\infty \tag{53}$$

A second, nearly two-component anisotropic source, was also implemented, to test the potential impact of free-stream anisotropy:

$$\mathcal{P}_{ij_\infty} = \frac{2}{3} \begin{bmatrix} \frac{9}{10} & 0 & 0 \\ 0 & \frac{9}{10} & 0 \\ 0 & 0 & \frac{2}{10} \end{bmatrix} \epsilon_\infty \tag{54}$$

Note that, as the dissipation of the scalar flux $\overline{su_i}$ is typically included in the pressure-scrambling model, and any of the flux-vector components may potentially be negative, such source terms for the flux-transport equations were not applied. Unless otherwise stated, in the simulations conducted for this work, the free-stream values k_∞ ϵ_∞ were chosen such that $\nu_{t_\infty} = C_\mu k_\infty^2 / \epsilon_\infty \approx 0.5\nu$ For the simulations presented in this work, these forms were employed. Further study may be needed to select forms which properly account for the stratification and background anisotropy.

2.9. Simulation Approach

As indicated in the introduction, the models were implemented for the open-source finite-volume code OpenFOAM. For all of the simulations detailed here, a second-order backward numerical differentiation scheme was applied for the temporal derivatives. Second-order linear schemes were applied for all spatial derivatives, with limiters applied as necessary to ensure convergence. The momentum Equation (1) and continuity Equation (2) were coupled using the widely employed hybrid PISO-SIMPLE (PIMPLE) algorithm, modified to include a Boussinesq body force after the fashion of Issa and Oliveira [39]. Note that, under the $2D + t$ approach employed, the axial component of the velocity U_1 given by (1) is actually the velocity difference:

$$U_1 = U_s = U_{1,total} - U_B \tag{55}$$

where U_B is the propagation velocity of the wake generator. The mean velocity and turbulence kinetic energy were initialized according to idealized models of drag and self-propelled (or net-zero momentum, NZM) wakes. The wakes were initially axially symmetric, and the Reynolds stress tensor was initially isotropic. For the drag wake, the expressions for velocity and TKE, in terms of radial position r, were:

$$U_s = U_{s,CL,0} \exp\left[-\frac{1}{2}\left(\frac{r}{D}\right)^2\right] \tag{56}$$

$$k = k_{CL,0}\left[1 + 4\left(\frac{r}{D}\right)^2\right] \exp\left[-2\left(\frac{r}{D}\right)^2\right] \tag{57}$$

where the initial centerline values are dependent on the case in question. The form of (57) produces a TKE distribution roughly in line with the sphere wake measurements of Uberoi and Freymuth [40]. For the self-propelled (NZM) wake, the expressions for velocity and TKE were:

$$U_s = U_{s,CL,0}\left[1 - 2\left(\frac{r}{D}\right)^2\right] \exp\left[-2\left(\frac{r}{D}\right)^2\right] \tag{58}$$

$$k = k_{CL,0} \exp\left[-2\left(\frac{r}{D}\right)^2\right] \tag{59}$$

The form of (59) produces a TKE distribution which roughly resembles the measurements in the wake of the disk/jet wake of Naudascher [41]. The expressions for both wake types are plotted as a function of radial position in Figure 1. Note that the expression for the self-propelled wake produces a smaller total amount of TKE than that of the drag wake. For all cases, the dissipation rate ϵ was set such that the turbulent Reynolds number $Re_T = k^2/\nu\epsilon$ had a constant value of 10,000 throughout the wake. In studies employing scale-resolving methods (e.g., Dommermuth et al. [14] or Brucker and Sarkar [16]), it was found that initializing fluctuations in the scalar value did not substantially change the behavior of the wake. In applying the RSMs to the same problem, algebraic expressions were tested to initialize $\overline{su_i}$ and $\overline{s^2}$, based on assuming equilibrium for the transport equations of those quantities. However, use of the algebraic initialization was found to have little effect on the RSM predictions outside of the initial stage of wake development. As the LES studies simply initialized these quantities to zero, the scalar-associated turbulence quantities were therefore likewise initialized to zero for this study.

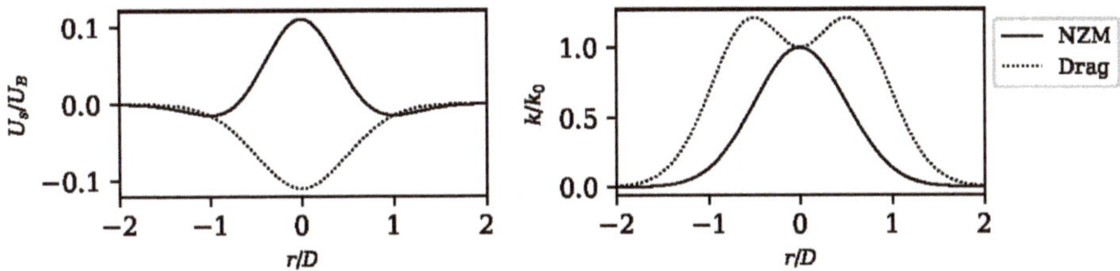

Figure 1. The initial radial distribution of axial velocity and TKE, given by Equations (56)–(59), for both a drag and a self-propelled wake.

The simulation domain consisted of a square two-dimensional grid, $120D$ in both vertical and horizontal extent. The pressure field p was given a Dirichlet boundary condition, with a fixed value of zero. The other flow variables were given mixed Dirichlet/Neumann boundary conditions, dependent on the flux of the quantity at the boundary.

As is readily seen in Equations (56)–(59), the scale of the initial mean velocity and TKE distribution is primarily set by the initial wake diameter D. The key time scales of the problem are associated with the mean velocity $((D/U_B)$, the turbulence time scale predicted by the RANS model (k/ϵ), and the oscillation period due to buoyant forcing (where the buoyancy frequency is given by $N = \sqrt{-(g/\rho_0)(\partial\rho/\partial x_3)}$).

For all of the simulations conducted in this study, the initial wake diameter was given a dimensional value of $D = 1$ m. The gravitational vector was aligned with the x_3 axis ($g_3 = -9.81$ m/s). The body propagation velocity U_B was chosen to obtain the desired Reynolds number $Re = U_B D/\nu$. The background salinity stratification $\partial S_B/\partial x_3$ was then set to provide the buoyancy frequency required to match a given internal Froude number $Fr = U_B/ND$.

3. Results and Discussion

The employed model variations and the associated equations are summarized in Table 1. RSM1 is a stress transport model employing the realizable, two-component limit pressure–strain model. The variant RSM1a employs an anisotropic model of the dissipation rate tensor ϵ_{ij}. The variant RSM1b employs an anisotropic free-stream turbulence source. RSM2 is a a stress transport model employing the simpler linear pressure–strain model, and EVM1 is the eddy–viscosity model.

The conditions simulated are summarized in Table 2. The scale-resolving simulation studies of Brucker and Sarkar [16] and Dommermuth et al. [14] were used for comparison due to use of the "temporal model", which is more analogous to the $2D + t$ approach than a body-inclusive simulation.

Table 2. Summary of the different conditions simulated, with the reference experiment or eddy-resolving simulation.

Tag	Type	$Re = \frac{U_B D}{\nu}$	$Fr = \frac{U_B}{ND}$	$100\frac{U_{s,0}}{U_B}$	$100\frac{\overline{u_{i}^2}_{CL,0}^{1/2}}{U_B}$	$100\frac{\overline{u_{i}^2}_{\infty}^{1/2}}{U_B}$	Compare With
LP	NZM	20,000	30	16	14	0	Lin and Pao [24] (LP1979)
BS1	NZM	50,000	4	11	8	0	Brucker and Sarkar [16] (BS2010)
BS1a	NZM	50,000	4	11	8	2	Brucker and Sarkar [16] (BS2010)
BS2	Drag	50,000	4	11	8	0	Brucker and Sarkar [16] (BS2010)
DOM	Drag	100,000	2	11	4.5	0	Dommermuth et al. [14] (DOM2002)

3.1. Turbulence Decay

The decay of the root-mean-square vertical-velocity fluctuations (the square-root of the vertical normal stress $\overline{u_3 u_3}$) and root-mean-square scalar fluctuations (the square-root of the scalar variance $\overline{s^2}$) along the wake centerline are depicted in Figures 2 and 3, respectively. The stress transport models achieve the correct decay rate, matching the $(Nt)^{-1}$ exponential decay measured for the experiments detailed by Lin and Pao [24]. The rate is captured for both $\overline{u_3 u_3}$ and $\overline{s^2}$. The differences between RSM1 and RSM2 are trivial for this case. The eddy–viscosity model was not paired with a transport equation for $\overline{s^2}$, and so is omitted from Figure 3.

While reproducing the decay rate, the RSMs systematically under-predict the magnitude of the scalar variance (both the peak value and the value during the decay). A partial explanation of the deficiency may be the uncertainty in the initial conditions for the problem. As noted in the previous section, the scalar-associated turbulence quantities were initially uniformly zero. While computationally convenient, this is clearly non-physical, as mixing of the scalar quantities begins at the onset of turbulence in the body boundary layer, not at a finite downstream distance.

The eddy–viscosity model (EVM1) predicts a much too rapid decay, likely indicating that the expression for the coefficient $c_{\epsilon 3}$ from Henkes et al. [37] is poorly tuned for this particular problem. The rapid extinction of turbulence quantities leads to an unrealistically high preserved wake momentum in the later stages of the wake, as will be discussed in Section 3.2.

Finally, the time-dependent behavior of the centerline value of ϵ is given for the self-propelled case under the same conditions as the LES of Brucker and Sarkar [16] as Figure 4. The RSMs predict a decay rate in keeping with the LES, while the EVM again predicts a much too rapid decay. The introduction of an anisotropic dissipation rate tensor for RSM1a does not produce a significant difference in behavior, despite the dependence of ϵ equation coefficients on the stress anisotropy for the RSMs.

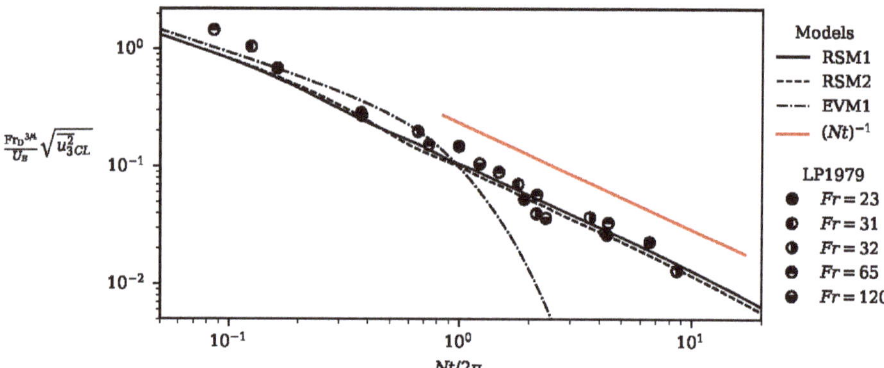

Figure 2. Time evolution of the centerline RMS fluctuating vertical velocity for $Re = 20,000$, $Fr = 30$ self-propelled stratified wake, with data from Lin and Pao [24] collected at various Froude numbers.

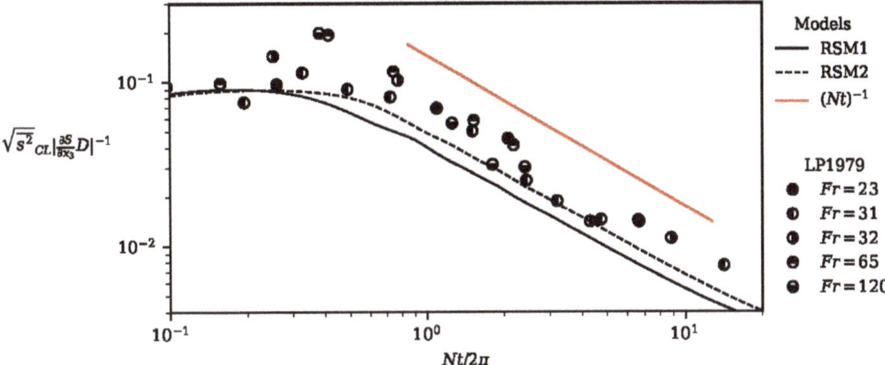

Figure 3. Time evolution of the centerline RMS fluctuating scalar for $Re = 20{,}000$, $Fr = 30$ self-propelled stratified wake, with data from Lin and Pao [24] collected at various Froude numbers.

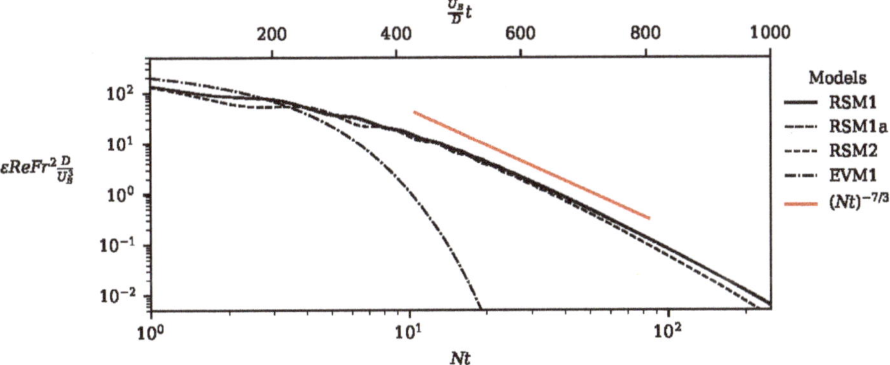

Figure 4. Time evolution of the centerline TKE dissipation rate for $Re = 50{,}000$, $Fr = 4$. The $(-7/3)$ exponential decay rate is the same as observed in the LES simulations of Brucker and Sarkar [16].

3.2. Wake Momentum Decay

Model predictions of the decay of the mean defect velocity are compared to the scale-resolving simulations of Brucker and Sarkar [16]. Figure 5 shows the comparison for the drag wake. The models all correctly predict the prolonged duration of the momentum wake due to the suppression of turbulent mixing by stratification. As noted in Section 3.1, the rapid extinction of turbulence quantities for EVM1 results in that model predicting a much larger sustained centerline defect velocity.

The RSMs in general reproduce the overall velocity decay well. The RANS models under-predict relative to the scale-resolving model in an approximate region between $Nt \approx 6$ and $Nt \approx 70$. This roughly corresponds to the NEQ region of the wake, according to the stage breakdown suggested by Spedding [18]. Further analysis is required to determine if the disagreement with the scale-resolving simulation can be explained by some physical mechanism occurring in this stage of the wake. As with the model predictions of the turbulence decay, there is not a substantial difference between RSM1 and RSM2 for this metric.

Figure 6 shows the comparison for a self-propelled (NZM) wake. The figure shows both the peak value of the momentum associated with the thrust portion of the wake (U_s^+) and the value associated with the drag portion of the wake (U_s^-). As with the drag wake, the preservation of the wake momentum to late Nt is reproduced by the RANS models. EVM1 again predicts a too-high preserved mean momentum.

The RSM predictions for both the thrust and drag portions of the wake are in fair agreement with the scale-resolving model for the NZM condition as well. The under-prediction in the NEQ stage is not present for the self-propelled case. The differences between RSM1 and RSM2 are somewhat more pronounced. RSM1 predicts a slight increase in U_s^- near $Nt = 5$; however, this ultimately puts that model's prediction more in line with the scale-resolved simulation predictions. Finally, it is notable that use of the anisotropic dissipation rate expression in model RSM1a does not produce any qualitative differences in behavior, and appears to reduce agreement with the scale-resolving simulation.

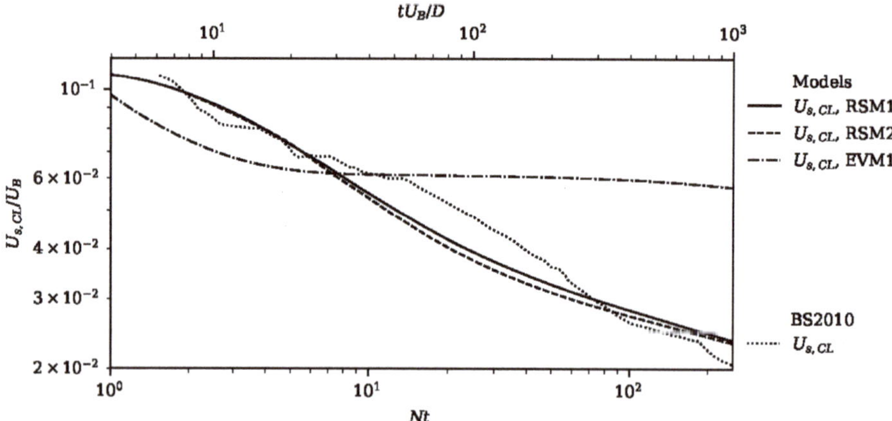

Figure 5. Time evolution of the wake velocity defect for the drag wake at $Re = 50{,}000$, $Fr = 4$. U_s^+ indicates the maximum thrust velocity, and U_s^- indicates the maximum drag velocity. With LES predictions from ?].

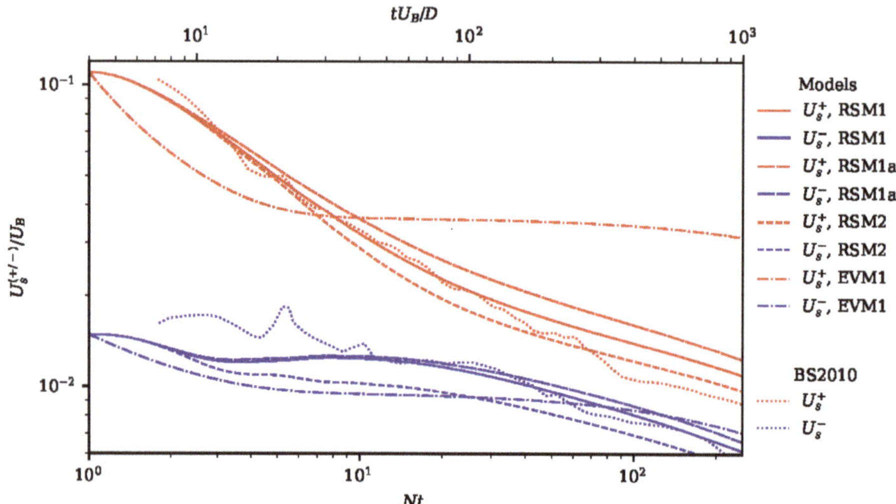

Figure 6. Time evolution of the wake velocity defect for the NZM wake at $Re = 50{,}000$, $Fr = 4$. U_s^+ indicates the maximum thrust velocity, and U_s^- indicates the maximum drag velocity. With LES predictions from Brucker and Sarkar [16].

3.3. Wake Dimensions

In evaluating model prediction of wake dimensions, the general definition of wake height/width suggested by Brucker and Sarkar [16] is adopted:

$$R_i = 2\frac{\iint U_1^2(x_i - x_i^c)^2 dx_2 dx_3}{\iint U_1^2 dx_2 dx_3}, \qquad x_i^c = \frac{\iint U_1^2 x_i dx_2 dx_3}{\iint U_1^2 dx_2 dx_3} \tag{60}$$

The integrated expression for the momentum width or height allows for direct comparison between the drag and self-propelled cases. Figure 7 shows the model predicted wake dimensions for the pure drag case. The RANS model predictions of the wake growth rate in a horizontal direction roughly agree with the scale-resolving simulation (disregarding the eddy–viscosity model). However, after approximately $Nt = 100$, the RSMs predict a slowing in horizontal growth, which is not observed in the LES results. However, there is substantial disagreement in the predictions of the vertical growth of the wake. Both the RANS and LES approaches predict a local peak in R_3 shortly after $Nt = 1$. However, the scale-resolving simulation of Brucker and Sarkar [16] predicts a wake which shrinks in the vertical axis over most of the wake's lifetime, while the RANS models predict a small but positive growth rate. The discrepancy is difficult to explain, and further study is needed to determine the cause of the qualitative difference in behavior.

Figure 8 shows the model predicted wake dimensions for the self-propelled case. The RSM models systematically under-predict the wake width of the self-propelled wake in comparison with the LES; however, the growth rate is in approximate agreement over a portion of the wake lifetime. In contrast with the drag wake, the wake height predictions agree fairly well with the LES, with both the RSMs and the LES indicating a wake which maintains a roughly constant height $R_3/D \approx 0.95$ at late Nt. The stronger agreement with LES suggests that the discrepancy seen in Figure 7 may be due to poor initial conditions in the drag wake simulations, rather than a deficiency in the RSMs themselves.

For both the drag and self-propelled cases, the differences between the predictions of RSM1, RSM1a, and RSM2 are again mostly trivial.

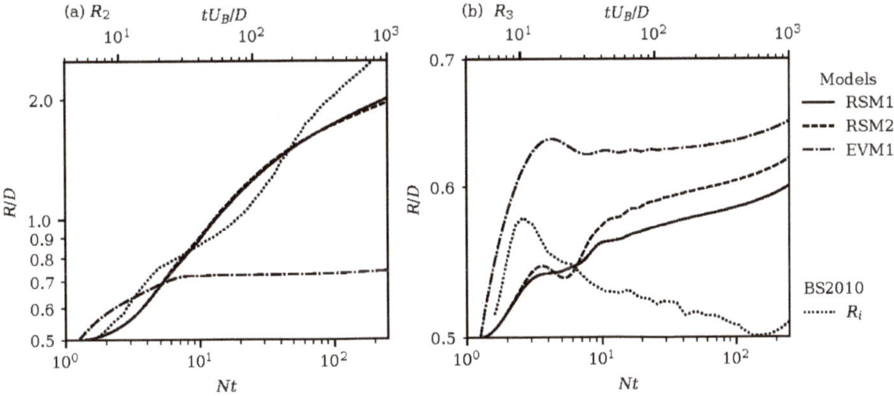

Figure 7. Time evolution of the wake width/height based on the integral of the axial momentum. For a self-propelled wake, $Re = 50{,}000$, $Fr = 4$. With LES predictions from Brucker and Sarkar [16]. (**a**) width (R_2), (**b**) height (R_3).

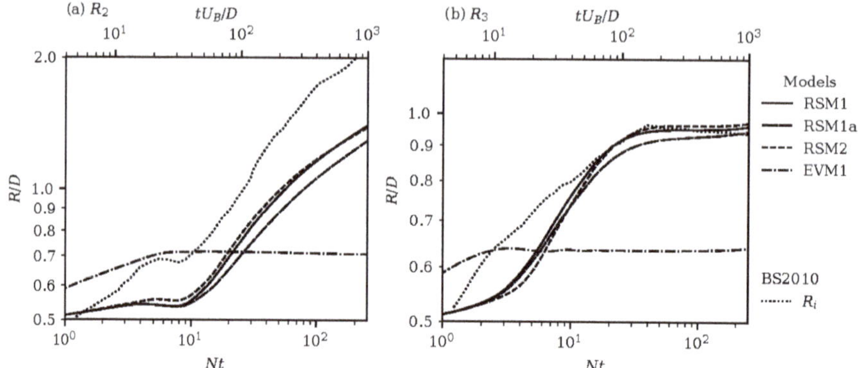

Figure 8. Time evolution of the wake width/height based on the integral of the axial momentum. For a self-propelled wake $Re = 50,000$, $Fr = 4$. With LES predictions from Brucker and Sarkar [16]. (**a**) width (R_2), (**b**) height (R_3).

3.4. Spatial Energy Distribution

The spatial distribution of the energy of a wake also be examined; again for comparison with the LES simulations of Brucker and Sarkar [16]. For the drag wake, Figure 9 shows a slice of the domain with the local mean kinetic energy (MKE), while Figure 10 supplies the same for the turbulent kinetic energy. Likewise, Figures 11 and 12 show the MKE and TKE distributions for the self-propelled wake. The predicted MKE and TKE distributions are vertically symmetric.

The drag wake MKE shows a wake which has grown in the horizontal direction, while growth in the vertical is suppressed, in keeping with the thicknesses measured in Section 3.3. Figure 10 shows that at late Nt the TKE has separated into two peaks with a saddle point on the centerline, which is broadly in agreement with the behavior predicted by the LES of Brucker and Sarkar [16]. The primary difference between RSM1 and RSM2 is a slightly larger peak TKE value for the former.

Examining Figure 11, the self-propelled wake possesses two distinct regions of mean kinetic energy. The thrust portion of the wake is still concentrated at the centerline, while the drag portion has been separated into two "lobes" roughly one diameter above and below the centerline. The distribution of MKE compares favorably with the LES of Brucker and Sarkar [16], which predicted similar lobes in similar locations. Figure 12 illustrates perhaps the most significant difference in behavior between RSM1 and RSM2 observed in this study. The former predicts a core of TKE for the self-propelled case, while the latter predicts two separate peaks (similar to the behavior of the drag wake). The prediction of RSM1 is qualitatively more similar to the distribution observed by Brucker and Sarkar [16] for the self-propelled case.

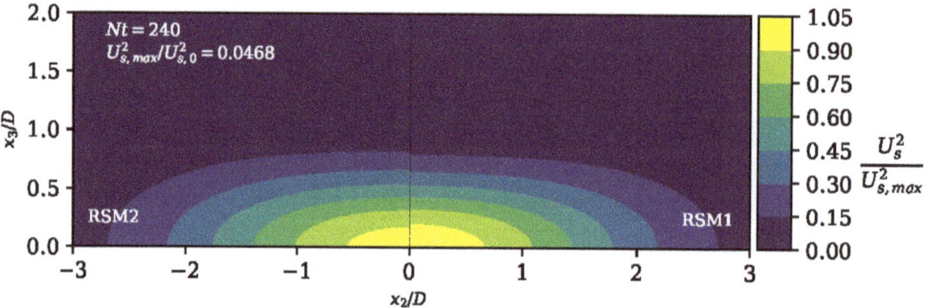

Figure 9. Distribution of wake MKE for a drag wake, $Re = 50,000$, $Fr = 4$, $Nt = 240$.

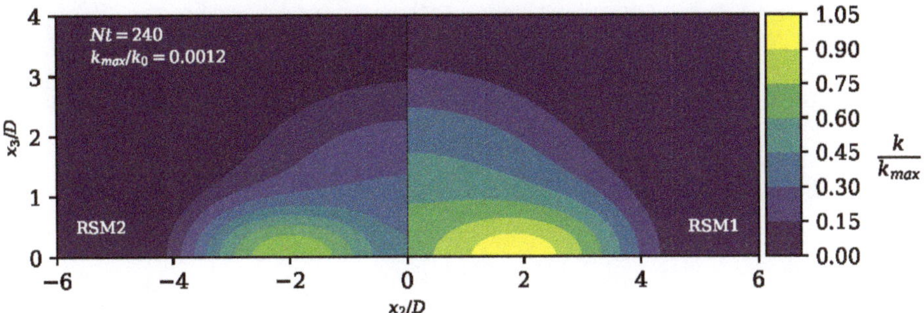

Figure 10. Distribution of wake TKE for a drag wake, $Re = 50{,}000$, $Fr = 4$, $Nt = 240$.

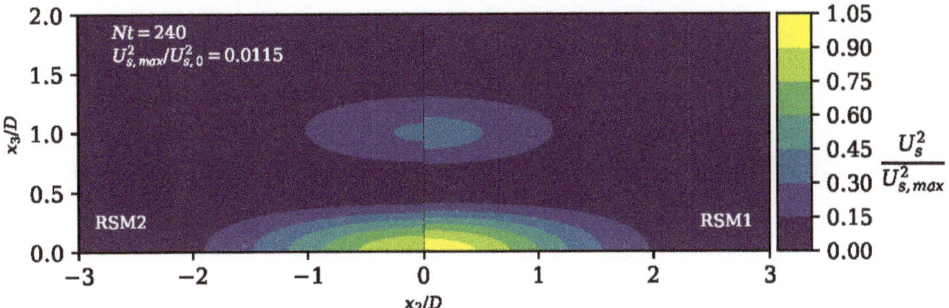

Figure 11. Distribution of wake MKE for a self-propelled wake, $Re = 50{,}000$, $Fr = 4$, $Nt = 240$.

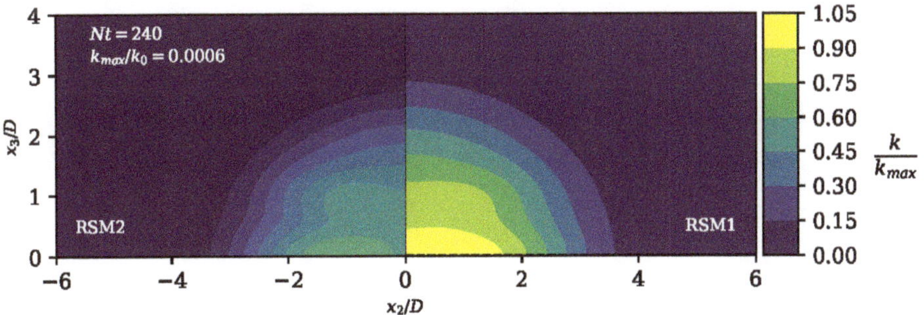

Figure 12. Distribution of wake TKE for a self-propelled wake, $Re = 50{,}000$, $Fr = 4$, $Nt = 240$.

3.5. Collapse-Induced Internal Gravity Waves

The internal gravity waves (IGWs) produced by the vertical collapse of the wake can be examined taking slices in the $(x_2 - x_3)$ and $(x_1 - x_3)$ planes (note that the x_1 direction is taken to be related to t by the body speed for the type of simulation conducted in this study, i.e., $x_1 = U_B t$). Figure 13 shows the perturbation to the salinity field (which, in this case, is equivalent to showing the density perturbation) as a function of time and vertical location. The wake produces waves which propagate upward and downward through the linear stratification, with an oscillation period roughly corresponding to the buoyancy period $2\pi/N$. Figures 14 and 15 show slices in the $x_2 - x_3$ plane, depicting waves which radiate from the wake centerline. The number of rays increases with increasing Nt, or equivalently, with downstream distance. It is important to note that both temporal model LES and $2D + t$ RANS models omit body-generated lee waves.

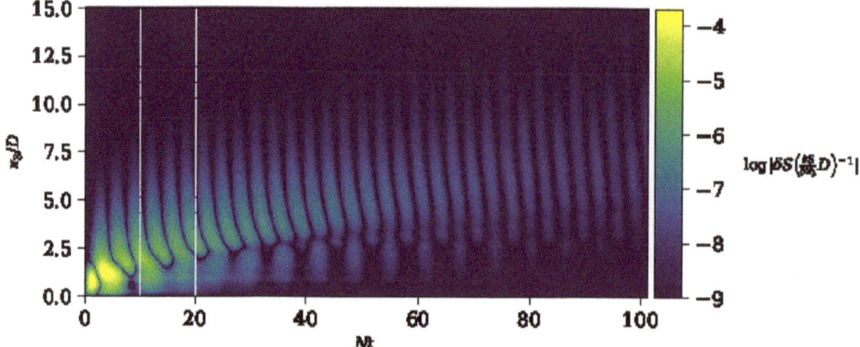

Figure 13. Model-predicted density perturbation for a self-propelled wake at $Re = 50{,}000$, $Fr = 4$, showing the collapse-generated IGWs. The vertical lines at $Nt = 10$ and $Nt = 20$ indicate the locations of the slices shown in Figures 14 and 15.

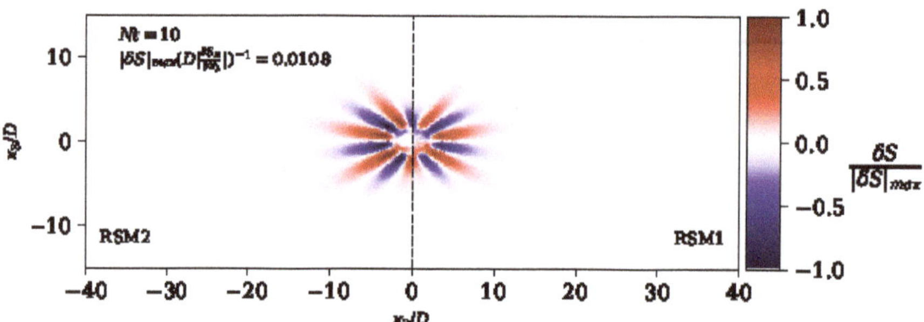

Figure 14. Model-predicted density perturbation for a self-propelled wake at $Re = 50{,}000$, $Fr = 4$, showing the collapse-generated IGWs. $Nt = 10$. There are 14 rays, spaced at roughly $25°$.

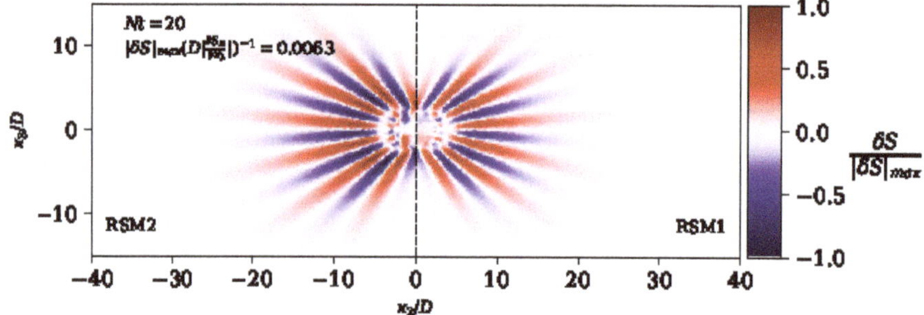

Figure 15. Model-predicted density perturbation for a self-propelled wake at $Re = 50{,}000$, $Fr = 4$, showing the collapse-generated IGWs. $Nt = 20$. There are 26 rays, spaced at roughly $11°$.

3.6. Integrated Energy Decay

Figure 16 shows the time variation of the turbulent kinetic and turbulent potential energy (TKE, TPE), integrated over the entire domain. The TKE is separated into vertical (VTKE) and horizontal (HTKE) components. The case is a drag wake under the same conditions as the LES of Dommermuth et al. [14], and the behavior depicted may be qualitatively compared with that study. The RSM predicts a decay rate roughly in line with the $(-2/3)$ exponential decay measured in the LES over a portion of the wake's lifetime. Additionally, the vertical TKE and TPE oscillate with a period roughly equally to the buoyancy period $2\pi/N$, exchanging energy as they do so. The oscillatory behavior is also in line with the behavior observed by Dommermuth et al. [14].

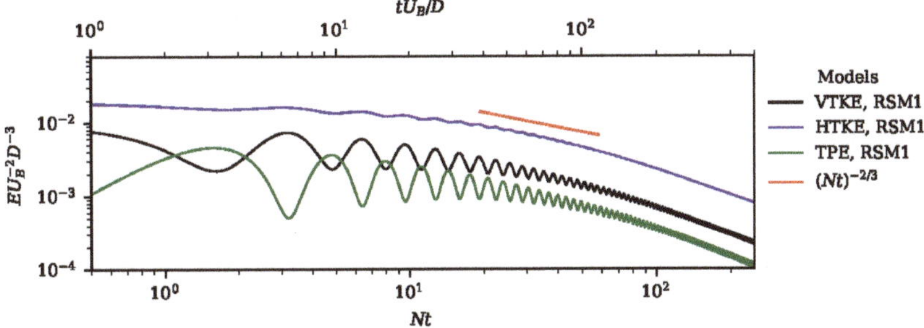

Figure 16. Model-predicted time evolution of integrated wake vertical and horizontal turbulent kinetic (VTKE, HTKE), and TPE for a drag wake at $Re = 10^5$, $Fr = 2$. The $(-2/3)$ decay is that predicted by the LES simulations of Dommermuth et al. [14].

3.7. Free-Stream Turbulence Effects

Finally, the effect of increased free-stream turbulence intensity may be briefly explored by increasing the strength of the sustaining sources included in the turbulence models. As indicated in Table 2, a second set of simulations was conducted with a high free-stream TI for the self-propelled case (case BS1a). The test was conducted for both an isotropic free-stream $\overline{u_i u_j}$ source, and the anisotropic source given by (54). Figure 17 shows the decay of the mean velocity. Strong background turbulence predictably increases the rate at which the mean velocity decays. The anisotropic source term appears to produce a stronger effect for the same free-stream TI.

Figure 18 depicts the predicted wake dimensions under the same conditions. In this case, the sustaining sources are strong enough to overcome the buoyancy effects, and the wake grows in both the vertical and horizontal direction. The growth continues past the point at which the wake height ceases growth in quiescent conditions. By late Nt, the wake turbulence has been reduced to the background value, and turbulent transport of the wake momentum is exclusively by the background turbulence.

While the predicted behavior shown in Figures 17 and 18 makes intuitive sense, further study is required to confirm the efficacy of the free-stream source approach employed.

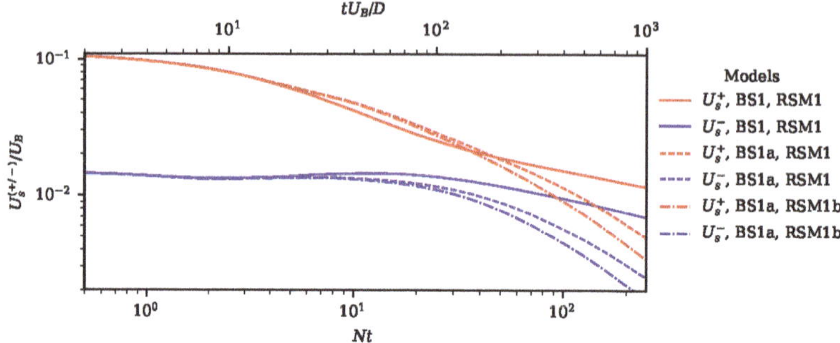

Figure 17. Time evolution of the wake velocity defect for the self-propelled wake at $Re = 50{,}000$, $Fr = 4$, with a free-stream turbulence intensity of 2%. U_s^+ indicates the maximum thrust velocity, and U_s^- indicates the maximum drag velocity.

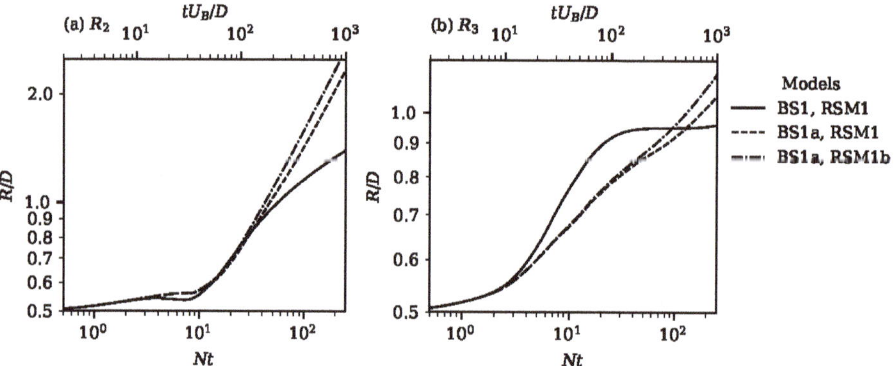

Figure 18. Time evolution of the wake dimensions for a self-propelled wake at $Re = 50{,}000$, $Fr = 4$, with a free-stream turbulence intensity of 2%.

4. Conclusions

In this work, we have demonstrated the use of a pair of anisotropic stress-transport RANS turbulence models intended to be used to simulate full-scale wakes in an active ocean environment. The models were found to reproduce a number of important stratified wake behaviors as observed in LES and laboratory studies. In particular, the models capture the decay rates of key turbulence quantities, the preservation of mean momentum to late Nt due to suppression of turbulence, and the wake collapse and internal gravity wave production. It was found that the more complex TCL pressure–strain based RSM did not differ significantly in behavior from the simpler linear model under the conditions simulated. Likewise, the use of an anisotropic dissipation-rate tensor for the stress transport equations did not substantially improve agreement with LES model predictions. Use of these more complex models required approximately 10–20% more computing times over the linear RSM for a given wake case. Further study at late Nt is needed to determine if the TCL model's additional cost is justified for low turbulence Froude numbers. Application of the models to late Nt will also likely require other modifications; the wake approaches both low turbulence Reynolds number and low Froude number conditions, likely beyond the range of validity for the models as implemented here (which were developed for high Reynolds number turbulence). Finally, the models were further modified with additional source terms to supply a nonzero background turbulence, which was found to increase the rate of wake decay and increase wake thickness growth. More tests are needed to carefully validate this capability.

Author Contributions: Methodology, software, writing—original draft preparation, D.W.; software, supervision, project administration, writing—review, and editing, E.P. All authors have read and agreed to the published version of the manuscript.

Funding: This research was funded via contract with a corporate partner.

Acknowledgments: Virginia Tech Advanced Research Computing and the Department of Defense High Performance Computing Modernization Program are acknowledged for providing supercomputing resources. The authors are also thankful to K.J. Moore, President of Cortana Corporation, for his continuing support.

Conflicts of Interest: The authors declare no conflict of interest.

References

1. Wall, D.; Paterson, E. Anisotropic RANS Turbulence Modeling for Ship Wakes in an Active Ocean Environment. In Proceedings of the 33rd Symposium on Naval Hydrodynamics, Osaka, Japan, 18–23 October 2020.
2. Mellor, G.L.; Yamada, T. A Hierarchy of Turbulence Closure Models for Planetary Boundary Layers. *J. Atmos. Sci.* **1974**, *31*, 1791–1806. [CrossRef]
3. Rodi, W. Examples of calculation methods for flow and mixing in stratified fluids. *J. Geophys. Res.* **1987**, *92*, 5305–5328. [CrossRef]
4. Hassid, S. Collapse of turbulent wakes in stably stratified media. *J. Hydronaut.* **1980**, *14*, 25–32. [CrossRef]
5. Voropaeva, O.F.; Ilyushin, B.B.; Chernykh, G.G. Numerical simulation of the far momentumless turbulent wake in a linearly stratified medium. *Dokl. Phys.* **2002**, *47*, 762–766. [CrossRef]
6. Voropaeva, O.F.; Druzhinin, O.A.; Chernykh, G.G. Numerical simulation of momentumless turbulent wake dynamics in linearly stratified medium. *J. Eng. Thermophys.* **2016**, *25*, 85–99. [CrossRef]
7. Lewellen, W.S.; Teske, M.; Donaldson, C.D. *Turbulent Wakes in a Stratified Fluid. Part 1: Model Development, Verification, and Sensitivity to Initial Conditions*; Technical Report; Defense Technical Information Center: Fort Belvoir, VA, USA, 1974. [CrossRef]
8. Voropaeva, O.F. Anisotropic turbulence decay in a far momentumless wake in a stratified medium. *Math. Model. Comput. Simul.* **2009**, *1*, 605–619. [CrossRef]
9. Craft, T.J.; Ince, N.Z.; Launder, B.E. Recent developments in second-moment closure for buoyancy-affected flows. *Dyn. Atmos. Ocean* **1996**, *23*, 99–114. [CrossRef]
10. Craft, T.J.; Launder, B.E. Application of TCL modelling to stratified flows. *Closure Strategies for Turbulent and Transitional Flows*; Cambridge University Press: Cambridge, UK, **2002**; pp. 407–423.
11. Suga, K. Predicting turbulence and heat transfer in 3-D curved ducts by near-wall second moment closures. *Int. J. Heat Mass Transf.* **2003**, *46*, 161–173. [CrossRef]
12. Craft, T.J.; Gerasimov, A.V.; Iacovides, H.; Kidger, J.W.; Launder, B.E. The negatively buoyant turbulent wall jet: Performance of alternative options in RANS modelling. *Int. J. Heat Fluid Flow* **2004**, *25*, 809–823. [CrossRef]
13. Armitage, C.A. Computational Modelling of Double-Diffusive Flows in Stratified Media. Ph.D. Thesis, University of Manchester, Manchester, UK, 2001.
14. Dommermuth, D.G.; Rottman, J.W.; Innis, G.E.; Novikov, E.A. Numerical simulation of the wake of a towed sphere in a weakly stratified fluid. *J. Fluid Mech.* **2002**, *473*, 83–101. [CrossRef]
15. Brucker, K.A. Numerical Investigation of Momentumless Wakes in Stratified Fluids. Ph.D. Thesis, UC San Diego, La Jolla, US, 2009.
16. Brucker, K.A.; Sarkar, S. A comparative study of self-propelled and towed wakes in a stratified fluid. *J. Fluid Mech.* **2010**, *652*, 373–404. [CrossRef]
17. Chongsiripinyo, K.; Sarkar, S. Decay of turbulent wakes behind a disk in homogeneous and stratified fluids. *J. Fluid Mech.* **2020**, *885*. [CrossRef]
18. Spedding, G.R. The evolution of initially turbulent bluff-body wakes at high internal Froude number. *J. Fluid Mech.* **1997**, *337*, 283–301. [CrossRef]

19. Somero, R.; Basovich, A.; Paterson, E.G. Structure and Persistence of Ship Wakes and the Role of Langmuir-Type Circulations. *J. Ship Res.* **2018**, *62*, 241–258. [CrossRef]
20. Amoura, Z.; Roig, V.; Risso, F.; Billet, A.M. Attenuation of the wake of a sphere in an intense incident turbulence with large length scales. *Phys. Fluids* **2010**, *22*, 055105. [CrossRef]
21. Rind, E.; Castro, I.P. Direct numerical simulation of axisymmetric wakes embedded in turbulence. *J. Fluid Mech.* **2012**, *710*, 482. [CrossRef]
22. Radko, T.; Lewis, D. The age of a wake. *Phys. Fluids* **2019**, *31*, 076601. [CrossRef]
23. Khani, S.; Waite, M. Buoyancy scale effects in large-eddy simulations of stratified turbulence. *J. Fluid Mech.* **2014**, *754*, 75. [CrossRef]
24. Lin, J.T.; Pao, Y.H. Wakes in stratified fluids. *Annu. Rev. Fluid Mech.* **1979**, *11*, 317–338. [CrossRef]
25. Lewellen, W.S.; Teske, M.; Donaldson, C.D. Application of turbulence model equations to axisymmetric wakes. *AIAA J.* **1974**, *12*, 620–625. [CrossRef]
26. Hanjalić, K.; Launder, B. *Modelling Turbulence in Engineering and the Environment: Second-Moment Routes to Closure*; Cambridge University Press: Cambridge, UK, 2011.
27. Daly, B.J.; Harlow, F.H. Transport Equations in Turbulence. *Phys. Fluids* **1970**, *13*, 2634–2649. [CrossRef]
28. Chou, P.Y. On velocity correlations and the solutions of the equations of turbulent fluctuation. *Q. Appl. Math.* **1945**, *3*, 38–54. [CrossRef]
29. Rotta, J. Statistische Theorie nichthomogener Turbulenz. *Z. Phys.* **1951**, *129*, 547–572. [CrossRef]
30. Launder, B.E.; Reece, G.J.; Rodi, W. Progress in the development of a Reynolds-stress turbulence closure. *J. Fluid Mech.* **1975**, *68*, 537–566. [CrossRef]
31. Gibson, M.M.; Launder, B.E. Ground effects on pressure fluctuations in the atmospheric boundary layer. *J. Fluid Mech.* **1978**, *86*, 491–511. [CrossRef]
32. Chongsiripinyo, K.; Sarkar, S. *Stratified Turbulence in Disk Wakes*; Department of Mechanical and Aerospace Engineering, University of California at San Diego, La Jolla, CA, 2019; p. 6.
33. Umlauf, L.; Burchard, H. A generic length-scale equation for geophysical turbulence models. *J. Mar. Res.* **2003**, *61*, 235–265. [CrossRef]
34. Speziale, C.G. Turbulence modeling: Present and future Comment 2. In *Whither Turbulence? Turbulence at the Crossroads*; Lecture Notes in Physics; Lumley, J.L., Ed.; Springer: Berlin/Heidelberg, Germany, 1990; pp. 490–512._64. [CrossRef]
35. Pereira, J.C.F.; Rocha, J.M.P. Prediction of stably stratified homogeneous shear flows with second-order turbulence models. *Fluid Dyn. Res.* **2010**, *42*, 045509. [CrossRef]
36. Hallbäck, M.; Groth, J.; Johansson, A.V. An algebraic model for nonisotropic turbulent dissipation rate in Reynolds stress closures. *Phys. Fluids A* **1990**, *2*, 1859–1866. [CrossRef]
37. Henkes, R.A.W.M.; Van Der Vlugt, F.F.; Hoogendoorn, C.J. Natural-convection flow in a square cavity calculated with low-Reynolds-number turbulence models. *Int. J. Heat Mass Transf.* **1991**, *34*, 377–388. [CrossRef]
38. Spalart, P.R.; Rumsey, C.L. Effective Inflow Conditions for Turbulence Models in Aerodynamic Calculations. *AIAA J.* **2007**, *45*, 2544–2553. [CrossRef]
39. Issa, R.I.; Oliveira, P.J. An Improved Piso Algorithm for the Computation of Buoyancy-Driven Flows. *Numer. Heat Transf. Part B* **2001**, *40*, 473–493. [CrossRef]
40. Uberoi, M.S.; Freymuth, P. Turbulent Energy Balance and Spectra of the Axisymmetric Wake. *Phys. Fluids* **1970**, *13*, 2205–2210. doi:10.1063/1.1693225. [CrossRef]
41. Naudascher, E. Flow in the wake of self-propelled bodies and related sources of turbulence. *J. Fluid Mech.* **1965**, *22*, 625–656. [CrossRef]

Publisher's Note: MDPI stays neutral with regard to jurisdictional claims in published maps and institutional affiliations.

 *fluids*

Article

Multi-Scale Localized Perturbation Method in OpenFOAM

Erik Higgins *, Jonathan Pitt and Eric Paterson

Department of Aerospace and Ocean Engineering, Virginia Polytechnic Institute and State University, Blacksburg, VA 24061, USA; jspitt@vt.edu (J.P.); egp@vt.edu (E.P.)
* Correspondence: erikth1@vt.edu

Received: 9 October 2020; Accepted: 15 December 2020; Published: 19 December 2020

Abstract: A modified set of governing differential equations for geophysical fluid flows is derived. All of the simulation fields are decomposed into a nominal large-scale background state and a small-scale perturbation from this background, and the new system is closed by the assumption that the perturbation is one-way coupled to the background. The decomposition method, termed the multi-scale localized perturbation method (MSLPM), is then applied to the governing equations of stratified fluid flows, implemented in OpenFOAM, and exercised in order to simulate the interaction of a vertically-varying background shear flow with an axisymmetric perturbation in a turbulent ocean environment. The results demonstrate that the MSLPM can be useful in visualizing the evolution of a perturbation within a complex background while retaining the complex physics that are associated with the original governing equations. The simulation setup may also be simplified under the MSLPM framework. Further applications of the MSLPM, especially to multi-scale simulations that encompass a large range of spatial and temporal scales, may be beneficial for researchers.

Keywords: multi-scale simulation; OpenFOAM; geophysical fluids; ocean physics

---

1. Introduction

Advancements in computational methods and hardware have enabled researchers and engineers to use computers in order to understand and model increasingly-complex phenomena. High-resolution, large-eddy simulations of vehicle bodies are being performed at increasingly-high grid resolution with hundreds of millions of cells, allowing for simulations to better approach the scales of experimental studies [1]. More complex computational models are also being developed, including those that utilize multi-physics modeling. Researchers now have the ability to calculate, for example, not just the hydrodynamics of the Earth's oceans over time, but also the distribution of temperature, salinity, mixed-layer depth, and sea-ice thickness, and then compare these quantities across different software and models [2].

The time and space resolution of a computational simulation limits the time- and length-scales, which may be resolved by any given simulation. When a wide range of scales must be resolved, computational resource limitations require that concessions be made in which scales receive adequate resolution, an example of this being turbulence modeling for modeling the smallest necessary scales. Multi-scale modeling and simulation often forgoes attempting to resolve all desired scales within a single simulation, and instead "nests" simulations of decreasing scale and increasing resolution within each other. Information can then be propagated between simulations, either from the large scales to the small scales or in both directions. This approach has seen success in weather simulations and simulations of wind around buildings [3,4]. A great degree of scale disparity is not necessarily required for multi-scale modeling and simulation. The Urban Multi-scale Environmental Predictor, for example, has models for a range of city-scale down to street-scale phenomena [5]. A novel application of multi-scale modeling include the coupling of a large-scale 1D pipe-flow simulation with a small-scale 3D fire-in-pipe simulation [6]. Groen et al. reviewed other applications of multi-scale modeling in science [7].

Perturbation methods are popular within science and engineering, due to their utility. Given a "base" solution to a differential equation, one can derive additional correction terms that are multiplied by a small perturbation parameter that accounts for a perturbation in the system [8]. This method assumes that the perturbation to a term is small when compared to the original term or other terms in the system. A variant of the perturbation method approach that is seen in fluid mechanics is the acoustic perturbation equations, where it is known *a priori* that sound waves are very small in amplitude when compared to a general incompressible flow, in which they propagate [9,10]. Such an approach produces additional source terms that are related to the evolution of the incompressible mean flow, which allows for the propagation of sound waves under their own dynamics as well as the influence of the incompressible mean flow.

The aim of this work is to simulate the dynamics of a perturbation to a geophysical flow while using modified forms of the governing equations of stratified fluid dynamics. The ability to study the evolution of a perturbation within a nominal "background" may be useful to geophysical flow studies, particularly environmental engineering or climatological ones. This approach, which is called the multi-scale localized perturbation method (MSLPM), is implemented in OpenFOAM, an open-source CFD software suite programmed in C++ [11]. The high-level language that OpenFOAM uses to build its solvers closely mirrors the mathematical expressions that one would find in the definition of a model. Its open-source nature also enables users to freely modify solvers for their own purposes and speed the development of new computational models. OpenFOAM has been used in order to successfully simulate incompressible flows under a variety of turbulence models [12–15].

2. Materials and Methods

The MSLPM is derived from the idea of scale separation between background fields and corresponding perturbations from the background. Beginning with a partial differential equation operating on an arbitrary simulation variable ϕ, such that

$$PDE\left(\phi\right) = 0 \tag{1}$$

One can then decompose ϕ into a background component ϕ_b and a perturbation component $\delta\phi$. Within this paper, the subscript b is added to a variable in order to indicate that it is a background field, and a prefix δ is added to a variable to indicate that a perturbation field. It is assumed that both background and perturbation components may vary in time and space.

$$\phi = \phi_b + \delta\phi \tag{2}$$

This substitution may be freely made into Equation (1), as it does not change the nature of the governing PDE. However, we may expand the governing PDE, as shown in Equation (3), where the original PDE in terms of background and perturbation components can be written as a sum of the original PDE in terms of background fields only, the original PDE in terms of perturbation fields only, and then the non-linear interaction between the background and perturbation fields. The latter terms, which are denoted by $NL\left(\phi_b, \delta\phi\right)$, arise due to any non-linearity in the governing PDE and they are assumed to go to zero if either the background or perturbation is identically zero. This process is shown for an example partial differential equation in Appendix A.

$$PDE\left(\phi_b + \delta\phi\right) = PDE\left(\phi_b\right) + PDE\left(\delta\phi\right) + NL\left(\phi_b, \delta\phi\right) = 0 \tag{3}$$

Closing the above equation now remains, and this is accomplished through assumptions regarding the evolution of the background and perturbation components. Geophysical flows exhibit a great degree of scale separation; an atmospheric front may evolve on the scale of days to weeks and span continents, while the wake of a wind turbine may evolve over the span of minutes to hours and over meters to kilometers. If it is assumed that the phenomena that are associated with the background are large-scale, long-timescale in nature, and that phenomena associated with the perturbation are much smaller in scale than those in the background, then one may assume that the perturbation will diffuse before it can impart any significant changes upon the background. In this way, the background can be assumed to evolve independently from the perturbation, due to a separation of scales. Mathematically, this is equivalent to saying that the background evolves, as if $\delta\phi \to 0$ everywhere for all time, and, therefore, Equation (3) becomes $PDE(\phi_b) = 0$. With this second equation, a set of governing equations forms:

$$PDE\left(\phi_b\right) = 0$$
$$PDE\left(\phi_b\right) + PDE\left(\delta\phi\right) + NL\left(\phi_b, \delta\phi\right) = 0$$

This simplifies to

$$PDE\left(\phi_b\right) = 0 \tag{4}$$
$$PDE\left(\delta\phi\right) + NL\left(\phi_b, \delta\phi\right) = 0 \tag{5}$$

The outcome of this separation of background and perturbation components is a set of governing equations that are derived from the original governing equation. This set of equations is one-way coupled, which means that the second equation depends on the solution of the first for the current timestep, but the first does not depend on the solution of the second. Equation (4) describes the evolution of the background as a function of background field only and it has the same form as Equation (1). Equation (5) describes the evolution of the perturbation field as a function the perturbation field as well as any non-linear interaction between the background and perturbation. The effect of this interaction on the background is assumed to be negligible, but it is potentially significant for the perturbation.

This assumption lends itself to many geophysical flows. For example, one might be able to determine the average wind velocity and turbulence profile for a given field of land. If a wind turbine was built upon this field, the average wind velocity and turbulence profiles were measured again, then a wake would be found, where the velocity immediately downstream of the turbine is lower and turbulence immediately downstream of the turbine is higher, but, outside of this wake, the average wind and turbulence profiles would not experience a great change. This is because the wind and turbulence at these locations is far more dependent on other factors than they are on the presence and operation of the wind turbine, and this holds even far downstream of the wind turbine when the wake has effectively dissipated. In the notation of the MSLPM, the wind velocity and turbulence profiles in the case without the wind turbine could be considered as the background conditions, and the deviations from these conditions that the wind turbine introduces are the perturbation to the background. The perturbations are localized to the vicinity of the wind turbine and they do not have an appreciable impact on the evolution of the background far away from the wind turbine when compared to factors, such as diurnal forcing and topography, including upstream mountain ranges. Therefore, we might say that the background evolution can be assumed to be independent of the perturbation.

The assumption of the evolution of the background independent of the perturbation may not hold for perturbations, which would, upon interacting with the background, affect the evolution of the background to the point where it evolves on the same timescale as the perturbation. An example of this may be the case where the background is marginally stable and the perturbation is enough to destabilize a large portion of the background. In the MSLPM, this perturbation-on-background interaction would not be resolved, and the background would continue to evolve as if the perturbation was not there, which is not reflective of the actual physics.

The derivation that is shown above is not explicitly restricted to stratified fluid flows, so it is possible for it to be applied to other sets of governing equations given that the aforementioned assumptions are also valid for the new governing equations. A more thorough derivation of the above equations may be seen in [16]. It should be noted that the perturbation that was introduced into this system can be made distinct from the background due to temporal- and spatial-scale separation, unlike traditional perturbation methods that assume the perturbation is small relative to the other terms in the governing equation.

Several different approaches may be taken in order to solve Equations (4) and (5). One may apply a time-marching scheme in order to solve Equation (4) for the background fields for each timestep, and then use this updated information to solve Equation (5) for the perturbation fields. For one-way coupled multi-scale simulations, in which a large-scale simulation is performed beforehand, one may also spatially and temporally interpolate the fields from the large-scale simulation onto the small-scale simulation as the background fields, and then simulate a perturbation on the small-scale domain. This method, as roughly shown in Figure 1, is reminiscent to the nested simulations that are seen in multi-scale simulations at present, and may be beneficial for these types of simulations. In this way, only Equation (5) needs to be solved. Another possible method is to prescribe the background fields from an analytic or empirical expression. When assuming the timescale of the simulation is very small when compared to the timescale of the background, one may also treat the background fields as constant in time and, therefore, treat them as a function of space only.

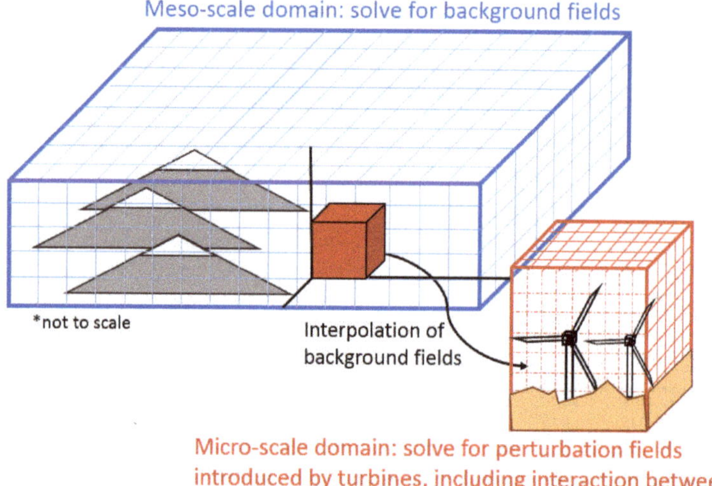

Figure 1. Example of an application of multi-scale localized perturbation method (MSLPM) using nested domains. Here, one could simulate meso-scale wind pattern, including the affects of a nearby mountain range, and then perform a micro-scale simulation of a wind turbine site localized within the meso-scale domain.

2.1. Governing Equations

The MSLPM is applied to the governing equations of stratified turbulent flow in order to determine the equations of motion for the background fields and perturbation fields. Beginning with the incompressible unsteady Reynolds-averaged Navier–Stokes equations and a scalar transport equation for temperature T, as given in Equations (6)–(8), below. Here, $\vec{U}$ is the velocity vector; ν, kinematic viscosity; $\overline{R}$, Reynolds stress tensor; p, deviation from the hydrostatic pressure; ρ_0, constant reference density; ρ, density of the fluid; $\vec{g}$, gravity vector; and, κ, molecular diffusivity of temperature. An eddy diffusivity model, such that turbulent diffusivity $\kappa_t \propto \nu_t$, where ν_t is the eddy viscosity, is used in order to close the turbulent fluxes for T. The divergence of the Reynolds stress tensor accounts for the average affect of turbulent fluctuations on the mean flow.

$$\nabla \cdot \vec{U} = 0 \tag{6}$$

$$\frac{\partial \vec{U}}{\partial t} + \vec{U} \cdot \nabla \vec{U} - \nu \nabla^2 \vec{U} = -\frac{1}{\rho_0} \nabla p - \nabla \cdot \overline{R} + \frac{\rho - \rho_0}{\rho_0} \vec{g} \tag{7}$$

$$\frac{\partial T}{\partial t} + \vec{U} \cdot \nabla T - \nabla \cdot ([\kappa + \kappa_t] \nabla T) = 0 \tag{8}$$

The total fields are then decomposed into their background and perturbation components following the MSLPM, which results in two sets of equations, with the first set describing the motion of the background fields given in Equations (9)–(11) and the second set describing the motion of the perturbation fields, including any non-linear interaction between the background and perturbation fields, in Equations (12)–(14). A linear equation of state, as defined in Equations (15)–(17), is used, where the thermal expansion coefficient of seawater β is taken as -2.1×10^{-4} °C/m. Equation (18) gives the Brunt–Väisälä frequency N in terms of both density and temperature.

$$\nabla \cdot \vec{U}_b = 0 \tag{9}$$

$$\frac{\partial \vec{U}_b}{\partial t} + \vec{U}_b \cdot \nabla \vec{U}_b - \nu \nabla^2 \vec{U}_b = -\frac{1}{\rho_0} \nabla p_b - \nabla \cdot \overline{R}_b + \frac{\rho_b - \rho_0}{\rho_0} \vec{g} \tag{10}$$

$$\frac{\partial T_b}{\partial t} + \vec{U}_b \cdot \nabla T_b - \nabla \cdot ([\kappa + \kappa_{tb}] \nabla T_b) = 0 \tag{11}$$

$$\nabla \cdot \delta\vec{U} = 0 \tag{12}$$

$$\frac{\partial \delta\vec{U}}{\partial t} + \left(\vec{U}_b + \delta\vec{U}\right) \cdot \nabla \delta\vec{U} - \nu \nabla^2 \delta\vec{U} = -\frac{1}{\rho_0} \nabla \delta p - \nabla \cdot \delta\overline{R} + \frac{\delta\rho}{\rho_0} \vec{g} - \delta\vec{U} \cdot \nabla \vec{U}_b \tag{13}$$

$$\frac{\partial \delta T}{\partial t} + \left(\vec{U}_b + \delta\vec{U}\right) \cdot \nabla \delta T - \nabla \cdot ([\kappa + \kappa_{tb} + \delta\kappa_t] \nabla \delta T) = -\delta\vec{U} \cdot \nabla T_b + \nabla \cdot (\delta\kappa_t \nabla T_b) \tag{14}$$

$$\rho_b(T_b) = \rho_0 + \rho_0 \beta (T_b - T_{ref}) \tag{15}$$

$$\rho(T) = \rho_0 + \rho_0 \beta (T - T_{ref}) \tag{16}$$

$$\delta\rho(\delta T) = \rho(T) - \rho_b(T_b) = \beta \delta T \tag{17}$$

$$N^2 = \frac{g}{\rho_0} \frac{\partial \rho}{\partial z} = g\beta \frac{\partial T}{\partial z} \tag{18}$$

The buoyant $k - \epsilon$ turbulence model is used in order to close the turbulent momentum fluxes [17]. This model is an extension to the popular $k - \epsilon$ model, and it accounts for the effects of buoyancy through a source term in both the k and ϵ equations [18]. While this model is not the most realistic for geophysical turbulence, it was chosen for its speed in the case presented here. The model equations are given in total form in Equations (19)–(21), with their equivalents for the background turbulence fields in Equations (22)–(24), and the governing equations for the perturbation turbulence variables in Equations (25)–(27). The following model coefficients are used: $C_\mu = 0.09$, $C_{1\epsilon} = 1.44$, $C_{2\epsilon} = 1.92$, $C_{3\epsilon} = 0.1$, $\sigma_k = 1$, $\sigma_\epsilon = 1.3$, and $\sigma_t = 1$.

$$\frac{\partial k}{\partial t} + \vec{U} \cdot \nabla k - \nabla \cdot \left(\frac{\nu_t}{\sigma_k} \nabla k \right) = P - G - \epsilon \tag{19}$$

$$\frac{\partial \epsilon}{\partial t} + \vec{U} \cdot \nabla \epsilon - \nabla \cdot \left(\frac{\nu_t}{\sigma_\epsilon} \nabla \epsilon \right) = \frac{\epsilon}{k} \left(C_{1\epsilon} P - C_{3\epsilon} G \right) - C_{2\epsilon} \frac{\epsilon^2}{k} \tag{20}$$

$$\nu_t = C_\mu \frac{k^2}{\epsilon} \tag{21}$$

where

$$P \equiv \nu_t \left(\nabla \vec{U} + \nabla \vec{U}^T \right) : \nabla \vec{U}$$

$$G \equiv \frac{\nu_t}{\sigma_t} N^2$$

$$\frac{\partial k_b}{\partial t} + \vec{U}_b \cdot \nabla k_b - \nabla \cdot \left(\frac{\nu_{tb}}{\sigma_k} \nabla k_b \right) = P_b - G_b - \epsilon_b \tag{22}$$

$$\frac{\partial \epsilon_b}{\partial t} + \vec{U}_b \cdot \nabla \epsilon_b - \nabla \cdot \left(\frac{\nu_{tb}}{\sigma_\epsilon} \nabla \epsilon_b \right) = \frac{\epsilon_b}{k_b} \left(C_{1\epsilon} P_b - C_{3\epsilon} G_b \right) - C_{2\epsilon} \frac{\epsilon_b^2}{k_b} \tag{23}$$

$$\nu_{tb} = C_\mu \frac{k_b^2}{\epsilon_b} \tag{24}$$

$$\frac{\partial \delta k}{\partial t} + \left(\vec{U}_b + \delta \vec{U} \right) \cdot \nabla \delta k - \nabla \cdot \left(\frac{\nu_{tb} + \delta \nu_t}{\sigma_k} \nabla \delta k \right) = \delta P - \delta G - \delta \epsilon - \delta \vec{U} \cdot \nabla k_b + \nabla \cdot \left(\frac{\delta \nu_t}{\sigma_k} \nabla k_b \right) \tag{25}$$

$$\frac{\partial \delta \epsilon}{\partial t} + \left(\vec{U}_b + \delta \vec{U} \right) \cdot \nabla \delta \epsilon - \nabla \cdot \left(\frac{\nu_{t_b} + \delta \nu_t}{\sigma_\epsilon} \nabla \delta \epsilon \right) = \frac{\epsilon}{k} \left(C_{1\epsilon} P - C_{3\epsilon} G \right) \dots$$

$$\dots - \frac{\epsilon_b}{k_b} \left(C_{1\epsilon} P_b - C_{3\epsilon} G_b \right) - C_{2\epsilon} \left(\frac{\epsilon^2}{k} - \frac{\epsilon_b^2}{k_b} \right) - \delta \vec{U} \cdot \nabla \epsilon_b + \nabla \cdot \left(\frac{\delta \nu_t}{\sigma_\epsilon} \nabla \epsilon_b \right) \tag{26}$$

$$\delta \nu_t = C_\mu \left(\frac{k^2}{\epsilon} - \frac{k_b^2}{\epsilon_b} \right) \tag{27}$$

2.2. Simulation Setup

A simulation case focusing on the application of the MSLPM to a geophysical fluid flow is defined, focusing on the interaction of a perturbation with the background current. A thermally-mixed axisymmetric current is initialized as a perturbation within the computational domain, where a vertically-varying background shear layer is present. The currents are situated, such that they flow perpendicular with respect to each other. The axisymmetric current is allowed to evolve in time and, as the simulation begins, this current will collapse from buoyancy forces, forming internal gravity waves. These internal gravity waves will radiate out, with some of them reaching the background shear layer and interacting with it. Such internal gravity wave-shear layer interactions have been theoretically studied by Eltayeb and McKenzie, and numerically studied by Galmiche et al. and Javam et al. [19–21]. It is assumed that the background shear layer is also slowly varying, such that it can be assumed to be constant in time over the span of the simulation. This case may be representative of a sewage outflow into a bay or estuary with a strong submerged current. This demonstration simulation is divided into two sections: a one-dimensional (1D) precursor simulation where turbulence fields are initialized and background fields are constructed, and then a 2D+t simulation where the axisymmetric current evolves under its own dynamics as well as influence from the shear layer. The axisymmetric current and its associated turbulence fields in the latter simulation are initialized to its respective perturbation fields, while the quantities that are associated with the shear layer are stored within the background fields. Because the background shear layer is held fixed in time, only Equations (12)–(14) and (25)–(27)

are solved each timestep. These equations are implemented into OpenFOAM while using the methods described in [16].

2.2.1. Precursor Simulation

The 1D precursor simulation is performed in order to generate the background fields for the 2D+t simulation. A 500 m tall 1D domain, stretching from $z = -250$ m to $z = 250$ m, is discretized into 751 cells, and the simulation itself takes place over 3200 s, such that it ends when the lowest Richardson number in the domain, as defined by Equation (28), is just above 0.25. The shear layer is initially 40 m wide and it is centered at $z = 50$ m, and it is initialized within this region using Equation (29), and elsewhere the fluid is initially quiescent.

$$Ri = \frac{N^2}{(dv_b/dz)^2} \tag{28}$$

$$v_b(z, t = 0) = -0.1 + 25N \sinh^{-1}\left(\frac{2z}{12\sqrt{3}}\right) \tag{29}$$

The remainder of the fields are initialized using the following equations:

$$\vec{U}_b(z, t = 0) = (0, v_b(z, 0), 0)^T \tag{30}$$

$$T_b(z, t = 0) = T_s + \frac{dT_b}{dz}z \tag{31}$$

$$k_b(z, t = 0) = 2.4 \times 10^{-9}\text{m}^2/\text{s}^2 \tag{32}$$

$$\epsilon_b(z, t = 0) = 1.9 \times 10^{-11}\text{m}^2/\text{s}^3 \tag{33}$$

N is the Brunt-Väisälä frequency and it is set at 5 cph for this simulation. dT_b/dz is taken as 0.00505 °C/m. The values for k_b and ϵ_b are bounded, such that they do not decrease below the values that they are initialized at. Free-slip conditions are used on the z-normal boundaries and periodic conditions are used on the lateral boundaries.

2.2.2. 2D+t Simulation

A rectangular 2D domain that is defined in Figure 2 is used for this demonstration case, similar to the one that was used by Hassid [22]. The 2D+t approach in this simulation can be used to approximate a steady, 3D domain while using far fewer cells than the 3D domain would nominally require, and the data periodically saved over the course of the unsteady simulation can be transformed into a third spatial dimension through a Galilean transform and a characteristic velocity U while using $x = Ut$. These data can then be assembled into a 3D domain with the initial conditions located at $x = 0$. The downside to the 2D+t approach is that gradients in the x direction are lost, and so this approximation only holds if the flow slowly varies in this direction. Here, we assume that the shear layer varies in only one dimension reflective of a large-scale current, while Hassid shows that the treatment of flow that is similar to the axisymmetric current through a 2D+t application matches the experimental data well.

The domain is periodic in the x dimension, and it has a damping beach that is placed on both sides of the domain that monotonically increases in strength as it approaches the lateral boundaries. Cell size is uniform within the central, undamped portion of the domain, but, within the beach, cells progressively stretch along the y axis as they approach the lateral boundaries. The domain is discretized into one cell in the x direction, 1125 cells in the y direction, and 751 cells in the z direction.

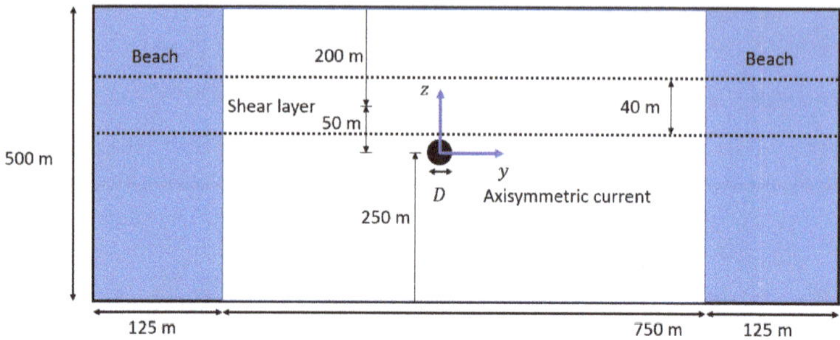

Figure 2. Simulation domain with the axisymmetric current and shear layer (not to scale).

Equations (34)–(38) are adapted from Hassid and used as initial conditions [22]. Here, $U_{D0} = 0.16\, U_0$, $k_0 = (0.06 U_0)^2$, $U_0 = 1$ m/s, and $D = 10$ m.

$$\delta u(y, z, t = 0) = U_{D0}\left(1 - \frac{8(y^2 + z^2)}{D^2}\right) \exp\left(\frac{-8(y^2 + z^2)}{D^2}\right) \tag{34}$$

$$\delta\vec{U}(y, z, t = 0) = (\delta u(y, z, 0), 0, 0)^T \tag{35}$$

$$\delta k(y, z, t = 0) = k_0 \exp\left(\frac{-4(y^2 + z^2)}{D^2}\right) \tag{36}$$

$$\delta\epsilon(y, z, t = 0) = \sqrt{12 k_0}\frac{k_0}{D} \exp\left(\frac{-6(y^2 + z^2)}{D^2}\right) \tag{37}$$

$$\delta T(y, z, t = 0) = -\frac{dT_b}{dz} z \exp\left(\frac{-4(y^2 + z^2)}{D^2}\right) \tag{38}$$

Background fields are set from the results of the precursor simulation. Because only the perturbation fields are solved over the course of this simulation, no boundary conditions are needed for the background fields and, instead, their boundary values are set from the precursor simulation results. The outflow boundary conditions are used on the y-normal boundaries, free-slip conditions on the z-normal boundaries, and periodic boundary conditions on the x-normal boundaries. This simulation takes place over 3600 s.

Overall, the following assumptions were made over the course of these simulations:

- the background shear current is assumed to evolve very slowly when compared to the axisymmetric current and can be modeled as evolving approximately independently of it;
- the background shear current is assumed to evolve over a timescale much longer than the 2D+t simulation length, so over the course of this simulation it may be assumed fixed in time;
- the background shear varies in only the vertical direction;
- the axisymmetric current and any resulting internal gravity waves vary slowly in the streamwise direction; and,
- the axisymmetric current and any results internal gravity waves do not have any significant impact on the evolution of the background shear current.

3. Results

3.1. Precursor Simulation

The 1D domain is simulated for nearly 3200 s to develop a final velocity and turbulence field. Figure 3 provides the vertical profile of the initial and final horizontal velocity normalized by the initial axisymmetric current centerline velocity U_{D0}, which is taken to be 0.16 m/s here. Figure 4 gives the vertical profile of eddy viscosity normalized by the molecular kinematic viscosity, here 10^{-6} m^2/s.

This field is determined algebraically from the k and ϵ fields of the precursor simulation, and these fields are mapped onto the 2D+t fields as well. The shear layer is shown to be highly turbulent, with eddy viscosity on the order of tens of thousands of times larger than molecular viscosity. This background turbulence may lead to interactions between itself and the perturbation flow, including any perturbation to turbulence. Finally, the vertical profile of $1/Ri$ is given by Figure 5. The line $1/Ri = 4$ denotes the classical Miles–Howard stability criterion, where regions in which $Ri < 1/4$ are dynamically unstable and prone to the formation of shear instabilities. The entire domain can be assumed to be stable because the entire profile has an inverse Richardson number of less than 4. Once the 1D precursor simulation is completed, its fields may be mapped to the 2D+t domain to be used as the time-constant background fields.

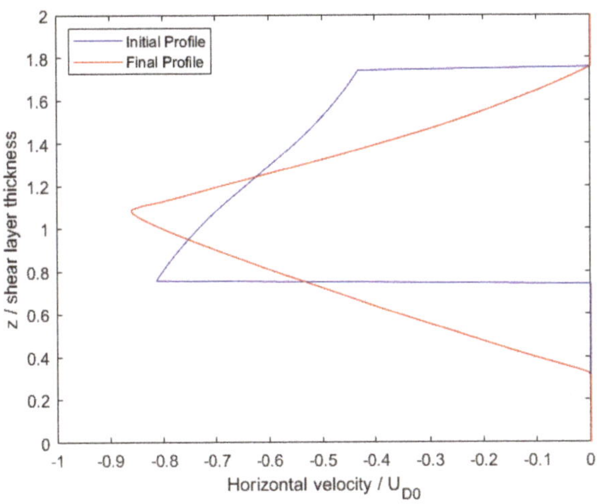

Figure 3. Initial and final velocity for the precursor simulation, normalized by the initial axisymmetric current centerline velocity. Taken from [16].

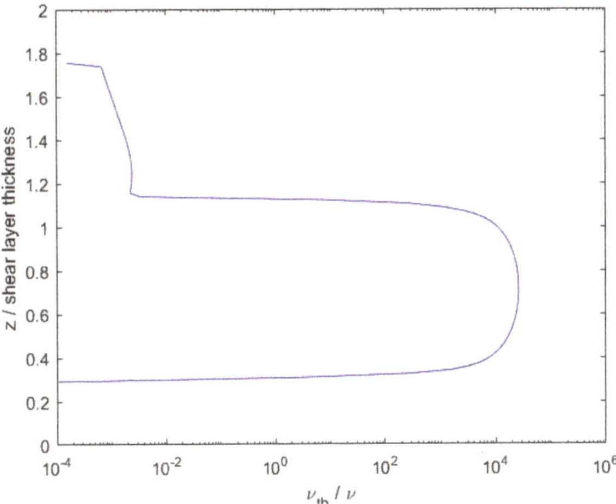

Figure 4. Final eddy viscosity ν_{tb} from precursor simulation, normalized by molecular kinematic viscosity. Taken from [16].

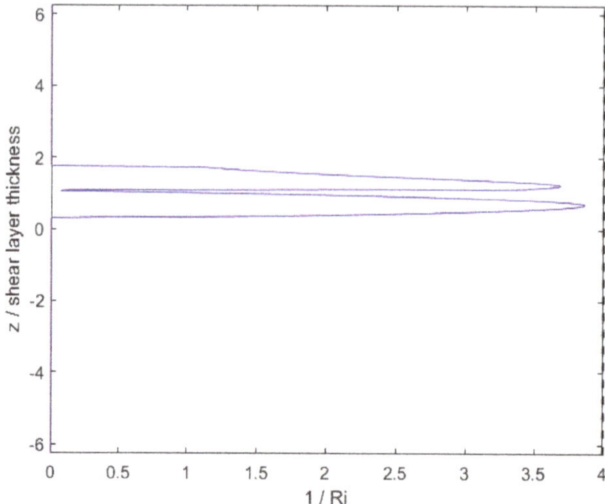

Figure 5. Final inverse gradient Richardson number profile from the precursor simulation. Taken from [16].

3.2. 2D+t Simulation

The 2D+t domain is simulated for one hour of simulation time with background fields being held constant in time. The results show that the axisymmetric current collapses due to density differences between it and the ambient stratification. This collapse generated internal gravity waves that radiate out in all directions, with some internal gravity waves entering the shear layer located above the axisymmetric current. The waves that enter are advected in the $-y$ direction by the strong shear current, and interactions between the internal gravity waves and the background shear layer may be seen.

Figures 6–8 show the perturbations to the x, y, and z components of velocity, respectively, at the final timestep. The axisymmetric current retains most of its original velocity distribution in the x direction, but internal gravity waves are seen in the δv and δw components. These internal gravity waves are seen entering the shear layer, and then dissipating due to the strong transverse current and high background turbulence within the shear layer, particularly seen in the close-up of the perturbation to vertical velocity in Figure 9. In spite of this, a wave-like pattern is still seen within the shear layer shown in Figures 7 and 8. The magnitude of the velocity perturbation is small, as seen in Figure 10, approximately 3% of the original centerline velocity defect. Figure 11 shows the perturbation to eddy viscosity $\delta \nu_t$ at the final timestep. This variable serves as a surrogate for the general state of turbulence that is introduced by the perturbation and the interaction between the perturbation and background. The highest perturbation to eddy viscosity is seen within the decaying axisymmetric current and within the shear layer just above the axisymmetric current. Turbulent viscosity is seen advected downstream by the shear current, and small quantities of $\delta \nu_t$ are also seen upstream within the shear layer. These latter quantities likely result from small perturbations to the shear layer that are caused by the axisymmetric perturbation and generated internal gravity waves.

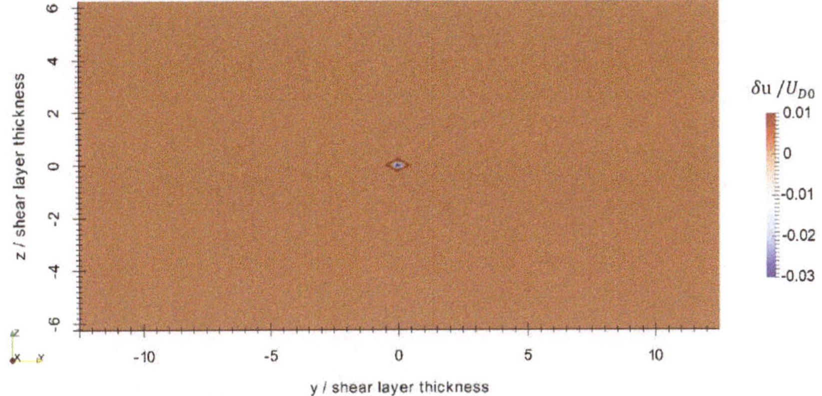

Figure 6. x velocity component perturbation at $t = 3600$ s. Adapted from [16].

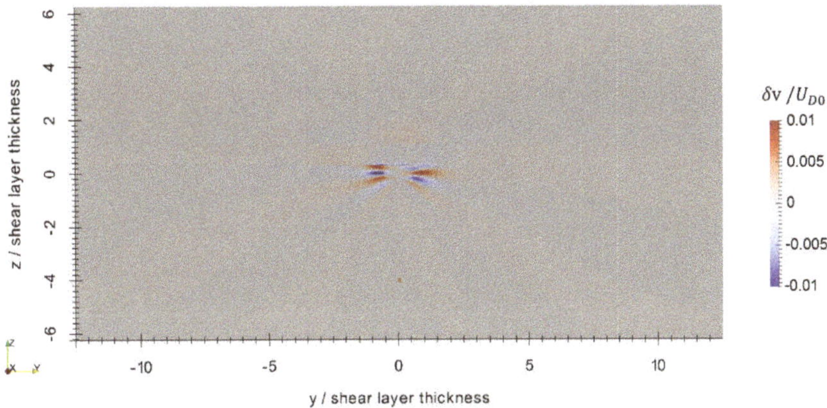

Figure 7. y velocity component perturbation at $t = 3600$ s. Adapted from [16].

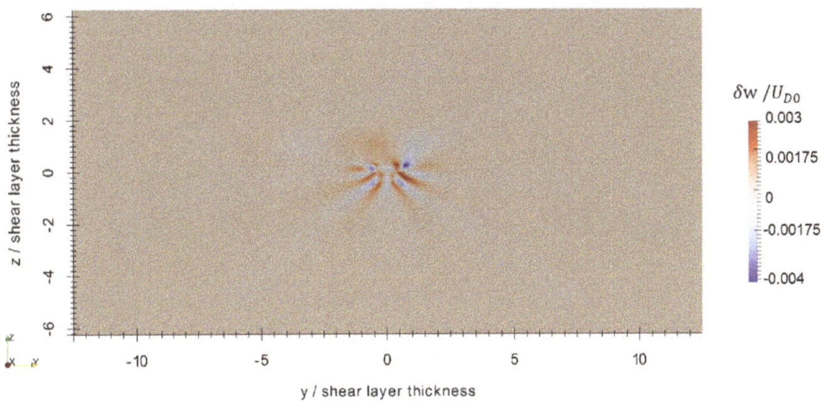

Figure 8. z velocity component perturbation at $t = 3600$ s. Adapted from [16].

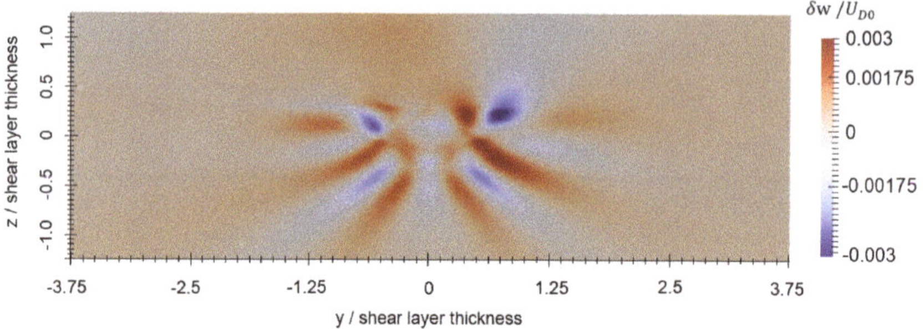

Figure 9. Close up of the z velocity component perturbation at $t = 3600$ s.

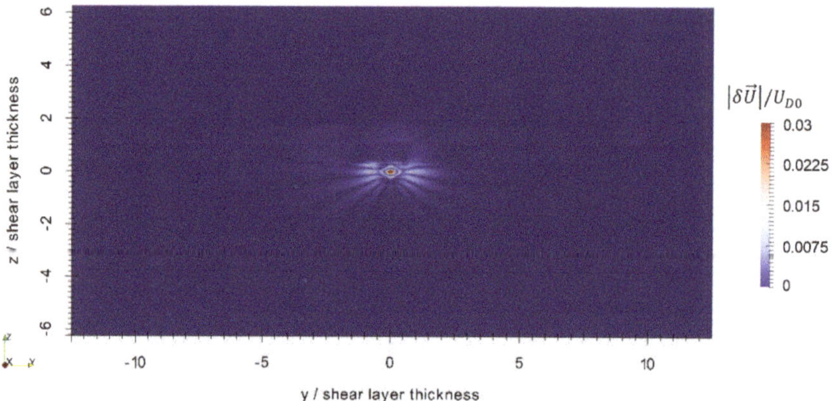

Figure 10. Magnitude of the velocity perturbation at $t = 3600$ s. Adapted from [16].

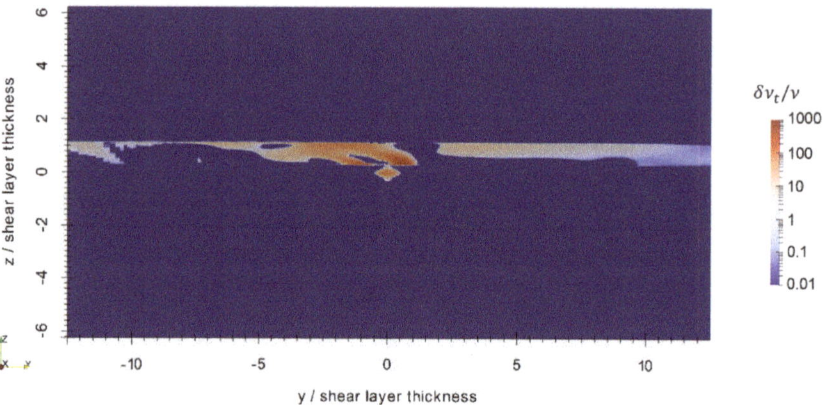

Figure 11. Perturbation to eddy viscosity at $t = 3600$ s. Adapted from [16].

4. Discussion

The ability to independently track the evolution of a perturbation within an evolving background is one of the most prominent benefits of the MSLPM. Using traditional methods, several simulations may be required in order to ascertain the time history of a simulation without perturbation so that the effects of a perturbation can be determined. However, with the MSLPM, this perturbation is intrinsically separated and tracked. This benefit becomes more apparent when one must perform a series of calculations as part of a parametric study, where one may compare the evolution of a perturbation across a range of different background conditions. The separation of a perturbation from a nominal background may also yield numerical benefits, as linear solvers may be determining a solution based around zero rather than the conditions of the background state.

In the simulation documented here, a time-constant background field is defined within the domain and the perturbation allowed to interact with it. This is done without the need for boundary conditions or body source terms, which may be of interest in stratified fluid cases. The generation of internal gravity waves that radiate in all directions may result in internal gravity waves impinging on the upstream boundary, and the use of an inlet boundary condition on this same surface may lead to the reflection of waves back into the domain. Localizing the internal gravity waves to the perturbation allows for the user to place an inlet boundary condition on the background flow, then a sponge layer and an open boundary condition may be placed upon the perturbation fields in order to prevent any waves from reflecting back into the domain.

Furthermore, as the background is imposed upon the domain while using data that were collected from a precursor simulation, the influences acting upon the background can be retained, even if they are not resolved within a small-scale simulation. While it is not seen here, one could ensure that the background evolves over a certain scale. This may be important for multi-scale simulation. For example, a large-scale simulation examining continental wind patterns may examine how an upstream mountain range affects a potential downstream site, which is the subject of a small-scale simulation. This smaller-scale simulation may utilize the large-scale simulation data as a set of initial conditions, but, once the information is localized to the small-scale domain, it loses any influences of far-upstream or far-downstream phenomena which may be important. A possible remedy for this is to couple the two simulations together, which is done by the MSLPM.

4.1. Limitations of Assumptions

The MSLPM is derived under the assumption that there is sufficient scale separation between the background and perturbation, and that the perturbation does not have a significant influence upon the background. This assumption limits the perturbations that may be studied, for example, a large temperature or velocity perturbation to the entire domain, as these perturbations may have a significant impact on the dynamics of the background. Because the one-way coupling of the background and perturbation is the result of this assumption, the implementation of two-way coupling may partially relax this assumption and permit a greater range of perturbations to be studied.

Certain solution methods may also limit the applicability of the MSLPM. The case that is described in Sections 2 and 3 above assumes that the time scale of the background processes is so large compared to the simulation duration that the background can be assumed to be constant in time. This requirement may be restrictive, depending on the background flow used, but two other solution methods are described in Section 2. The simultaneous solution of background and perturbation fields on the same domain may also be used for this case and, in particular, may have been more appropriate given the nature of the background shear flow. Interpolating the background fields temporally and spatially from a separate simulation could also be used. This method may be of interest to nested multi-scale simulations; however, one must ensure that proper interpolation is used.

The 1D and 2D+t simulations were also conducted under the assumption that the flow evolves slowly in certain directions. The 1D simulation was performed in order to provide initial conditions for the background shear in the 2D+t simulation, and as such it loses any evolution in other directions, particularly the y direction in which it flows. Within the 2D+t simulation, any significant changes in the direction in which the axisymmetric current flows is lost. While this mainly affects the evolution of the axisymmetric current, the interactions between internal gravity waves and the background shear layer may lose any 3D effects that may be important in the interactions. Similarly, the choice of a $k - \epsilon$ type model may also affect the quantitative values that are seen in Figures 6–10. In this case, the results seen here may only be qualitatively correct. As a final note, this case is demonstrative of the capabilities of the MSLPM, so no quantitative conclusions should be drawn from these results.

4.2. Future Work

Further cases with more diverse phenomena and physics may be useful in order to better quantify the capabilities of the MSLPM relative to traditional simulation methods. In particular, the different solution methods that are described in Section 2 may be tested, as these will give researchers the ability to fine-tune their simulation for the given background flows of interest. Further benchmark cases may be useful in understanding which background solution method is most practical for different flows and phenomena. Two-way coupling of the background and perturbation fields may also be of interest in order to allow for the perturbation to affect the evolution of the background. This is particularly relevant to turbulence quantities, as the great range in temporal and spatial scales of turbulent motion means that scale overlap is nearly guaranteed. While the methods that are described in this paper are in their infancy, they show great promise for geophysical flows, including multi-scale simulation of geophysical flows and, potentially, other realms of computational physics.

Author Contributions: Conceptualization, E.H., J.P. and E.P.; methodology, E.H.; software, E.H.; validation, E.H.; formal analysis, E.H.; investigation, E.H.; resources, E.P.; data curation, E.H.; writing—original draft preparation, E.H.; writing—review and editing, J.P. and E.P.; visualization, E.H.; supervision, J.P. and E.P.; project administration, E.P.; funding acquisition, J.P. and E.P. All authors have read and agreed to the published version of the manuscript.

Funding: This research received no external funding.

Acknowledgments: The authors acknowledge Advanced Research Computing at Virginia Tech for providing computational resources and technical support that have contributed to the results reported within this paper. URL: arc.vt.edu. The authors would also like to thank the anonymous referees for their constructive feedback.

Conflicts of Interest: The authors declare no conflict of interest.

Appendix A

The expansion shown in Equation (3) can be demonstrated with an example case for the following partial differential equation in Equation (A1), but this process is also identical for other partial differential equations, including multi-variable ones, as well ordinary differential equations.

$$PDE(\phi) = \frac{\partial \phi}{\partial t} - \phi \frac{\partial \phi}{\partial x} = 0 \tag{A1}$$

The variable ϕ is accordingly decomposed into a background component ϕ_b and a perturbation component $\delta\phi$ such that $\phi = \phi_b + \delta\phi$. This yields Equation (A2).

$$PDE(\phi) = PDE(\phi_b + \delta\phi) = \frac{\partial(\phi_b + \delta\phi)}{\partial t} - (\phi_b + \delta\phi)\frac{\partial(\phi_b + \delta\phi)}{\partial x} = 0 \tag{A2}$$

Equation (A2) can then be manipulated into the form shown in Equation (A3).

$$PDE(\phi_b + \delta\phi) = \frac{\partial \phi_b}{\partial t} - \phi_b\frac{\partial \phi_b}{\partial x} + \frac{\partial \delta\phi}{\partial t} - \delta\phi\frac{\partial \delta\phi}{\partial x} - \delta\phi\frac{\partial \phi_b}{\partial x} - \phi_b\frac{\partial \delta\phi}{\partial x} = 0 \tag{A3}$$

One may notice the first two terms in the middle equality of Equation (A3) are identical in form to the original $PDE(\phi)$ but only depending on the background component ϕ_b, and the next two terms are identical in form to the original PDE but only depending on the perturbation component $\delta\phi$. The final two terms arise from the non-linearity of $PDE(\phi)$ and would be absent if this differential equation were linear. Similarly, if either ϕ_b or $\delta\phi$ were identically zero for all time, these terms would be equal to zero. We may, therefore, make the following definitions in Equations (A4)–(A6).

$$PDE(\phi_b) \equiv \frac{\partial \phi_b}{\partial t} - \phi_b \frac{\partial \phi_b}{\partial x} \tag{A4}$$

$$PDE(\delta\phi) \equiv \frac{\partial \delta\phi}{\partial t} - \delta\phi \frac{\partial \delta\phi}{\partial x} \tag{A5}$$

$$NL(\phi_b, \delta\phi) \equiv -\delta\phi \frac{\partial \phi_b}{\partial x} - \phi_b \frac{\partial \delta\phi}{\partial x} \tag{A6}$$

which, when substituted into Equation (A3), yield Equation (3).

References

1. Fureby, C.; Anderson, B.; Clarke, D.; Erm, L.; Henbest, S.; Giacobello, M.; Jones, D.; Nguyen, M.; Johansson, M.; Jones, M.; et al. Experimental and numerical study of a generic conventional submarine at 10° yaw. *Ocean. Eng.* **2016**, *116*, 1–20. [CrossRef]
2. Danabasoglu, G.; Yeager, S.G.; Bailey, D.; Behrens, E.; Bentsen, M.; Bi, D.; Biastoch, A.; Böning, C.; Bozec, A.; Canuto, V.M.; et al. North Atlantic simulations in Coordinated Ocean-ice Reference Experiments phase II (CORE-II). Part I: Mean states. *Ocean. Model.* **2014**, *73*, 76–107. [CrossRef]
3. Talbot, C.; Bou-Zeid, E.; Smith, J. Nested Mesoscale Large-Eddy Simulations with WRF: Performance in Real Test Cases. *J. Hydrometeorol.* **2012**, *13*, 1421–1441. [CrossRef]
4. Wiersema, D.J.; Lundquist, K.A.; Chow, F.K. Mesoscale to Microscale Simulations over Complex Terrain with the Immersed Boundary Method in the Weather Research and Forecasting Model. *Mon. Weather. Rev.* **2020**, *148*, 577–595. [CrossRef]
5. Lindberg, F.; Grimmond, C.; Gabey, A.; Huang, B.; Kent, C.W.; Sun, T.; Theeuwes, N.E.; Järvi, L.; Ward, H.C.; Capel-Timms, I.; et al. Urban Multi-scale Environmental Predictor (UMEP): An integrated tool for city-based climate services. *Environ. Model. Softw.* **2018**, *99*, 70–87. [CrossRef]
6. Colella, F.; Rein, G.; Verda, V.; Borchiellini, R. Multiscale modeling of transient flows from fire and ventilation in long tunnels. *Comput. Fluids* **2011**, *51*, 16–29. [CrossRef]
7. Groen, D.; Zasada, S.J.; Coveney, P.V. Survey of Multiscale and Multiphysics Applications and Communities. *Comput. Sci. Eng.* **2014**, *16*, 34–43. [CrossRef]
8. Lin, C.C.; Segel, L.A. *Mathematics Applied to Deterministic Problems in the Natural Sciences*, 1st ed.; Macmillan: New York, NY, USA, 1974.
9. Hardin, J.C.; Pope, D.S. An acoustic/viscous splitting technique for computational aeroacoustics. *Theor. Comput. Fluid Dyn.* **1994**, *6*, 323–340. [CrossRef]
10. Ewert, R.; Schröder, W. Acoustic perturbation equations based on flow decomposition via source filtering. *J. Comput. Phys.* **2003**, *188*, 365–398. [CrossRef]
11. Jasak, H.; Jemcov, A.; Tuković, Ž. OpenFOAM: A C++ Library for Complex Physics Simulations. In Proceedings of the International Workshop on Coupled Methods in Numerical Dynamics, Zagreb, Croatia, 19–21 September 2007; pp. 47–66.
12. Alfonsi, G.; Ferraro, D.; Lauria, A.; Gaudio, R. Numerical Investigation of Natural Rough-Bed Flow. In *Numerical Computations: Theory and Algorithms*; Sergeyev, Y.D., Kvasov, D.E., Eds.; Springer International Publishing: Cham, Switzerland, 2020; pp. 280–288.
13. Tafarojnoruz, A.; Lauria, A. Large eddy simulation of the turbulent flow field around a submerged pile within a scour hole under current condition. *Coast. Eng. J.* **2020**, *62*, 489–503. [CrossRef]
14. Lauria, A.; Alfonsi, G.; Tafarojnoruz, A. Flow Pressure Behavior Downstream of Ski Jumps. *Fluids* **2020**, *5*, 168. [CrossRef]
15. Vuorinen, V.; Chaudhari, A.; Keskinen, J.P. Large-eddy simulation in a complex hill terrain enabled by a compact fractional step OpenFOAM® solver. *Adv. Eng. Softw.* **2015**, *79*, 70–80. [CrossRef]
16. Higgins, E.T. Multi-Scale Localized Perturbation Method for Geophysical Fluid Flows. Master's Thesis, Virginia Polytechnic Institute and State University, Blacksburg, VA, USA, 2020.
17. Rodi, W. Examples of calculation methods for flow and mixing in stratified fluids. *J. Geophys. Res.* **1987**, *92*, 5305. [CrossRef]
18. Launder, B.E.; Spalding, D.B. The Numerical Computation of Turbulent Flows. *Comput. Methods Appl. Mech. Eng.* **1974**, *3*, 21. [CrossRef]

19. Eltayeb, I.A.; McKenzie, J.F. Critical-level behaviour and wave amplification of a gravity wave incident upon a shear layer. *J. Fluid Mech.* **1975**, *72*, 661–671. [CrossRef]
20. Galmiche, M.; Thual, O.; Bonneton, P. Direct Numerical Simulation of Turbulence in a Stably Stratified Fluid and Wave-Shear Interaction. *Appl. Sci. Res.* **1997**, *59*, 15. [CrossRef]
21. Javam, A.; Imberger, J.; Armfield, S.W. Numerical study of internal wave–caustic and internal wave–shear interactions in a stratified fluid. *J. Fluid Mech.* **2000**, *415*, 89–116. [CrossRef]
22. Hassid, S. Collapse of turbulent wakes in stably stratified media. *J. Hydronautics* **1980**, *14*, 25–32. [CrossRef]

Publisher's Note: MDPI stays neutral with regard to jurisdictional claims in published maps and institutional affiliations.

Article

A Coupled OpenFOAM-WRF Study on Atmosphere-Wake-Ocean Interaction

John Gilbert [1,*] and Jonathan Pitt [1,2]

[1] Hume Center for National Security and Technology, Virginia Tech, Blacksburg, VA 24061, USA
[2] Crofton Department of Aerospace and Ocean Engineering, Virginia Tech, Arlington, VA 22203, USA; jspitt@vt.edu
* Correspondence: johng12@vt.edu

Abstract: This work aims to better understand how small scale disturbances that are generated at the air-sea interface propagate into the surrounding atmosphere under realistic environmental conditions. To that end, a one-way coupled atmosphere-ocean model is presented, in which predictions of sea surface currents and sea surface temperatures from a microscale ocean model are used as constant boundary conditions in a larger atmospheric model. The coupled model consists of an ocean component implemented while using the open source CFD software OpenFOAM, an atmospheric component solved using the Weather Research and Forecast (WRF) model, and a Python-based utility `foamToWRF`, which is responsible for mapping field data between the ocean and atmospheric domains. The results are presented for two demonstration cases, which indicate that the proposed coupled model is able to capture the propagation of small scale sea surface disturbances in the atmosphere, although a more thorough study is required in order to properly validate the model.

Keywords: atmosphere-ocean coupling; air-sea interface; OpenFOAM; Weather Research and Forecast (WRF)

Citation: Gilbert, J.; Pitt, J. A Coupled OpenFOAM-WRF Study on Atmosphere-Wake-Ocean Interaction. *Fluids* **2021**, *6*, 12. https://doi.org/10.3390/fluids6010012

Received: 13 November 2020
Accepted: 23 December 2020
Published: 30 December 2020

Publisher's Note: MDPI stays neutral with regard to jurisdictional claims in published maps and institutional affiliations.

1. Introduction

Large scale oceanic disturbances, such as earthquakes or tsunamis, are known to generate gravity waves that propagate through the atmosphere up to the ionosphere, where they produce electron density variations that are detectable by global navigation satellite systems (GNSS) [1,2]. Smaller scale disturbances, such as the passage of a surface ship, also clearly manifest in the atmosphere in a number of ways. For example, Yuan et al. [3] trained a deep neural network to detect surface ship tracks that are present in satellite imagery resulting from aerosol-cloud interactions in the atmosphere. Characterizing these types of disturbances and distinguishing them from the normal variability in the ambient environment requires a deep understanding of the dynamics within the atmosphere, ocean, and air-sea interface.

This effort is made all the more challenging, due to the wide range of relevant spatial and temporal scales that are involved in this problem. Characteristic lengths in the atmospheric boundary layer can range from 10 s of meters, in the case of water vapor variability [4,5], to 10 s or 100 s of kilometers, in the case of horizontal rolls and convective cells. On the other hand, the relevant length scales for ship wakes range from <1 m in the vicinity of the ship to 10 s of kilometers behind the vessel, in the case of surface current perturbations. Modeling and simulation solutions exist for the disparate domains, length, and time scales, but a fully coupled solution does not yet exist.

Previous efforts to model the atmosphere-ocean system can be broadly categorized as one-way or two-way coupled. In a one-way coupled model, one model (typically the atmosphere) is used in order to define boundary conditions that drive the response of the other model (typically the ocean). The data are assumed to flow in one direction only, and there is no feedback between the two models. In a two-way model, information passes

Fluids **2021**, *6*, 12. https://doi.org/10.3390/fluids6010012

between the both models during the simulation. The majority of one-way coupled models have primarily focused on initializing microscale CFD analyses with more realistic environmental conditions from mesoscale atmospheric simulations. For example, Boutanios et al. [6] conducted atmospheric mesoscale simulations of the Grosse Isle Manitoba storm whlie using the Weather Research and Forecasting (WRF) model. Boundary conditions that were extracted from this simulation were then used in a microscale CFD analysis in order to investigate the flow around steel transmission towers. Several other authors have taken a similar mesoscale-to-microscale, one-way coupled approach to study the flow around buildings in urban environments [7,8], and around wind farms [9–14]. Previous two-way coupled atmosphere-ocean models have largely focused on improved regional and global weather forecasting (e.g., [15–17]) and do not possess the necessary spatial resolution in the ocean model in order to resolve small scale disturbances.

This paper presents a one-way coupled atmosphere-ocean model, in which sea surface currents and sea surface temperatures that are predicted by the ocean model are used as boundary conditions for the atmospheric model. This work is distinct from previous works in two ways: (1) atmospheric boundary conditions are set via a microscale CFD ocean model and (2) the CFD ocean model is developed in order to resolve small scale disturbances at the air-sea interface. The remainder of the paper is organized, as follows: Section 2 provides details on the computational approach and discusses some of the limitations of the current coupled model, Section 3 presents results for two demonstration cases that are intended to exercise the one-way coupling approach, and Section 4 provides concluding remarks and a discussion of future work.

2. Computational Approach

A partitioned approach is used in this work in order to study the coupled problem, in which the atmosphere and ocean domains are solved by separate numerical approaches that are best suited to their different domains. Data (i.e., surface currents, sea surface temperatures, etc.) are exchanged at the air-sea interface via an appropriate coupling scheme. In this case, one-way coupling is accomplished by extracting constant boundary field information from ocean model predictions at the sea surface. Details on each of the component models, as well as the coupling, are provided in the following sections.

2.1. Atmospheric Model

The atmospheric subsystem is solved while using the well-known numerical weather prediction software Weather Research and Forecasting model (WRF). WRF is a fully-compressible, Eulerian non-hydrostatic equations solver that uses terrain-following hydrostatic-pressure vertical coordinates. WRF includes several boundary layer physics schemes and sub-grid scale turbulence formulations. Predictions of three-dimensional winds, pressure, precipitation, air temperature, surface sensible, and latent heat fluxes, as well as many more, are available. WRF is known to be highly scalable and it supports grid nesting to improve resolution over specific areas of interest. Table 1 provides a summary of the atmospheric model configuration used in the current work. The interested reader is referred to Skamarock et al. [18] for a complete description of WRF.

Table 1. A list of the key physics schemes and meteorological data sources that are used in this work.

Atmospheric Model	Configuration
Software	WRF Version 4.2.1
Integration Domain	Gulf of Mexico
Grid	Arakawa semi-staggered C-grid
Initial and boundary conditions	NCEP GDAS/FNL Reanalysis, 0.25° × 0.25°
	6-hour update of boundary conditions
SST	NCEP GDAS/FNL Reanalysis
Surface Layer	Revised MM5 Monin-Obukhov scheme [19]
Planetary Boundary Layer	YSU [20]
Microphysics	Thompson scheme [21]
Cumulus	None
Land surface	Noah [22]
Radiation	RRTMG scheme used for longwave and shortwave radiation [23]

2.2. Ocean Model

The primary focus of this effort is to characterize the propagation of small scale sea surface perturbations into the atmosphere in realistic environments. To that end, the proposed ocean model is designed in roder to allow for a specification of realistic initial conditions and a variety of geophysical forcings. The equations governing the behavior of the ocean domain are solved while using the open-source CFD software, OpenFOAM™. This work uses OpenFOAM v2006 that is distributed by ESI-OpenCFD (Bracknell, UK).

2.2.1. Governing Equations

The ocean domain is assumed to satisfy the incompressible unsteady Reynolds-averaged Navier–Stokes (URANS) equations for a Boussinesq fluid, such that the balance of mass and momentum can be expressed as

$$\nabla \cdot \mathbf{U} = 0, \tag{1}$$

$$\frac{\partial \mathbf{U}}{\partial t} + \mathbf{U} \cdot \nabla \mathbf{U} - \nu \nabla^2 \mathbf{U} = -\nabla \hat{p} + \nabla \cdot \overline{u_i' u_j'} - \frac{\Delta \rho}{\rho_0} \mathbf{g} + \mathbf{F_b}. \tag{2}$$

Here, $\mathbf{U} = [U, V, W]^T$ is the mean velocity field, u_i' is the fluctuating component of velocity due to turbulence, ν is the kinematic viscosity, $\Delta \rho = \rho - \rho_b(z)$ indicates a perturbation from the background density profile, $\rho_b(z)$, and $\rho_0 = \rho_b(z = 0)$ is a reference density. This work also makes use of the piezometric pressure, $\hat{p} = (p - \rho_0 \mathbf{g} \cdot \mathbf{x})/\rho_0$, where $\mathbf{x}$ is the position vector, $\mathbf{g}$ is the gravity vector, and p is the total pressure. The final term on the right-hand side of Equation (2), $\mathbf{F_b}$, represents external body forces other than gravity (e.g., Coriolis force, etc.). Those forcings are neglected in the current work, but they can be easily included in the future.

The evolution of temperature and salinity are directly modeled in this work through the following transport-diffusion equations

$$\frac{\partial T}{\partial t} + \nabla \cdot (\mathbf{U}T) = \nabla \cdot (\kappa_T \nabla T) + \nabla \cdot \overline{u_i' t'}, \tag{3}$$

$$\frac{\partial S}{\partial t} + \nabla \cdot (\mathbf{U}S) = \nabla \cdot (\kappa_S \nabla S) + \nabla \cdot \overline{u_i' s'}. \tag{4}$$

where t' and s' are the unsteady fluctuations in temperature, T, and salinity, S, respectively, and κ_T and κ_S are the molecular diffusivity of heat and salinity. Modeling temperature and salinity, in this way, allows for the use of realistic environmental profiles that were recorded by oceanic buoys or Regional Ocean Modeling System (ROMS) forecasts. Equations (1)–(4) are closed by an equation of state (EOS) relating pressure, temperature, salinity, and density. In this work, the density is computed while using the TEOS-10 seawater EOS [24].

2.2.2. Turbulence Modeling

A standard $k - \epsilon$ model, modified to include buoyancy production effects, is used in order to compute the Reynolds stresses, $\overline{u'_i u'_j}$, and the turbulent fluxes $\overline{u'_i t'}$ and $\overline{u'_i s'}$. The Reynolds stress and turbulent fluxes are given by

$$-\overline{u'_i u'_j} = 2\nu_t S_{ij} - \frac{2}{3} k \delta_{ij}, \tag{5}$$

$$-\overline{u'_i t'} = \frac{\nu_t}{\sigma_T} \frac{\partial T}{\partial x_i}, \tag{6}$$

$$-\overline{u'_i s'} = \frac{\nu_t}{\sigma_S} \frac{\partial S}{\partial x_i}, \tag{7}$$

where $\nu_t = C_\mu \frac{k^2}{\epsilon}$ is the eddy viscosity and $S_{ij} = \frac{1}{2}\left(\frac{\partial U_i}{\partial x_j} + \frac{\partial U_j}{\partial x_i}\right)$ is the mean rate of strain. The following evolution equations describe the turbulent kinetic energy $k = \frac{1}{2}\left(\overline{u'_i u'_j}\right)$ and dissipation ϵ:

$$\frac{\partial k}{\partial t} + \nabla \cdot (\mathbf{U}k) = \nabla \left[\left(\nu + \frac{\nu_t}{\sigma_k}\right)\nabla k\right] + P_k + P_b - \epsilon, \tag{8}$$

$$\frac{\partial \epsilon}{\partial t} + \nabla \cdot (\mathbf{U}\epsilon) = \nabla \left[\left(\nu + \frac{\nu_t}{\sigma_\epsilon}\right)\nabla \epsilon\right] + C_{1\epsilon}\frac{\epsilon}{k}(P_k + C_{3\epsilon}P_b) - C_{2\epsilon}\frac{\epsilon^2}{k}, \tag{9}$$

$$P_k = -\overline{u'_i u'_j}\frac{\partial U_i}{\partial x_j} = \nu_t S^2, \tag{10}$$

$$P_b = \rho' \mathbf{g} \frac{\nu_t}{Pr_t}. \tag{11}$$

Here, P_k is the standard turbulent kinetic energy production term, P_b is the turbulent energy production due to buoyancy effects, and $C_{1\epsilon}$, $C_{2\epsilon}$, $C_{3\epsilon}$, σ_k, and σ_ϵ are the model constants.

2.3. One-Way Coupling Approach

WRF employs a surface layer scheme that is based on the Revised Monin–Obukhov similarity theory [19]. Within WRF, the surface layer is assumed to be the first vertical layer. Following Varlas et al. [17], the air-sea fluxes of momentum, sensible heat, and latent heat can be expressed as

$$\tau = \rho_a u_*^2 = \rho_a C_d (U - U_s)^2 \tag{12}$$

$$H = -\rho_a C_p u_* \theta_* = \rho_a C_p C_h U(\theta_s - \theta_a) \tag{13}$$

$$LH = L_v \rho_a u_* q_* = L_v \rho_a C_q U(q_s - q_a) \tag{14}$$

where ρ_a is the density of air in the surface layer, u_* is the friction velocity, U_s, is the sea surface current velocity, and U is the wind speed at the lower layer that is modified by convective velocity and a sub-grid velocity following Beljaars [25] and Mahrt and Sun [26], respectively. Additionally, θ_* and q_* are the temperature and moisture scales, respectively, C_p is the specific heat capacity at constant pressure, and L_v is the latent heat of vaporization. θ_a and θ_s are the air and sea surface potential temperatures, respectively, q_s is the specific humidity at the sea surface, q_a is the specific humidity of air at the lower level, and C_d, C_h, and C_q are the dimensionless bulk transfer coefficients for momentum, sensible heat, and moisture. For details on the parameterization of the bulk transfer coefficients, the reader is referred to Jiménez et al. [19].

In order to couple the atmosphere and ocean domains, predictions for sea surface currents, U_s, V_s, and sea surface temperature (SST), θ_s, from the ocean model are applied as the bottom surface boundary conditions and used in the surface momentum, heat, and moisture flux calculations of the atmospheric model, as given by Equations (12)–(14). For the results that are presented in this paper, the ocean domain is assumed to evolve much more slowly in time than the atmospheric domain. Consequently, predictions of surface currents and SST from the ocean model are taken to be constant relative to the atmosphere and result in a one-way coupled system. This greatly simplifies the coupling, as no data exchange is required during execution of either the ocean or atmosphere model. Figure 1a illustrates the proposed one-way coupling algorithm and it can be summarized, as follows:

1. initialize the ocean and atmospheric models with a consistent ambient environment;
2. run the ocean model and extract relevant surface fields (U_s, V_s, T_s);
3. map surface fields from the ocean domain onto the atmospheric domain;
4. overwrite initial atmospheric environment with mapped ocean fields; and,
5. run the atmospheric simulation as normal.

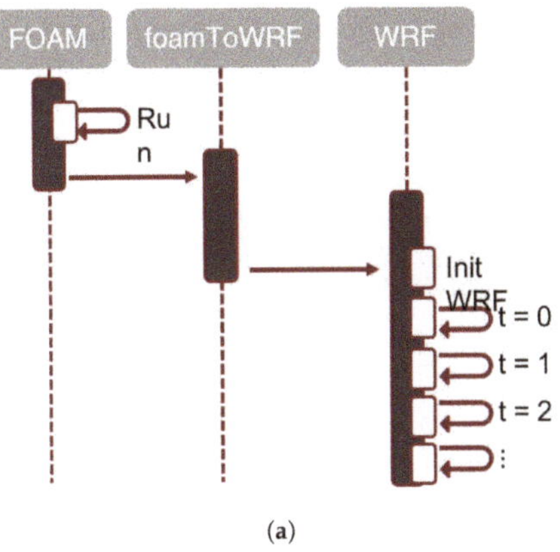

(a)

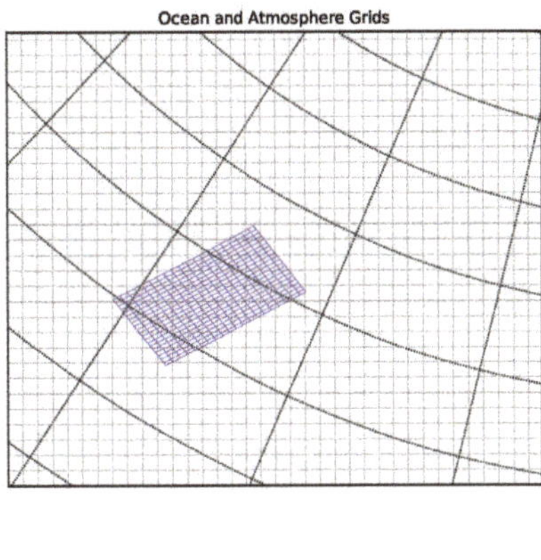

(b)

Figure 1. (a) Illustration of the one-way coupling algorithm. The ocean model (FOAM) is run prior to the atmospheric model to generate the required bottom boundary conditions. (b) Depiction of the possible disparate meshes that can exist between ocean and atmospheric model components. The ocean domain, drawn in blue, at a finer grid resolution, and not aligned with the underlying atmospheric domain.

2.3.1. Mapping Data Fields

In general, the atmosphere and ocean models will have different domain projections, different resolution requirements, and different reference frames. Consequently, the computational domains will often be mismatched at the air-sea interface, as illustrated in Figure 1b. The process of interpolating sea surface predictions from the ocean model onto the atmospheric domain involves four steps: (1) aligning the atmospheric and ocean domains, (2) transform the atmospheric domain geographic coordinates into Cartesian coordinates, (3) locate the atmospheric points in the ocean domain and interpolate field data from the ocean domain onto the specified interpolation points, and (4) output the new interpolated fields in an appropriate format.

In general, some amount of transformation (rotation and/or translation) is required in order to align the atmospheric and ocean reference frames. Often, the origin of the ocean domain is chosen in such a way as to simplify initialization of the model or to satisfy some other modeling consideration. For example, the surface ship wake application that is presented in Section 3 requires one coordinate (the x-coordinate, in this case) be aligned with the direction of ship motion in order to satisfy the 2D + t assumption. Predictions of sea surface currents must then be transformed in order to align with the ship's heading relative to the atmospheric domains reference frame.

For the applications under consideration in this work, the computational domain of the ocean model is much smaller than the atmospheric domain. Therefore, it is assumed that the x-coordinate direction is parallel to lines of longitude and the y-coordinate direction is parallel to the lines of latitude. Thus, to a reasonable degree of accuracy, the geographic and Cartesian coordinates can be related via the expressions:

$$\text{lat} = \frac{(y - y_{\min})(\text{lat}_{\max} - \text{lat}_{\min})}{(y_{\max} - y_{\min})} + \text{lat}_{\min}, \tag{15}$$

$$\text{lon} = \frac{(x - x_{\min})(\text{lon}_{\max} - \text{lon}_{\min})}{(x_{\max} - x_{\min})} + \text{lon}_{\min}, \tag{16}$$

where the min and max subscripts refer domain extents [10].

After the coordinate transformations are complete, the required data field must be interpolated from the ocean domain onto the atmospheric domain. In this work, both linear and cubic two-dimensional (2-D) interpolation schemes were tested. The bi-linear interpolation scheme proved to be the most robust at minimizing errors near domain boundaries and when the resolutions were significantly different between the ocean and atmospheric domains.

Finally, the interpolated boundary data must be stored in a format that is suitable for ingestion in the atmospheric model. In this case, the WRF input files are written in NetCDF format. This step, along with the mapping approach that is described above, has been implemented in a Python-based utility, called `foamToWRF`. This utility automates the extraction of surface field data from OpenFOAM result files, the alignment of the OpenFOAM and WRF domains, the interpolation of the OpenFOAM data onto the WRF grid, and the export of the resultant bottom boundary conditions in the NetCDF format that is required by WRF. This work uses NetCDF version 4.3.3.1.

3. Results and Discussion

The following section describes two examples of the proposed one-way coupled atmosphere-ocean model. In the first example, the surface currents and SST are analytically generated without the use of an ocean model. In the second example, the ocean model is used in order to predict sea surface current and SST modifications that are caused by the passage of a surface ship. The first example does not use the ocean model and it is intended to demonstrate the ability to modify the WRF boundary conditions, while the second example is illustrative of a more realistic scenario. In both cases, the focus of the investigation is on the microscale Large Eddy Simulation (LES) domain predictions.

3.1. Application to Surface Jets

This test case was conducted as a demonstration of the one-way coupling process. In this problem, modified sea surface temperatures and currents are analytically described by Gaussian jets to represent an exaggerated surface signature. In this case, two Gaussian jets describe the initial sea surface temperature and sea surface currents:

$$F(x, y) = \sum_i F_i(x, y), i = 1, 2, \tag{17}$$

where each jet is given by

$$F(x, y) = A(x) \exp\left[-\alpha(y - y_0)^2\right], \tag{18}$$

and

$$A(x) = A_0 \exp{-\beta(x)x^2}. \tag{19}$$

Here, $\beta(x)$ is given by

$$\beta(x) = \begin{cases} 0, & \text{if } \bar{x} < 0 \\ 10, & \text{if } \bar{x} > 0 \end{cases} \tag{20}$$

where $-\frac{1}{2} \le \bar{x} \le \frac{1}{2}$ is a normalized horizontal location. Table 2 lists the values of the Gaussian jet parameters.

Table 2. Gaussian jet parameters that were used in this study.

Parameter	Value for SST	Value for U	Value for V
A_0	20 K	0 m/s	10 m/s
y_{0_1}	8500 m	-	8500
y_{0_2}	6500 m	-	6500
α	1×10^{-6} m^2	-	1×10^{-6} m^2

The meteorological environment is chosen, so that surface winds are nearly perpendicular to the direction of the surface currents, as seen in Figure 2b, and the background SST is nearly uniform over the region of interest. This is constructed in such a way as to maximize the induced atmospheric perturbations. NCEP GDAS Final global analysis and forecast data provide meteorological initial and boundary conditions ($0.25° \times 0.25°$ resolution every 6 h). The total simulation length is 10 h, which includes 6 h of spin up time. The overall elevation in the atmospheric domain is approximately 20,500 m, and the simulation utilized a total of four nested grids, which are shown in Figure 2a. Table 3 lists the details of each grid.

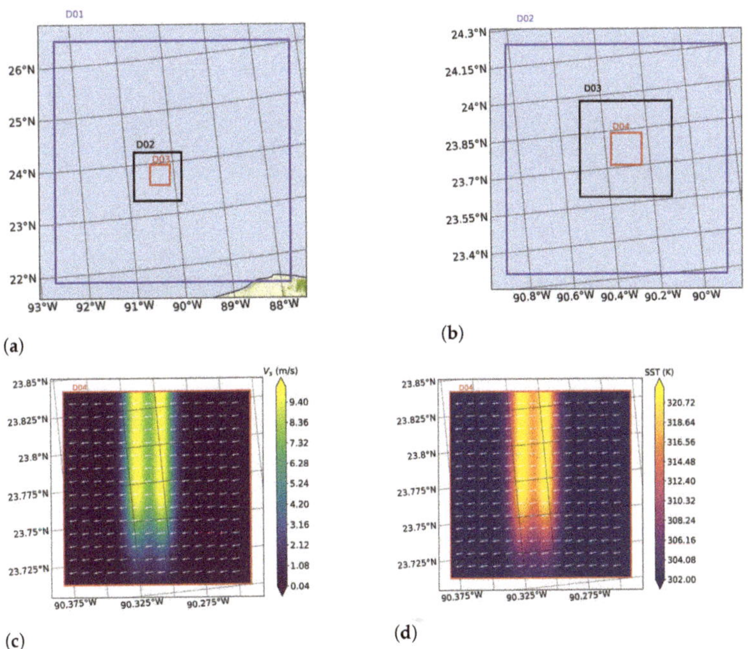

Figure 2. (**a**) Overview of nested domains D01-D03, (**b**) overview of nested domains D02–D03, (**c**) contours of sea surface meridonal currents (V_s) with white arrows showing 10 m wind direction, and (**d**) contours of sea surface temperature (SST).

Table 3. The Yonsel University Planetary Boundary Layer scheme (YSU PBL) is used for mesoscale simulations [20].

Simulation Type	PBL Treatment	Δx (m)	Height of 1st Grid Point (m)	Grid Points	Δt (s)
D01 mesoscale	YSU PBL	4500	5	$120 \times 120 \times 81$	27.0
D02 mesoscale	YSU PBL	900	5	$121 \times 121 \times 81$	5.4
D03 mesoscale	YSU PBL	300	5	$151 \times 151 \times 81$	1.8
D04 microscale	LES	100	5	$151 \times 151 \times 81$	0.6

Figure 2c,d show the contours of sea surface zonal currents and SST, respectfully, inside grid D04. As shown, the modified surface fields that are specified in Equations (17)–(20) are completely contained within the most refined grid (D04) in order to avoid introducing spurious behavior at the nested grid boundaries. This is consistent with the WRF published best practice recommendations, which suggest that nested grid boundaries be placed far away from regions of interest [27]. Figure 3 qualitatively shows the impact of the prescribed sea surface modifications, which show contours of the U-component of wind at 10 m elevation and vertical heat flux at the sea surface at the end of the 10-hour simulation time. In this case, three separate simulations are run: (1) a baseline case, in which no modification is introduced to the bottom boundary condition, (2) a SST-only case, in which the the SST is modified, and (3) a current-only case, in which only the meridonal currents are modified. This allows for independent evaluation of each modification on the atmospheric boundary layer. The signature of the prescribed jets is clearly visible in Figure 3, which indicates that the modified boundary conditions are being correctly ingested by the atmospheric model. However, both the 10 m wind and vertical heat flux are prognostic quantities that are computed directly from the imposed boundary conditions and are not necessarily indicative of the sensitivity of the model to the perturbations at the air-sea interface.

Figure 4 provides a more quantitative comparison, which show mean vertical profiles of potential temperature, water vapor mixing ratio, and vertical wind velocity evaluated at the center of grid D04. Again, deviations from the baseline predictions are readily apparent. Potential temperature and water vapor mixing ratio appear to be relatively insensitive to the imposed perturbations over the elevation ranges of interest. On the other hand, the vertical wind velocity that is shown in Figure 4c appears to be much more sensitive to modifications of the air-sea interface, which suggests that vertical wind measurements may be a key indicator in detecting sea surface perturbations in the atmosphere. However, a more rigorous parameter study is required before firm conclusions can be drawn.

3.2. Application to Surface Ship Wake

This section considers sea surface disturbances due to the passage of a surface ship. As the ship moves through the ocean, it generates a wake that perturbs the sea surface for many kilometers behind the traveling vessel. These perturbations propagate into the atmospheric boundary layer, although to what extent is not yet well understood.

A full, unsteady, three-dimensional simulation of the evolution of a ship wake is not computationally tractable, because of the range of length and time scales that are needed to resolve all of the relevant physics. Instead, a "two-dimensions plus time" (2D + t) approach is used, in which the problem domain is reduced to two spatial dimensions under the assumption that changes in the axial direction are negligible. Under this assumption, the wake is evolved in time, t, and in the (y, z) plane. Assuming that the ship is moving with constant speed, U_0, the x-location in the wake can be determined from the simulation time as $x = U_0 t$.

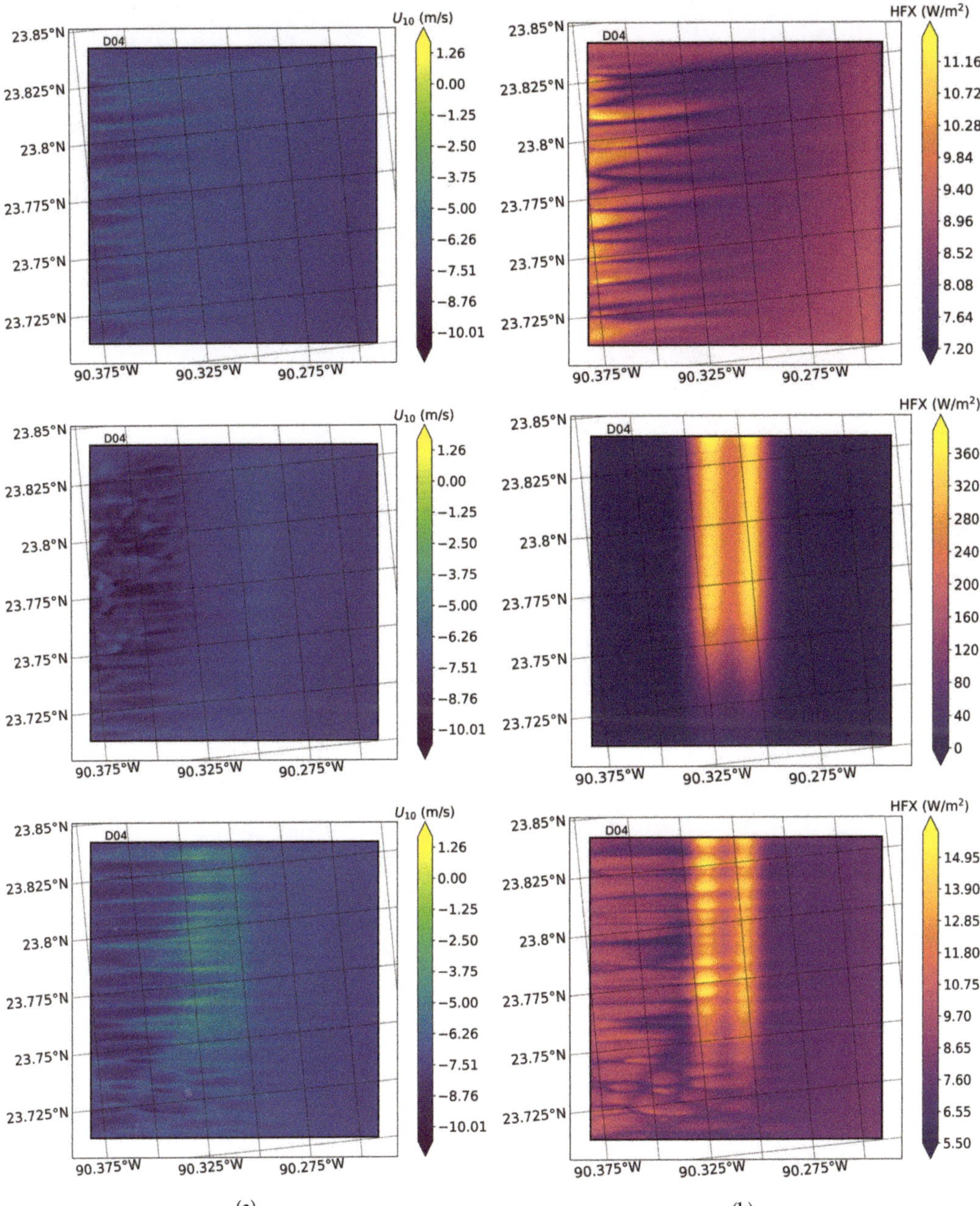

Figure 3. Predicted contours of (**a**) 10 m U-wind and (**b**) Upward heat flux. The top row contains predictions for the baseline case, the middle row contains predictions for SST-only modifications, and bottom most row shows predictions for sea surface current-only modifications.

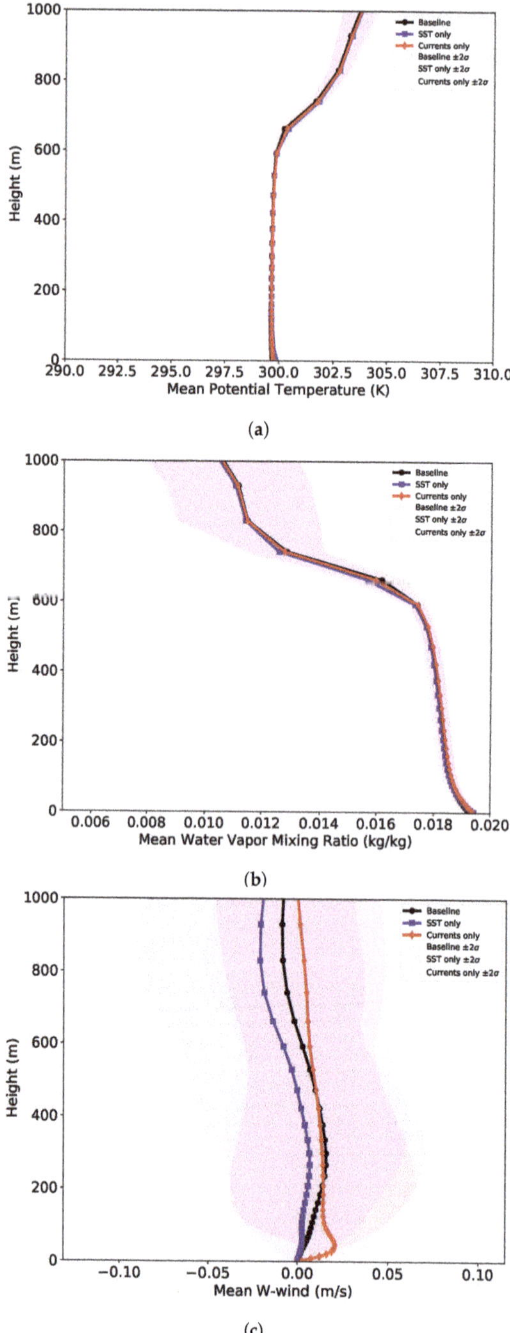

Figure 4. Mean vertical profiles of (**a**) potential temperature, (**b**) water vapor mixing ratio, and (**c**) vertical wind velocity. Shaded regions represent $\pm 2\sigma$ from the mean value.

The computational domain of the ocean model has the dimensions of -500 m $< y <$ 500 m by -500 m $< z < 0$ m and a nominal cell size of 0.3 m. The time step size is set to $\Delta t = 1$ s and the total simulation duration is 1 h. The $x = 0$ m location is set to be approximately one-half ship length downstream of the stern of the vessel, so as not to violate the 2D+t assumptions. In this study, the initial conditions at $t = 0$ s are generated via a semi-empirical formulation that is similar to Miner et al. [28]. Appendix A of Somero et al. [29] procides the complete details on the semi-empirical wake formulation.

Table 4 lists the details of the surface ship used in order to generate the initial conditions at $t = 0$ s. The ship's heading was specifically chosen to be aligned with the atmospheric domain's reference frame in order to assess relative resolution requirements of the atmospheric model.

Table 4. Surface ship parameters.

Parameter	Value
Forward speed (m/s)	13
Draft (m)	6.16
Beam (m)	18.9
Heading (deg)	5.52

Field data (e.g., currents, temperature, etc.) from the 2D + t predictions are first sampled along the free-surface ($z = 0$ m) in order to generate the bottom boundary conditions for the atmospheric model. The sampled fields are then mapped onto the atmospheric domain, as described in Section 2.3.1. Figure 5 shows contours of the sampled surface currents predicted by the ocean model and the corresponding mapped currents.

NCEP GDAS Final global analysis and forecast data provide the meteorological initial and lateral boundary conditions, as in Section 3.1. The total simulated time is 1 h and a total of five nested grid is used in this analysis in order to reach a sufficient level of refinement in the atmospheric domain in order to resolve enough of the wake. It should be noted that a simulation time of 1 h does not, in general, provide sufficient "spin up" time for a valid physical state to develop in the WRF model. As a result, specific model predictions during this time period are very likely inaccurate. Despite this limitation, this analysis is still useful for a qualitative assessment of the one-way coupled approach. Figure 6 shows the nested grids. Table 5 lists the relevant details for each grid used in this analysis.

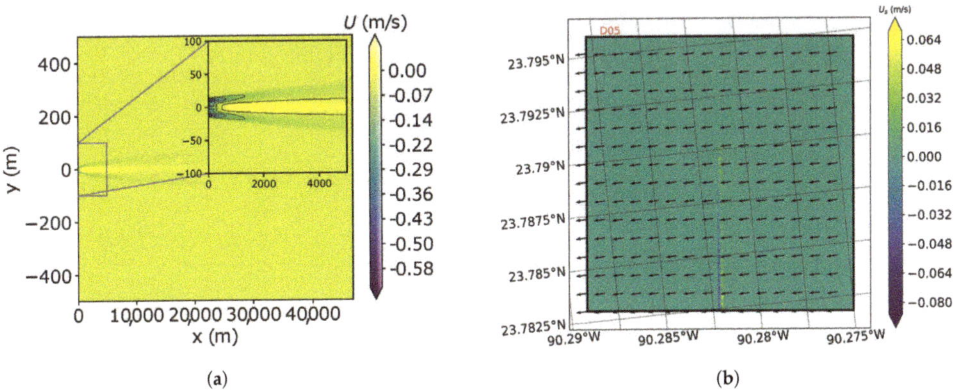

(a) (b)

Figure 5. *Cont.*

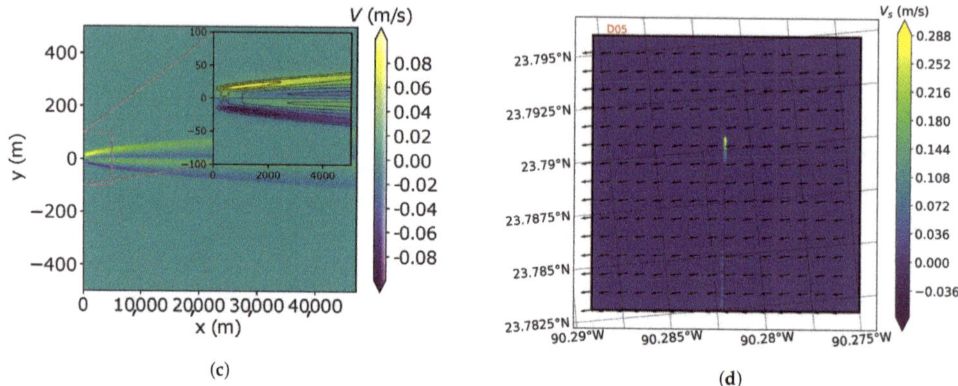

Figure 5. (**a**) U-component of surface currents predicted by the OpenFOAM model, (**b**) V-component of surface currents predicted by the OpenFOAM model, (**c**) U-component of surface currents mapped onto WRF domain, and (**d**) V-component of surface currents mapped onto WRF domain. The black arrows indicate the direction of the wind at 10 m elevation.

Table 5. The Yonsel University Planetary Boundary Layer scheme (YSU PBL) is used for mesoscale simulations [20].

Simulation Type	PBL Treatment	Δx (m)	Height of 1st Grid Point (m)	Grid Points	Δt (s)
D01 mesoscale	YSU PBL	2500	5	$120 \times 120 \times 81$	13.0
D02 mesoscale	YSU PBL	450	5	$121 \times 121 \times 81$	2.6
D03 mesoscale	YSU PBL	90	5	$151 \times 151 \times 81$	0.52
D04 microscale	LES	30	5	$151 \times 151 \times 81$	0.17
D05 microscale	LES	10	5	$151 \times 151 \times 81$	0.06

Figure 7 shows atmospheric model predictions of 10 m horizontal winds and sea surface vertical heat flux for the case with and without a ship wake. Larger fluctuations are predicted by the atmospheric model across the entire D05 domain for the ship wake case, and evidence of the ship wake is barely discernible in Figure 7f, around 23.79° N, 90.28125° W.

Figure 8 shows the vertical profiles of potential temperature, water vapor mixing ratio, and vertical wind taken from the center of domain D05 . Predictions of potential temperature and water vapor mixing ratio are not strongly affected by the presence of the modified surface currents caused by the ship wake, as was observed in Section 3.1. Conversely, differences in vertical wind, as shown in Figure 8c, are observed much higher into the air column, consistent with the results that are presented in Section 3.1. However, it should be noted that the fluctuations in the vertical wind are approximately an order of magnitude greater than those that are observed in Figure 4, due to the influence of transient pressure waves that are present during the spin up time.

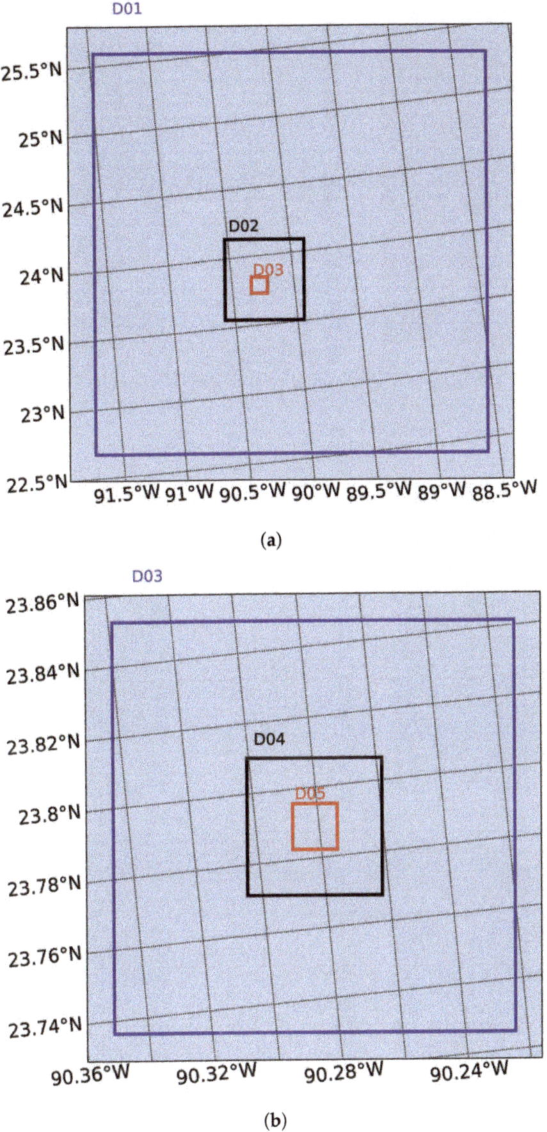

Figure 6. Overview of the atmospheric domain extents of (**a**) D01–D03, and (**b**) D03–D05 used in the surface ship wake analysis.

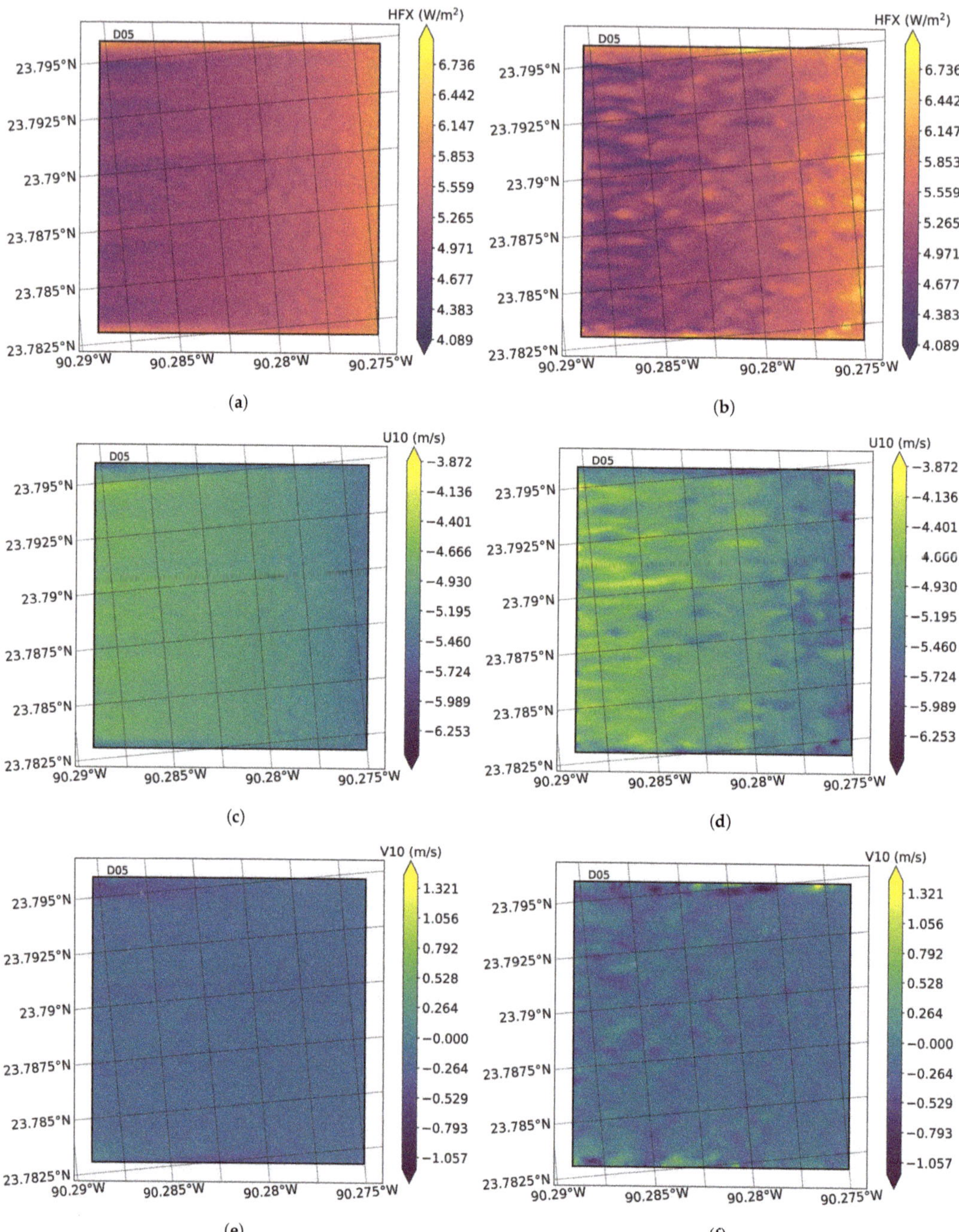

Figure 7. Predicted contours of upward heat flux (**a**,**b**), 10 m *U*-wind (**c**,**d**), and 10 m *V*-wind (**e**,**f**). Baseline case predictions without a ship wake are shown in the left column and the right column shows predictions with the ship wake present.

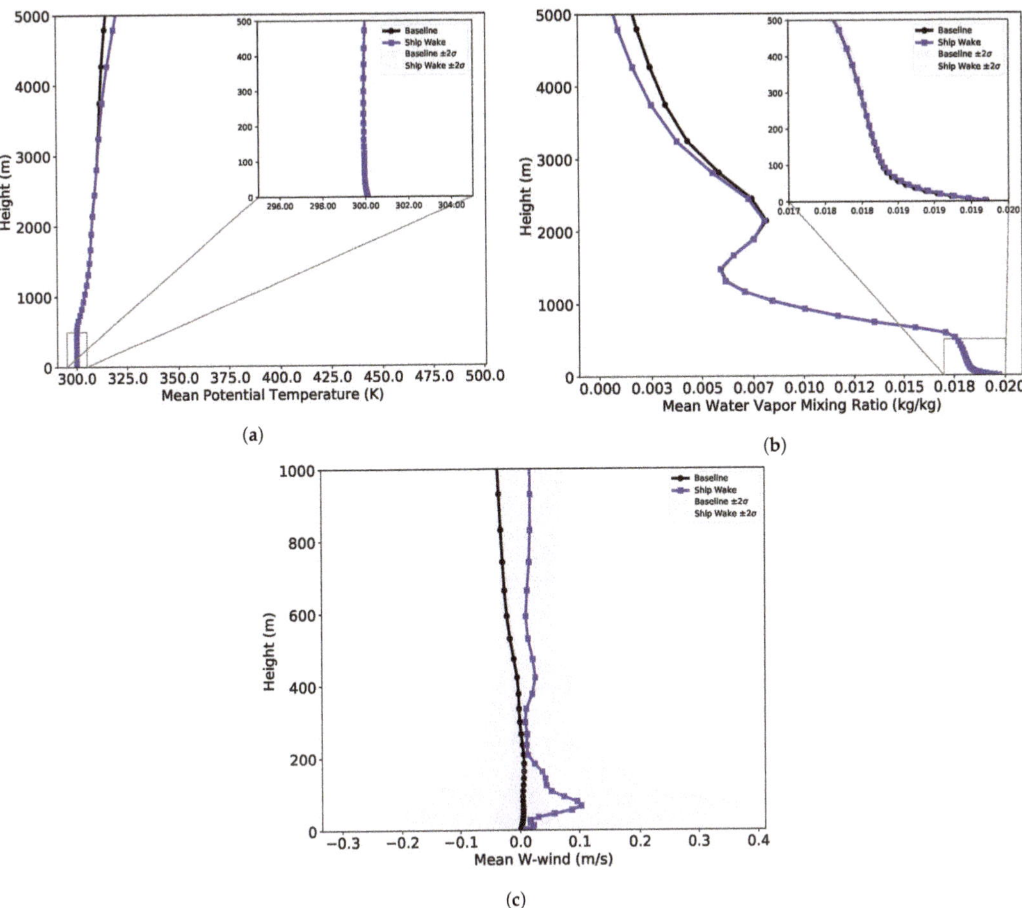

Figure 8. Vertical profiles of (**a**) potential temperature, (**b**) water vapor mixing ratio, and (**c**) vertical wind velocity. Shaded regions represent $\pm 2\sigma$ from the mean value.

4. Summary and Conclusions

A one-way coupled atmosphere-ocean model was presented in order to study the effects of air-sea interface perturbations on the atmospheric boundary layer. A partitioned approach is used in order to solve the coupled problem, in which the atmospheric domain is modeled while using the numerical weather prediction tool WRF, while the ocean domain is modeled using an unsteady RANS finite volume method implemented in OpenFOAM. One-way coupling is achieved by treating ocean model predictions of sea surface currents and sea surface temperature as a constant bottom boundary condition in the atmospheric model.

Two demonstration cases are presented in order to illustrate the performance of the proposed one-way coupled approach. The first case employs analytical functions in order to specify the perturbed currents and temperatures at the sea surface. The second case uses a semi-empirical relationship to describe the initial conditions of a surface ship wake, which is then evolved in time while using the ocean model. Predicted surface fields from the ocean model are extracted and applied as bottom boundary conditions in the atmospheric model. The modified boundary conditions were clearly visible in contours of 10 m U-wind and vertical heat flux predicted by the WRF model. In both demonstration cases, the vertical profiles of temperature and relative humidity that were predicted by the coupled model were relatively unaffected by the modified boundary conditions, while vertical wind velocity predictions were much more sensitive to the disturbed sea surface conditions.

It is important to note that the results that are presented herein should be considered preliminary and more work is required in order to improve and validate the proposed approach. Techniques, such as Digital Filter Initialization (DFI), need to be explored for reducing atmospheric model "spin up" time [30]. A careful grid convergence study needs to be performed in order to assess the impact of mesh resolution on the model predictions and determine appropriate resolution requirements for this application. Additionally, a parametric study should be conducted to assess which atmospheric parameters are the most sensitive to sea surface disturbances. Techniques for differentiating the relevant atmospheric fluctuations from the naturally occurring variability in the ambient environment is another key aspect of this problem that needs to be investigated. Finally, it will be important to assess whether the assumption of constant bottom boundary conditions is appropriate.

Author Contributions: Conceptualization, J.P.; Formal analysis, J.G.; Investigation, J.G. and J.P.; Methodology, J.G. and J.P.; Software, J.G.; Supervision, J.P.; Visualization, J.G.; Writing—original draft, J.G. and J.P. All authors have read and agreed to the published version of the manuscript.

Funding: This research was funded by the Cortana Corporation.

Institutional Review Board Statement: Not applicable.

Informed Consent Statement: Not applicable.

Data Availability Statement: The data presented in this study are available on request from the corresponding author.

Conflicts of Interest: The authors declare no conflict of interest.

Abbreviations

The following abbreviations are used in this manuscript:

WRF	Weather Research and Forecasting model
FVM	Finite Volume Method
RANS	Reynolds Averaged Navier Stokes
URANS	Unsteady RANS
2D + t	Two dimensions plus time
OpenFOAM	Open Field Operation And Manipulation
ROMS	Regional Ocean Modeling System

References

1. Komjathy, A.; Yang, Y.M.; Meng, X.; Verkhoglyadova, O.; Mannucci, A.J.; Langley, R.B. Review and perspectives: Understanding natural-hazards-generated ionospheric perturbations using GPS measurements and coupled modeling: NATURAL-HAZARDS-CAUSED TEC PERTURBATIONS. *Radio Sci.* **2016**, *51*, 951–961. [CrossRef]
2. Savastano, G.; Komjathy, A.; Verkhoglyadova, O.; Mazzoni, A.; Crespi, M.; Wei, Y.; Mannucci, A.J. Real-Time Detection of Tsunami Ionospheric Disturbances with a Stand-Alone GNSS Receiver: A Preliminary Feasibility Demonstration. *Sci. Rep.* **2017**, *7*, 46607. [CrossRef] [PubMed]
3. Yuan, T.; Wang, C.; Song, H.; Platnick, S.; Meyer, K.; Oreopoulos, L. Automatically Finding Ship Tracks to Enable Large-Scale Analysis of Aerosol-Cloud Interactions. *Geophys. Res. Lett.* **2019**, *46*, 7726–7733. [CrossRef]
4. Cooper, D.; Eichinger, W.; Barr, S.; Cottingame, W.; Hynes, V.; Keller, C.; Lebbda, C.; Poling, D.A. High-Resolution properties of the Equatorial Pacific marine atmospheric boundary layer from lidar and radiosonde observations. *J. Atmos. Sci.* **1996**, *53*, 2054–2074. [CrossRef]
5. Hagelberg, C.; Cooper, D.; Winter, C.; Eichinger, W. Scale properities of microscale convection in the marine surface layer. *J. Geophys. Res.* **1998**, *103*, 16897–16907. [CrossRef]
6. Boutanios, Z.; Miller, C.; Hangan, H. Computational Analysis of the Manitoba September 5 1996 Storm: Mesoscale WRF-ARW Simulations Coupled with Microscale OpenFOAM CFD Simulations. In Proceedings of the Fifth International Symposium on Computational Wind Engineering, Chapel Hill, NC, USA, 23–27 May 2010; p. 8.
7. Wyszogrodzki, A.A.; Miao, S.; Chen, F. Evaluation of the coupling between mesoscale-WRF and LES-EULAG models for simulating fine-scale urban dispersion. *Atmos. Res.* **2012**, *118*, 324–345. [CrossRef]

8. Miao, Y.; Liu, S.; Chen, B.; Zhang, B.; Wang, S.; Li, S. Simulating urban flow and dispersion in Beijing by coupling a CFD model with the WRF model. *Adv. Atmos. Sci.* **2013**, *30*, 1663–1678. [CrossRef]
9. Leblebici, E.; Tuncer, I.H. Coupled Unsteady Openfoam and WRF Solutions for an Accurate Estimation of Wind Energy Potential. In Proceedings of the VII European Congress on Computational Methods in Applied Sciences and Engineering, Crete Island, Greece, 5–10 June 2016; p. 8.
10. Leblebici, E.; Tuncer, D.I.H. Development of OpenFOAM—WRF Coupling Methodolgy for Wind Power Production Estimations. In Proceedings of the 4th Symposium on OpenFOAM in Wind Energy, Delft, The Netherlands, 2–4 May 2016; p. 3.
11. Haupt, S.E.; Berg, L.; Anderson, A.; Brown, B.; Churchfield, M.; Draxl, C.; Ennis, B.; Feng, Y.; Kosovic, B.; Kotamarthi, R.; et al. *First Year Report of the A2e Mesoscale to Microscale Coupling Project*; Technical Report; Pacific Northwest National Laboratory: Richland, WA, USA, 2015, [CrossRef]
12. Haupt, S.E.; Kotamarthi, R.; Feng, Y.; Mirocha, J.D.; Koo, E.; Linn, R.; Kosovic, B.; Brown, B.; Anderson, A.; Churchfield, M.J.; et al. *Second Year Report of the Atmosphere to Electrons Mesoscale to Microscale Coupling Project: Nonstationary Modeling Techniques and Assessment*; Technical Report PNNL-26267, 1573811; Pacific Northwest National Laboratory: Richland, WA, USA, 2017, [CrossRef]
13. Temel, O.; Bricteux, L.; van Beeck, J. Coupled WRF-OpenFOAM study of wind flow over complex terrain. *J. Wind Eng. Ind. Aerodyn.* **2018**, *174*, 152–169. [CrossRef]
14. Mughal, M.O.; Lynch, M.; Yu, F.; Sutton, J. Forecasting and verification of winds in an East African complex terrain using coupled mesoscale—And micro-scale models. *J. Wind Eng. Ind. Aerodyn.* **2018**, *176*, 13–20. [CrossRef]
15. Warner, J.C.; Armstrong, B.; He, R.; Zambon, J.B. Development of a Coupled Ocean–Atmosphere–Wave–Sediment Transport (COAWST) Modeling System. *Ocean Model.* **2010**, *35*, 230–244. [CrossRef]
16. Samson, G.; Masson, S.; Lengaigne, M.; Keerthi, M.G.; Vialard, J.; Pous, S.; Madec, G.; Jourdain, N.C.; Jullien, S.; Menkes, C.; Marchesiello, P. The NOW regional coupled model: Application to the tropical Indian Ocean climate and tropical cyclone activity. *J. Adv. Model. Earth Syst.* **2014**, *6*, 700–722. [CrossRef]
17. Varlas, G.; Katsafados, P.; Papadopoulos, A.; Korres, G. Implementation of a two-way coupled atmosphere-ocean wave modeling system for assessing air-sea interaction over the Mediterranean Sea. *Atmos. Res.* **2018**, *208*, 201–217. [CrossRef]
18. Skamarock, W.C.; Klemp, J.B.; Dudhia, J.; Gill, D.O.; Liu, Z.; Berner, J.; Wang, W.; Powers, J.G.; Duda, M.G.; Barker, D.M.; Huang, X.Y. *A Description of the Advanced Research WRF Model Version 4*; NCAR Technical Notes NCAR/TN-556+STR; National Center for Atmospheric Research: Boulder, CO,USA, 2019; p. 162.
19. Jiménez, P.A.; Dudhia, J.; González-Rouco, J.F.; Navarro, J.; Montávez, J.P.; García-Bustamante, E. A Revised Scheme for the WRF Surface Layer Formulation. *Mon. Weather Rev.* **2012**, *140*, 898–918. [CrossRef]
20. Hong, S.Y.; Noh, Y.; Dudhia, J. A New Vertical Diffusion Package with an Explicit Treatment of Entrainment Processes. *Mon. Weather Rev.* **2006**, *134*, 2318–2341. [CrossRef]
21. Thompson, G.; Field, P.R.; Rasmussen, R.M.; Hall, W.D. Explicit Forecasts of Winter Precipitation Using an Improved Bulk Microphysics Scheme. Part II: Implementation of a New Snow Parameterization. *Mon. Weather Rev.* **2008**, *136*, 5095–5115. [CrossRef]
22. Niu, G.Y.; Yang, Z.L.; Mitchell, K.E.; Chen, F.; Ek, M.B.; Barlage, M.; Kumar, A.; Manning, K.; Niyogi, D.; Rosero, E.; Tewari, M.; Xia, Y. The community Noah land surface model with multiparameterization options (Noah-MP): 1. Model description and evaluation with local-scale measurements. *J. Geophys. Res. Atmos.* **2011**, *116*. doi:10.1029/2010JD015139. [CrossRef]
23. Iacono, M.J.; Delamere, J.S.; Mlawer, E.J.; Shephard, M.W.; Clough, S.A.; Collins, W.D. Radiative forcing by long-lived greenhouse gases: Calculations with the AER radiative transfer models. *J. Geophys. Res. Atmos.* **2008**, *113*. doi:10.1029/2008JD009944. [CrossRef]
24. IOC; SCOR; IAPSO. *The International Thermodynamic Equation of Seawater—2010: Calculation and Use of Thermodynamic Properties*; Intergovernmental Oceanographic Commission, Manuals and Guides No. 56; UNESCO: Paris, France, 2010.
25. Beljaars, A.C.M. The parametrization of surface fluxes in large-scale models under free convection. *Q. J. R. Meteorol. Soc.* **1995**, *121*, 255–270. [CrossRef]
26. Mahrt, L.; Sun, J. The Subgrid Velocity Scale in the Bulk Aerodynamic Relationship for Spatially Averaged Scalar Fluxes. *Mon. Weather Rev.* **1995**, *123*, 3032–3041. [CrossRef]
27. Werner, K. Nesting in WRF. In Proceedings of the Basic WRF Tutorial, Boulder, CO, USA, 27–31 January 2020.
28. Miner, E.W.; Ramberg, S.E.; Swean, T.J. *A Method for Approximating the Initial Data Plane for Surface Ship Wake Simulations*; Technical Report NRL Memorandum Report 6376; Naval Research Laboratory: Washington, DC, USA, 1988.
29. Somero, J.R.; Basovich, A.; Paterson, E.G. Structure and Persistence of Ship Wakes and the Role of Langmuir-Type Circulations. *arXiv* **2018**, arXiv:1807.00441.
30. LYNCH, P.; HUANG, X.Y. Diabatic initialization using recursive filters. *Tellus A* **1994**, *46*, 583–597. doi:10.1034/j.1600-0870.1994.t01-4-00003.x. [CrossRef]

Article

On the Evaluation of Mesh Resolution for Large-Eddy Simulation of Internal Flows Using Openfoam [†]

Zahra Seifollahi Moghadam *, **François Guibault** and **André Garon**

Polytechnique Montréal, Mechanical Engineering Department, Montréal, QC H3T 1J4, Canada;
francois.guibault@polymtl.ca (F.G.); andre.garon@polymtl.ca (A.G.)
*   Correspondence: zahra.seifollahi-moghadam@polymtl.ca
†   This paper is an extended version of our paper published in The 15th OpenFOAM Workshop (OFW15), Washington, DC, USA, 22–25 June 2020.

Abstract: The central aim of this paper is to use OpenFOAM for the assessment of mesh resolution requirements for large-eddy simulation (LES) of flows similar to the ones which occur inside the draft-tube of hydraulic turbines at off-design operating conditions. The importance of this study is related to the fact that hydraulic turbines often need to be operated over an extended range of operating conditions, which makes the investigation of fluctuating stresses crucial. Scale-resolving simulation (SRS) approaches, such as LES and detached-eddy simulation (DES), have received more interests in the recent decade for understanding and mitigating unsteady operational behavior of hydro turbines. This interest is due to their ability to resolve a larger part of turbulent flows. However, verification studies in LES are very challenging, since errors in numerical discretization, but also subgrid-scale (SGS) models, are both influenced by grid resolution. A comprehensive examination of the literature shows that SRS for different operating conditions of hydraulic turbines is still quite limited and that there is no consensus on mesh resolution requirement for SRS studies. Therefore, the goal of this research is to develop a reliable framework for the validation and verification of SRS, especially LES, so that it can be applied for the investigation of flow phenomena inside hydraulic turbine draft-tube and runner at their off-design operating conditions. Two academic test cases are considered in this research, a turbulent channel flow and a case of sudden expansion. The sudden expansion test case resembles the flow inside the draft-tube of hydraulic turbines at part load. In this study, we concentrate on these academic test cases, but it is expected that hydraulic turbine flow simulations will eventually benefit from the results of the current research. The results show that two-point autocorrelation is more sensitive to mesh resolution than energy spectra. In addition, for the case of sudden expansion, the mesh resolution has a tremendous effect on the results, and, so far, we have not capture an asymptotic converging behavior in the results of Root Mean Square (RMS) of velocity fluctuations and two-point autocorrelation. This case, which represents complex flow behavior, needs further mesh resolution studies.

Keywords: mesh resolution; large-eddy simulation; OpenFOAM; channel flow; sudden expansion

Citation: Seifollahi Moghadam, Z.; Guibault, F.; Garon, A. On the Evaluation of Mesh Resolution for Large-Eddy Simulation of Internal Flows Using Openfoam. *Fluids* **2021**, *6*, 24. https://doi.org/10.3390/fluids6010024

Received: 1 November 2020
Accepted: 31 December 2020
Published: 5 January 2021

1. Introduction

Hydraulic-turbines are considered a highly reliable power source that can cover an extensive range of operating conditions in response to electricity demand. The importance of the large-eddy simulation (LES) for off-design operating conditions is due to the fact that RANS (Reynolds-averaged Navier–Stokes) studies are mainly capable of accurate flow simulations near the Best Efficiency Point, since optimal swirling flow occurs and the level of turbulence is low. However, at off-design operating conditions, swirling vortices are the main phenomena inside hydraulic turbine draft-tubes and the flow is highly turbulent. Therefore, it is necessary to use high fidelity methods, such as large-eddy simulation, to capture the turbulent fluctuations in the flow.

Analysis of flow phenomena for different components of hydraulic turbines has been performed by a large number of researchers in the last few decades. The URANS (Unsteady Reynolds-averaged Navier–Stokes) is the most common approach used by researchers for numerical analysis of the flow field in hydraulic turbines [1–3]. The limitation of URANS in predicting self-induced vortex rope was reported by Foroutan and Yavuzkurt (2012), and the importance of time-dependent boundary conditions with turbulent fluctuations at the inlet of the draft tube was shown [4]. SAS (Scale Adaptive Simulation), due to its capability to dynamically adjust the length scale in the turbulent flow and hence provide more detailed predictions of turbulent flow structures, is often employed for draft tube flow field behavior investigations [5,6]. For instance, in a study by Neto et al. (2012), SAS-SST (Shear Stress Transport) results of velocity components inside a draft tube were validated against experimental data for a Francis turbine at part load [7]. DES (Detached Eddy Simulation) as a Hybrid RANS/LES approach, applies the RANS approach for near wall modeling of turbulence and resolves the core flow region using LES. This approach was employed by Sentyabov et al. (2014) for the analysis of vortex core precession of a Kaplan turbine [8]. A comparison of URANS and DES potentials to predict turbulence statistics for the draft tube was presented by Paik et al. (2005). Although both methods agree in mean velocity field, the results of turbulence statistics show significant discrepancies for URANS [9]. DES was also used for the prediction of pressure pulsation in high-head Francis turbines by Minakov et al. (2015), and an accuracy of 10% was reported for the simulations [10].

A comparison of LES and SAS results for a sudden-expansion test case which resembles the vortex rope of a draft tube at part load was performed by Javadi and Nilsson (2014) [11]. Rotating stall mechanism of a pump-turbine was investigated by Pacot et al. (2014) using LES [12]. The results of efficiency prediction of a Kaplan turbine using Zonal LES (ZLES) is compared to SAS in a study by Morgut et al. (2015) [13]. LES with a cavitation model for Francis turbine is also performed at part load and high load in a study by Yang et al. (2016). In this study, however, the LES mesh requirements were not well-satisfied [14]. Comparison of LES results with DES and URANS results for the Francis-99 draft tube at part load, Best Efficiency Point (BEP), and high load was performed by Minakov et al. (2017) for mean velocity profiles and pressure fluctuations [15]. Compressible LES is performed by Trivedi and Dahlhaug (2018) for a whole configuration of the Francis-99 turbine to compare the high-amplitude stochastic fluctuations at speed no-load [16] and runaway [17].

In the papers featuring LES studies for hydraulic turbines, the effect of the mesh has barely been studied. Moreover, there is no consensus on the space and time resolution requirements for the LES or DES studies. The validation studies in most of the references are less than adequate, since most of the studies avoid to validate the turbulent characteristics of the flow. This points to a choice of academic test cases with proper Direct Numerical Simulation (DNS) or experimental validation data. In order to circumvent the complexities related to the geometry and other requirements for the boundary conditions of the flow simulation inside hydraulic turbine draft-tube, two academic test cases including internal non-swirling and swirling flows with available DNS and experimental data were chosen. First, a case of channel flow which has a simple geometry and includes complex near wall turbulent behavior was selected. Second, a sudden-expansion test case which resembles the swirling flow at part load operating condition of a hydraulic turbine draft-tube was investigated. In this paper, LES results are presented and analyzed for both test cases.

2. Simulation and Modeling of Turbulent Flow

In LES, the velocity field (U_i) is decomposed into a filtered (or resolved) component ($\overline{U_i}$) and a residual (or sub-grid scale, SGS) component (u'). By formally applying the filtering operation to the continuity equation and Navier–Stokes equations, it is possible to derive conservation laws for the filtered flow variables. Due to the linearity of the continuity equation, applying the filtering is straightforward. The form of the equation remains unchanged.

$$\frac{\partial \overline{U}_i}{\partial x_i} = 0.$$

(1)

The resulting Navier–Stokes equations are of the same form as the original Navier–Stokes equations and also a residual stress tensor or SGS stress tensor (τ_{ij}^r) appears:

$$\frac{\partial \overline{U}_j}{\partial t} + \overline{U}_i \frac{\partial \overline{U}_j}{\partial x_i} = \nu \frac{\partial^2 \overline{U}_j}{\partial x_i \partial x_i} - \frac{1}{\rho}\frac{\partial \overline{p}}{\partial x_j} - \frac{\partial \tau_{ij}^r}{\partial x_i}.$$

(2)

The equation of motions must be closed by modeling the SGS stress tensor. This is usually performed using an eddy viscosity model. A one-equation model and the WALE (Wall-Adapting Local Eddy-viscosity) model are used in the present research to model eddy-viscosity. In the WALE model, the eddy-viscosity reads as:

$$\nu_r = (C_w \Delta)^2 \frac{(S_{ij}^d S_{ij}^d)^{3/2}}{(\overline{S}_{ij}\overline{S}_{ij})^{5/2} + (S_{ij}^d S_{ij}^d)^{5/4}},$$

(3)

where

$$S_{ij}^d = \frac{1}{2}(\overline{g}_{ij}^2 + \overline{g}_{ji}^2) - \frac{1}{3}\delta_{ij}\overline{g}_{kk}^2$$

$$\overline{g}_{ij} = \frac{\partial \overline{U}_i}{\partial x_j} \quad ,$$

$$\overline{g}_{ij}^2 = \overline{g}_{ik}\overline{g}_{kj}$$

(4)

and the constant is in the range $0.55 \le C_w \le 0.60$. A one-equation model can be used to model SGS turbulent kinetic energy:

$$\frac{\partial k_{sgs}}{\partial t} + \frac{\partial (\overline{U}_j k_{sgs})}{\partial x_j} = \frac{\partial}{\partial x_j}\left[(\nu + \nu_{sgs})\frac{\partial k_{sgs}}{\partial x_j}\right] + P_{k_{sgs}} - \epsilon,$$

(5)

$$\nu_{sgs} = c_k \Delta k_{sgs}^{1/2},$$

(6)

$$\epsilon = C_\epsilon \frac{k_{sgs}^{3/2}}{\Delta},$$

(7)

$$P_{k_{sgs}} = 2\nu_{sgs}\overline{S}_{ij}\overline{S}_{ij}.$$

(8)

Very close to the wall, the van Driest damping function is used to correct the behavior for the ν_{sgs}. This function has the following form:

$$D = 1 - e^{\frac{-y^+}{A^+}},$$

(9)

where $A^+ = 26$. The final length scale is given by:

$$\Delta = min(\frac{\kappa y}{C_S}D, \Delta_g),$$

(10)

where Δ_g is a geometric-based filter length, such as the element cube-root volume delta.

Several approaches are presented in the literature to evaluate the resolution of LES. Following Gant (2010) [18] and Celik (2006) [19], some of these methods can be used by a prior RANS simulation or one LES calculation. For example, a good practice is to obtain the integral length scale L_{int} from a prior RANS simulation as $L_{int} = \frac{k^{3/2}}{\epsilon}$ and, therefore, to estimate the initial grid resolution for LES [20]. Therefore, they can also be called single-grid estimators, which means that, by running one LES calculation, they can be estimated. The multi-grid estimator methods require a number of LES calculations and some sort of Richardson Extrapolation. In this research, two-point autocorrelation is used for the evaluation of mesh resolutions in LES.

Correlation means the tendency of two values to change together, either in the same or opposite way. If we think of $u(x)$ as a component of the velocity along a line in statistically homogeneous turbulence, the autocovariance or two-point correlation in space is:

$$R(r) \equiv \langle u(x+r)u(x) \rangle, \tag{11}$$

where r is the separation distance between two points. As the two points move closer to each other, R increases. If the points move further away, R will go to zero. $R(r)$ is often normalized by the root-mean-square of velocity. Integral length scale L_{int} can be computed from the autocovariance which is the integral of $R(r)$ over the separation distance r, i.e.,

$$L_{int} = \frac{1}{\langle u^2 \rangle} \int_0^\infty R(r)dr. \tag{12}$$

The two point correlations are calculated by Davidson (2009, 2011) [20,21]. In this study, the separation distance is varied with the grid resolution. The number of cells needed for the autocovariance to approach zero is verified, thereby checking the grid resolution. It is recommended that in a good LES, the integral length scale should cover 8–16 cells. In this study, it is reported that the energy spectra and the ratio of the resolved turbulent kinetic energy to the total one is not a good measure of LES resolution. Two-point correlations are suggested as the best measures for estimating LES resolution.

The ways to create turbulent inlet boundary conditions vary and include the two main categories of synthesis inlets and precursor simulation methods which are briefly described in the following sections.

Divergence-free synthetic eddy method (DFSEM) was proposed by Poletto et al. (2013) [22] to reduce near-inlet pressure fluctuations and the development length in the original SEM method by Reference [23]. The steps in DFSEM are the following:

1. User selection of inlet surface Ω.
2. User definition of average velocity $u(x)$, Reynolds stresses and turbulence length-scales $\sigma(x)$, for $x \subset \Omega$.
3. Eddy bounding box taken as: $\max\{x+\sigma\}, \min\{x-\sigma\}$ for $x \subset \Omega$.
4. Definition of the number of eddies.
5. Assigning random positions x^k and intensities α^k to all the eddies.
6. Eddies being convected through the eddy box, $x^k = x^k + U_b * \Delta t$, where $U_b = \int_\Omega u ds / \int_\Omega ds$ is the bulk velocity.
7. $u'(x)$ calculated and superimposed to u to generate the inlet condition.
8. Repeat steps 6–7 for all the subsequent time steps.

The fluctuating velocity field is:

$$u'(x) = \sqrt{\frac{1}{N} \sum_{k=1}^{N} \frac{q_\sigma(|r^k|)}{|r^k|}} \times \alpha^k, \tag{13}$$

where:

- N: the number of eddies introduced into the DFSEM;
- $r^k = \frac{x-x^k}{\sigma^k}$ with σ^k being the eddy length scale for the kth eddy;
- $q_\sigma(|r|^k)$ is a suitable shape function; and
- α_i^k are random numbers with zero averages which represent eddy intensity.

The fluctuating velocity field, taking into account the turbulence anisotropy, is:

$$u'_\beta(x) = \sqrt{\frac{1}{N} \sum_{k=1}^{N} \sigma_\beta^k \left[1 - (d^k)^2\right] \epsilon_{\beta j l} r_j^k \alpha_l^k}, \tag{14}$$

where:

- $d^k = \sqrt{(r_j^k)^2}$; and
- $\epsilon_{\beta j l}$ is the Levi-Civita symbol.

The expression for the Reynolds stresses is also expressed as follows:

$$\langle u'_\beta u'_\gamma \rangle = \frac{1}{N} \sum_{k=1}^{N} \sigma_\beta^k \sigma_\gamma^k \epsilon_{\beta j l} \epsilon_{\gamma m n} \left\langle \left\{ \left[1 - (d^k)^2\right]^2 r_j^k r_m^k \right\} \right\rangle \langle (\alpha_l^k)(\alpha_n^k) \rangle. \tag{15}$$

Therefore, σ_β^k and σ_γ^k are the eddy length scales for the kth eddy in β and γ directions.

This method in comparison with original SEM methods is able to provide a divergence-free velocity field and also to reproduce all possible state of Reynolds stress anisotropy as a function of the characteristic ellipsoid eddy shapes described by the aspect ratio $\Gamma = \frac{\sigma_x}{\sigma_y} = \frac{\sigma_x}{\sigma_z}$.

3. Verification and Validation Results

3.1. Channel Flow Test Case

The turbulent flow in the channel is simulated at a moderate Reynolds number. The geometry of the channel is shown in Figure 1. The flow in the channel is considered statistically homogeneous in the spanwise direction (z) and statistically developing in the streamwise (x) and channel-height (y) directions. Domain size for the channel is $(20\pi, 2, \pi)$ [m]. The physical modeling is done based on friction Reynolds number which is $Re_{u_\tau} = \frac{u_\tau \delta}{\nu} = 395$, where $\delta = 1$ [m] is a characteristic length (Channel half-height) and friction velocity, $u_\tau = \sqrt{\frac{\tau_w}{\rho}}$ is assumed to be equal to 1 $[\frac{m}{s}]$. This research took advantage of OpenFOAM built-in mesh generators. Capabilities and meshing strategies of OpenFOAM are presented by Concli et al. (2020) [24]. The initial number of nodes for the mesh was chosen as (500, 46, 82) with a total of 1,866,000 cells. Therefore, the wall units which are the normalized distance of the first mesh vertex next to the wall are $(x^+, y^+, z^+) = (49.36, 1.1, 15.1)$. This orthogonal mesh results in a maximum aspect ratio of 23. Two sets of boundary conditions are considered for the generation of the inflow turbulence, periodic and Divergence Free Synthetic Eddy Method (DFSEM). The boundaries in the spanwise direction are considered periodic. k-equation is used as the SGS model.

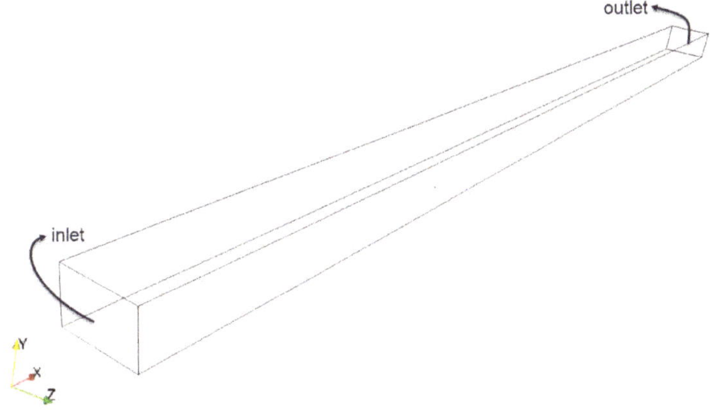

Figure 1. The geometry of the channel.

In the initial verification study, we want to investigate the influence of the inlet eddy sizes on the development of the flow inside the channel, and we want to verify that we can obtain the mean velocity and Reynolds stress field independent of the size of eddies generated at the inlet and transported into the domain. Therefore, the effect of the size of the inlet eddies will be evaluated based on the development length of the mean streamwise velocity profile inside the channel. Our hypothesis is that further increment in the size of the inlet eddies will not further exhibit changes in the flow development length. Starting the simulation by considering 1 cell per eddy simulation, and then increasing to 3 and 5, we expect that simulating each eddy with an adequate number of 3 cells would allow capturing non-decaying turbulence fluctuations in the domain.

The results of this verification study are provided in Figures 2 and 3. In these figures, the downstream development of the profiles of the mean streamwise velocity and $\langle u'u' \rangle$ component of the Reynolds stress is presented, respectively, for several cross sections and for different values of the parameter `nCellsPerEddy` representing the eddy sizes at the inlet. Moreover, the results for the periodic boundary condition case is also provided in the same figure, as a reference for the comparison.

The results of these simulations show enormous changes in the profiles of the mean streamwise velocity and $\langle u'u' \rangle$ component of the Reynolds stress, which are shown in Figures 2 and 3. In these figures, we can see how the flow develops across the channel using different conditions for the inlet eddies. Changing the value of the number of mesh cells per eddy to a value of 3, which means that each eddy is represented by at least 3 mesh cells in the domain, although requiring smaller time steps, improves the results of the development of the flow in the channel. The influence of further increasing this value to 5 does not exhibit a further improvement in the flow characteristics and only imposes more expensive computations in terms of time steps. In the aforementioned figures, the results at each streamwise section is averaged in the spanwise direction and time; therefore, the notation $\langle \rangle$ and $\{YZ\}$ index in the axis titles imply averaging in time and in the $\{YZ\}$ plane, respectively.

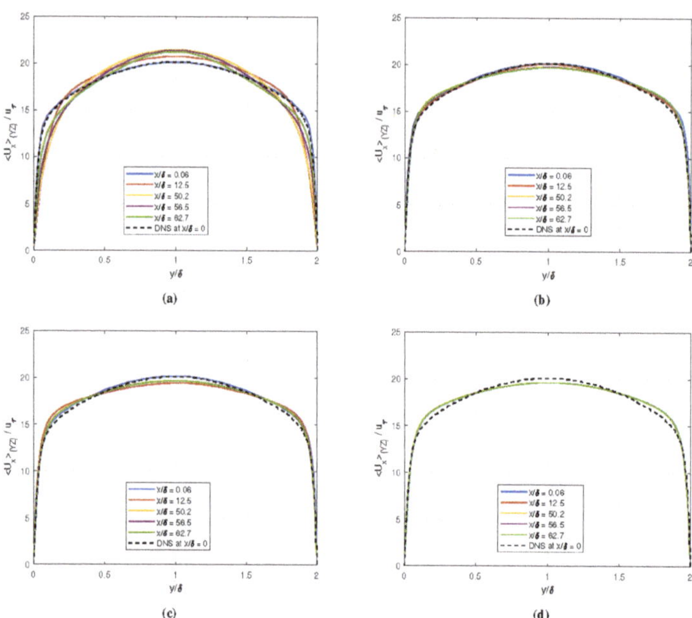

Figure 2. Mean velocity profiles at selected streamwise locations using various inlet conditionsas (**a**) 1 cell Per Eddy (**b**) 3 cells Per Eddy (**c**) 5 cells Per Eddy (**d**) Periodic.

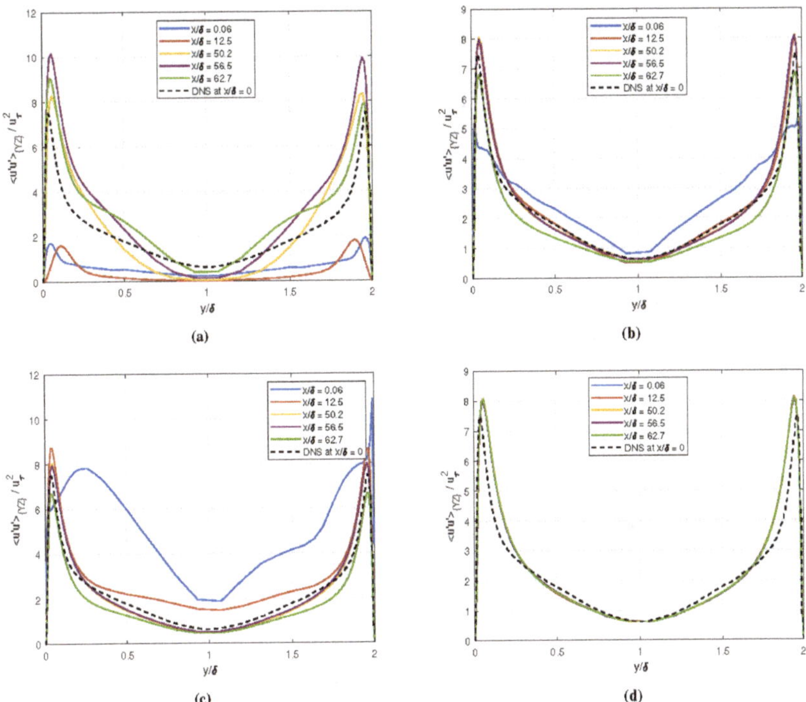

Figure 3. $\langle u'u' \rangle$ profiles at selected streamwise locations using various inlet conditions as (**a**) 1 cell Per Eddy (**b**) 3 cells Per Eddy (**c**) 5 cells Per Eddy (**d**) Periodic.

In Figure 3, the cross sections with the values of $x/\delta = 0.06$ and $x/\delta = 62.7$ represent the center of the first cell immediately after and before the inlet and outlet boundaries. It seems that a minimum number of cells in the streamwise direction and away from the boundaries is required for the attenuation of the effect of boundary conditions. Moreover, it seems that, although both boundary conditions with 3 and 5 mesh cells are representing shorter development length in comparison with the one with one mesh cell per eddy, which is evident by the results at downstream locations of x/δ of 12.5, 50.2, and 56.5 being almost on top of each other, the results of the 5 cells per eddy is worse than with 3 cells per eddy. A possible explanation could be that there should be an optimum number of cells per eddies for a given mesh, and increasing this value does not necessarily provide more accurate results. In other words, the optimum value for this parameter could be mesh-dependent. This hypothesis needs further investigation and analysis which is beyond the scope of this research.

For mesh resolution analysis, two other meshes are generated. The details of these meshes are presented in Table 1.

Table 1. Details of the numerical grids.

Name	Number of Cells	Total Size	Δx^+	Δz^+	y^+
Mesh 1	$397 \times 38 \times 65$	980,590	62.51	19.09	1.34
Mesh 2	$500 \times 46 \times 82$	1,886,000	49.63	15.13	1.11
Mesh 3	$630 \times 58 \times 103$	3,763,620	39.39	12.05	0.91

The results of the verification analysis for the three mesh resolutions including components of mean velocity and RMS of velocity fluctuations are presented in Figure 4. In this figure, the results are averaged in both $\{YZ\}$ and $\{XY\}$ planes, meaning both span-wise and stream-wise directions. The results show different flow behavior when moving farther from the wall in the outer layer ($y^+ > 50$). The finest mesh gives results that are in very good agreement with DNS across the whole region. We can conclude that grid refinement has a positive effect which is reflected in the gradual increase in the accuracy of the obtained results. The results of the RMS of velocity fluctuations show that all three components converge towards the DNS data as the mesh gets refined. Closer to the core region of the channel ($y/\delta > 0.5$), mesh 2 shows generally better prediction for the three components of RMS of velocities.

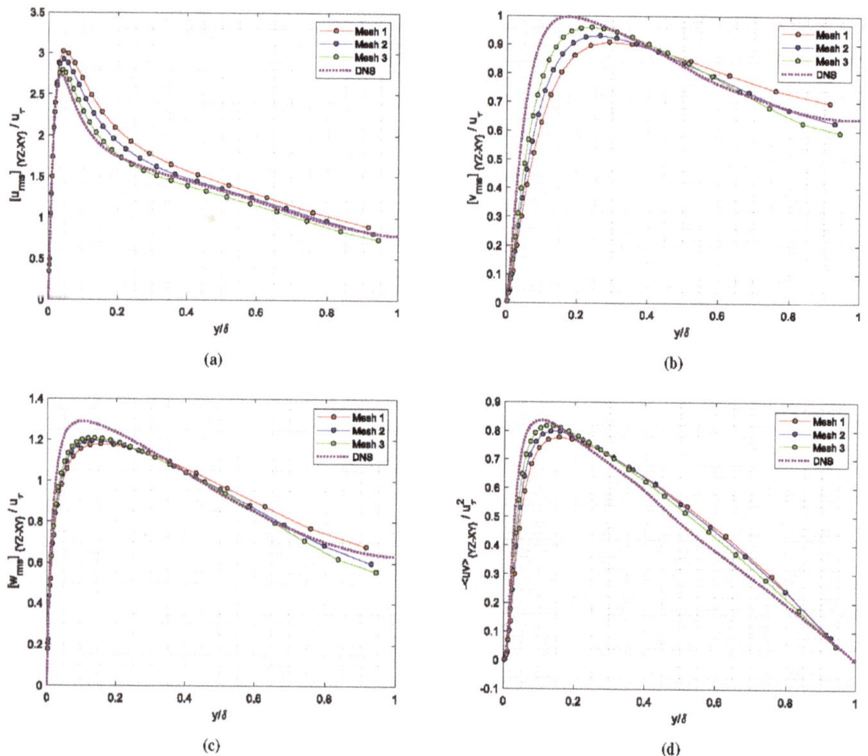

Figure 4. Comparison of RMS of velocity fluctuations for different mesh resolutions for (**a**) stream-wise, (**b**) wall-normal, (**c**) span-wise and (**d**) shear stress.

The second verification study for the mesh resolution is done for the two-point autocorrelation of velocity fluctuations. The results of this analysis are presented in Figure 5 for $y^+ = 5$. It should be noted that the results of autocorrelation in the x direction are averaged in a cross section plane (z-plane) and the results of autocorrelation in the z direction are averaged in the stream-wise plane (x-plane). In general, the profiles indicate that the chosen size of the domain is large enough to accommodate all the relevant turbulent structures, since the profiles reach near zero value or an asymptotic value in each computed direction. These profiles show higher sensitivity of this indicator to the mesh resolution, especially in the x-direction. The results computed with DFSEM boundary conditions, overall, show better performance than the periodic case, almost for all the cases, even for the coarsest mesh. In the integral length scale results in the Table 2, we do not observe a consistent trend in the results. These behaviors in the results show that further mesh refinement is still required to get a converged value for the integral length scales, especially in the streamwise direction, since the difference in the results for Mesh 3 and 2 is still big.

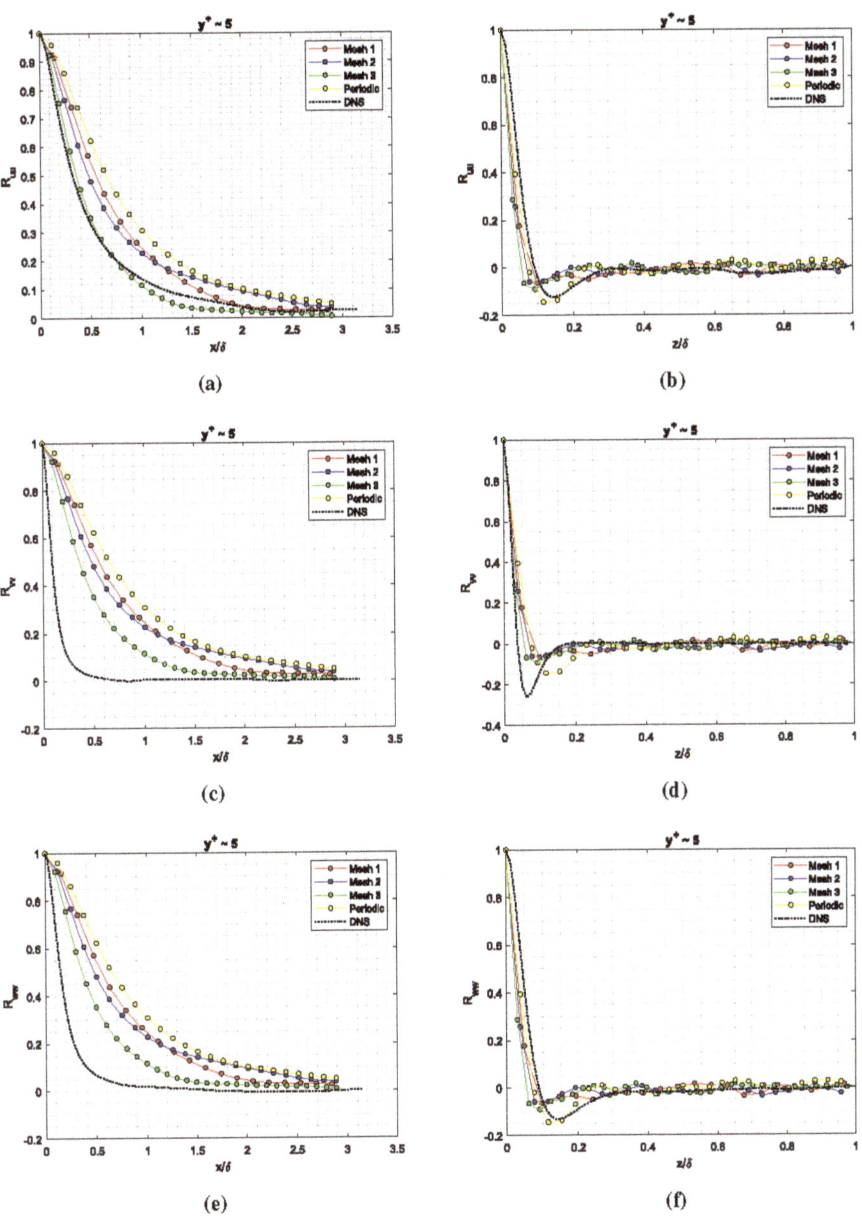

Figure 5. Spatial autocorrelation of velocity components at $y^+ = 5$: stream-wise velocity along (**a**) x and (**b**) z directions, wall-normal velocity along (**c**) x and (**d**) z directions, span-wise velocity along (**e**) x and (**f**) z directions.

Table 2. Integral length scales.

y^+		L_x					L_z				
		Mesh: 1	2	3	Periodic	DNS	Mesh: 1	2	3	Periodic	DNS
uu	5	0.74	0.75	0.51	0.90	0.53	0.02	0.01	0.01	0.02	0.02
uu	150	0.56	0.38	0.30	0.50	0.80	0.04	0.04	0.03	0.05	0.08
vv	5	0.74	0.75	0.51	0.90	0.13	0.02	0.01	0.01	0.02	0.01
vv	150	0.56	0.38	0.30	0.50	0.15	0.04	0.04	0.03	0.05	0.07
ww	5	0.74	0.75	0.51	0.90	0.20	0.02	0.01	0.01	0.02	0.02
ww	150	0.56	0.38	0.30	0.50	0.08	0.04	0.04	0.03	0.05	0.16

The third verification study for the mesh resolution is done based on the energy spectra. The objective of this verification analysis is to evaluate the influence of the mesh resolution on the energy spectra and to compare the spectra in the inertial subrange with the Kolmogorov spectrum. The results of this verification analysis are presented in Figure 6. These results show a much less sensitive behavior of the energy spectra to the mesh resolution; therefore, this parameter does not prove to be useful as an indicator of the LES mesh resolution.

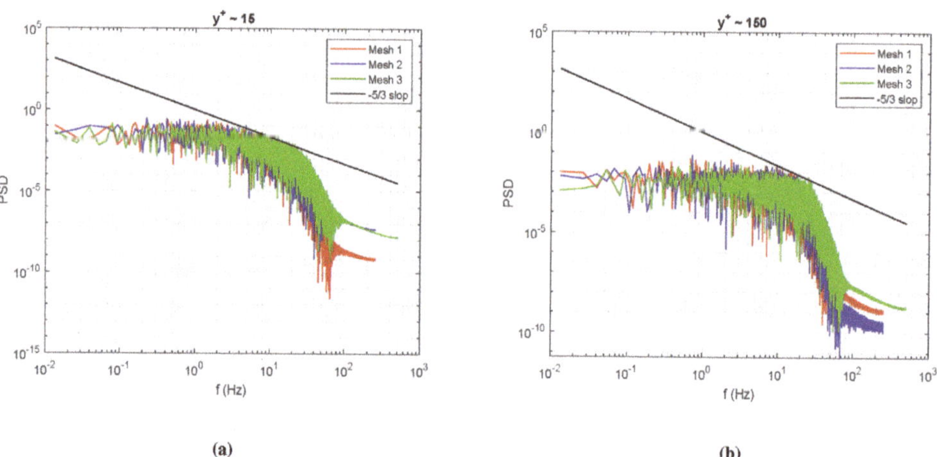

Figure 6. Energy spectra at two y^+ values of (**a**) 15 and (**b**) 150.

3.2. Sudden-Expansion Test Case

The Dellenback Abrupt Expansion test case resembles the swirling flow inside the draft-tube of hydraulic turbines at part load condition. Measurements of mean and fluctuating velocities were performed in a water flow with a laser Doppler anemometer [25]. The measurements were taken at the cross-sections shown in Figure 7. The simulation are performed at Reynolds number 30,000 and swirl number 0.6. The swirl number may be physically interpreted as the ratio of axial fluxes of swirl to linear momentum divided by a characteristic radius.

$$S = \frac{\int_0^{R_{in}} V_\theta V_z r^2 dr}{R_{in} \int_0^{R_{in}} V_z^2 r dr}\bigg|_{z/D_{in}=-2.00}.$$ (16)

The largest uncertainties in the experiments were reported to be about 2% in Reynolds number, 8% in swirl number, and 1% in probe volume positioning. Uncertainties in mean and RMS velocities stemming from the many possible biases and broadening errors were estimated to be about ±3% and ±10%, respectively.

The mesh is shown in Figure 8. This mesh has 1,567,944 cells. This mesh corresponds to a maximum aspect ratio of 21, maximum non-orthogonality of 28, average non-orthogonality of 5, and and maximum skewness of 0.7, which are within the acceptable range. For inlet boundary conditions, velocity profiles from the measurements are specified as axisymmetric profiles using the `profile1DfixedValue` boundary condition in OpenFOAM. The $k - \omega$ SST model is used as the turbulence model for the RANS and WALE model is used as the SGS model for the LES. For the case of simulations at swirl number 0.6, no inlet turbulence generation method is used, as it is stated in Reference [26], and the inlet turbulence is only important when the swirl in the flow is low.

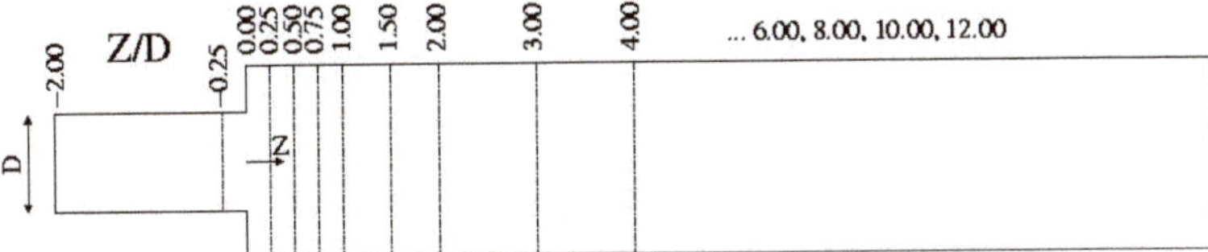

Figure 7. Measurement cross-sections. Numbers refer to Z/D, where D is the inlet diameter, and Z = 0 at the abrupt expansion.

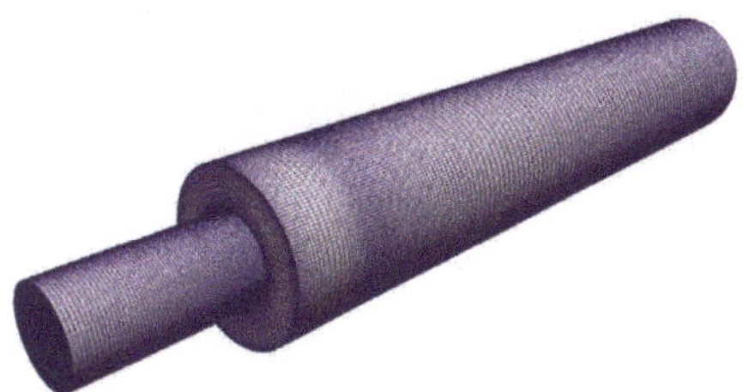

Figure 8. Numerical mesh for abrupt expansion test case.

The validation study is performed to compare the results of the mean velocity field and RMS of velocity fluctuations against experimental data. These results for the axial and tangential components of the mean velocity at the aforementioned cross sections are presented and compared with experimental data in Figures 9 to 10. The same comparison is also drawn for the axial and tangential components of the RMS of the velocity fluctuations in Figures 11 to 12. The results of the LES WALE model show great improvement over the results of the RANS k-ω SST model. These results show a huge improvement in the mean velocity results compared to the RANS model, especially immediately after the expansion where $z/D = 0.25$ to $z/D = 1.5$. LES also captured the RMS of velocity components with a good agreement with experimental data.

The verification analysis is performed to investigate the impact of the mesh resolution on the LES results. To this aim, two other meshes one coarser and one finer than the one for which results were presented in the previous paragraph were generated. The details about these meshes are given in Table 3. The results of these simulations are provided in Figures 13 and 14 for the mean velocities for axial and tangential components and in Figures 15 and 16 for RMS of velocity fluctuations, again for axial and tangential components. These results show higher sensitivity of the velocity profiles to mesh resolution for both mean and RMS values. A higher mesh resolution does not necessarily give less deviation from experimental data, due to the complex nature of the flow, especially for RMS of velocity fluctuations. However, in general, results for Mesh 3 appear to be slightly more consistent with experimental data.

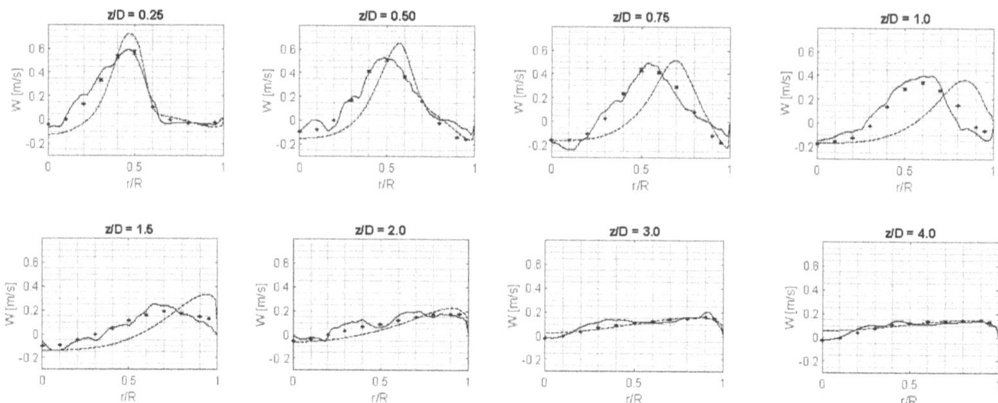

Figure 9. Mean axial velocity at several cross sections (dashed line: k-ω Shear Stress Transport (SST) Reynolds-averaged Navier–Stokes (RANS), solid line: Wall-Adapting Local Eddy-viscosity (WALE) large-eddy simulation (LES), dots: experiments).

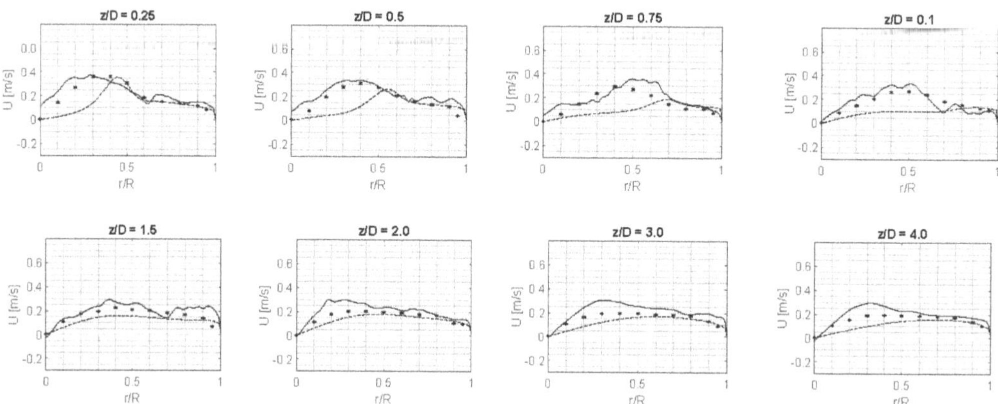

Figure 10. Mean tangential velocity at several cross sections (dashed line: k-ω SST RANS, solid line: WALE LES, dots: experiments).

The results of the two-point correlations of the components of the velocity fluctuations at the centerline of the cylinder downstream of the expansion are calculated for the three meshes. These results are presented in Figure 17. This figure shows that the complex vortical structures also represent a complex interactions which does not linearly go to zero as it changes randomly further downstream of the expansion.

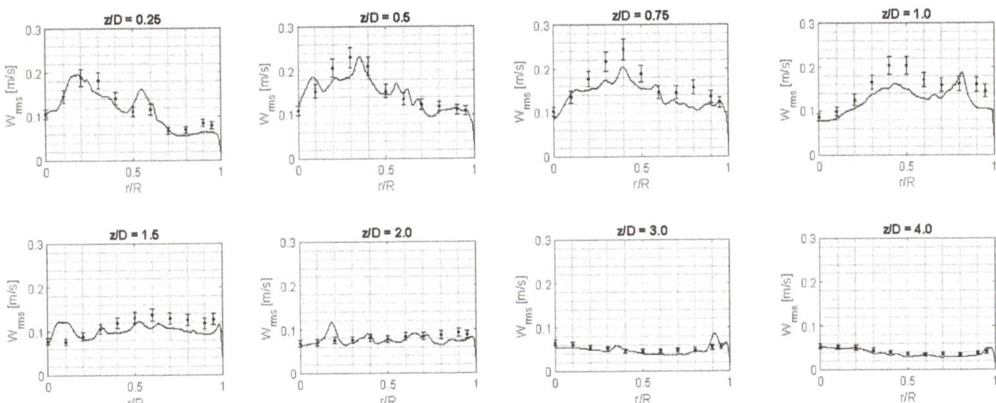

Figure 11. RMS of axial velocity at several cross sections (solid line: WALE LES, dots: experiments).

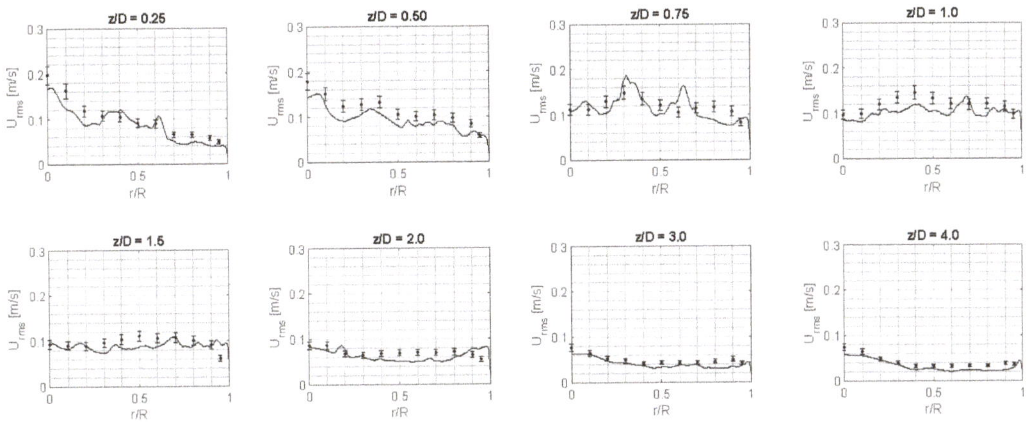

Figure 12. RMS of tangential velocity at several cross sections (solid line: WALE LES, dots: experiments).

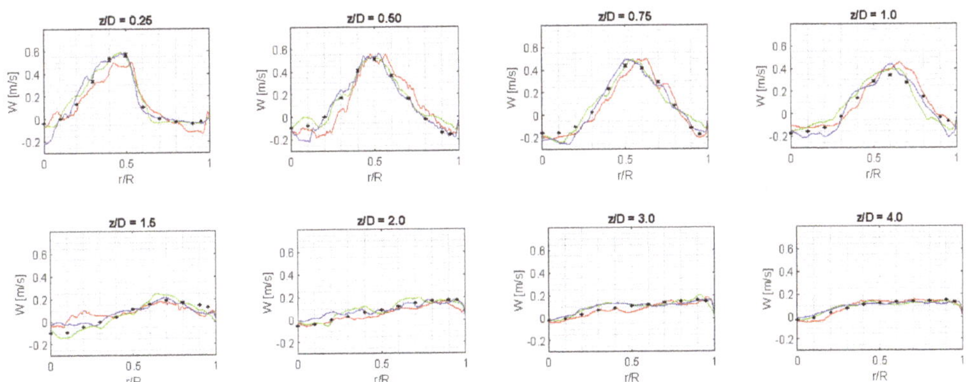

Figure 13. Mean axial velocity at several cross sections for different mesh resolutions (red: Mesh 1, green: Mesh 2, blue: Mesh 3, dots: experiments).

Table 3. Details of the numerical grids for sudden expansion test case.

Name	Number of Cells	Δt	y^+_{min}	y^+_{max}	$y^+_{average}$
Mesh 1	787,512	0.0005	0.195	35.42	5.29
Mesh 2	1,567,944	0.0005	0.360	28.70	4.21
Mesh 3	3,108,397	0.0001	0.412	54.17	6.38

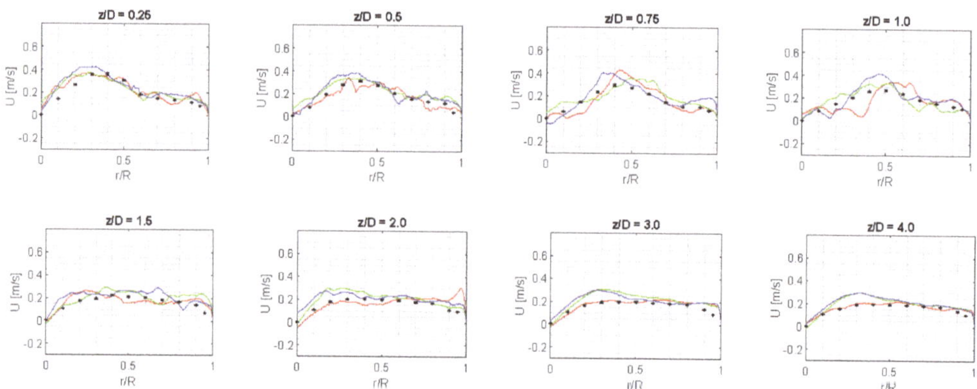

Figure 14. Mean tangential velocity at several cross sections for different mesh resolutions (red: Mesh 1, green: Mesh 2, blue: Mesh 3, dots: experiments).

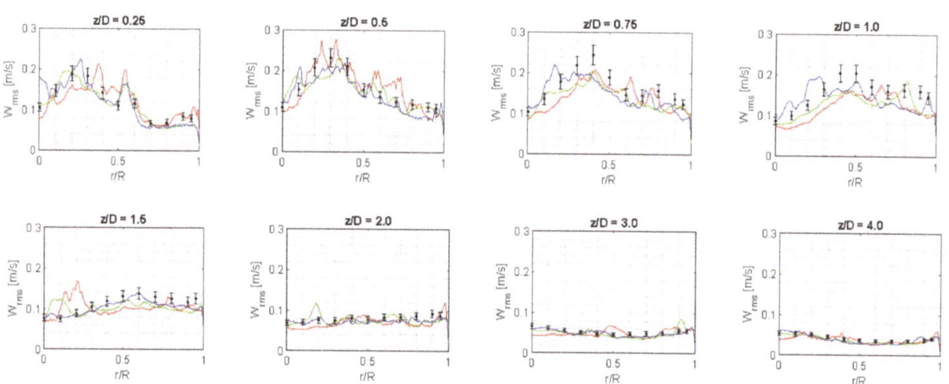

Figure 15. RMS of axial velocity at several cross sections for different mesh resolutions (red: Mesh 1, green: Mesh 2, blue: Mesh 3, dots: experiments).

The third verification analysis of the mesh resolution is performed for the results of the energy spectra of pressure fluctuations. The objective of this analysis, as for the case for the channel flow, is to investigate the effect of the mesh resolution on the convergence behavior of the energy spectra and also to compare the spectra in the inertial subrange with the Kolmogorov spectrum. Time history of the pressure fluctuations at several points at the wall is recorded and monitored during the simulations. Power spectral density of the pressure fluctuations at one point at $z/D = 2$ is calculated for the three meshes, and results are presented in Figure 18. This case also shows less sensitivity of the results to the mesh resolution, and this criteria can again not be used as an indicator of the mesh resolution adequacy.

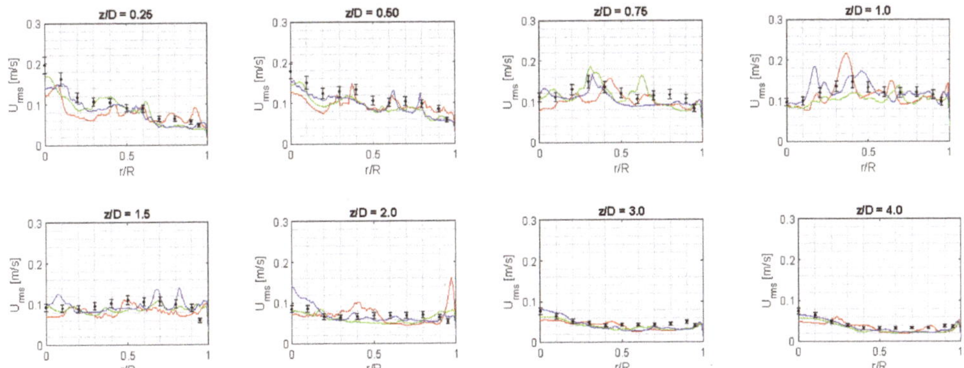

Figure 16. RMS of tangential velocity at several cross sections for different mesh resolutions (red: Mesh 1, green: Mesh 2, blue: Mesh 3, dots: experiments).

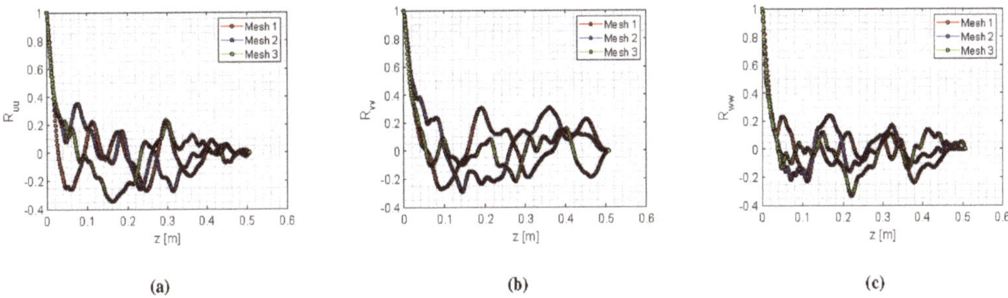

(a) (b) (c)

Figure 17. Two-point auto-correlation of the (**a**) stream-wise, (**b**) wall-normal, (**c**) span-wise components of the velocity fluctuations along the centerline.

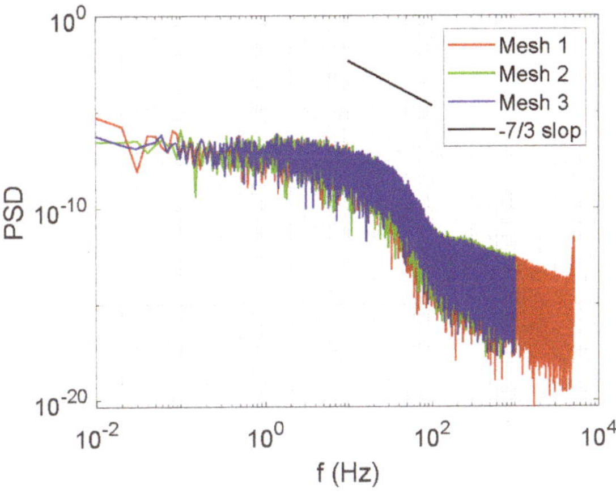

Figure 18. Power spectral density of the pressure fluctuations at one point on the wall at $z/D = 2$.

4. Conclusions

This work is dedicated to developing an expertise on how to use, validate, and verify the OpenFOAM CFD code for the large-eddy simulation of the flow types similar to the ones which occur at off-design operating condition, such as part load inside hydraulic turbines. Two main test cases, a case of turbulent channel flow and a case of sudden-expansion, were considered in this study. The first one brings the classical problem of near-wall turbulence complexities, and the second one resembles the swirling flow inside a hydraulic turbine draft-tube at part load off-design operating condition. The calculated profiles of two-point autocorrelation and hence the integral length scales showed that, even for a simple case of channel flow, obtaining mesh-independent results is still challenging, and further mesh-dependency analysis is required. The verified framework for LES on the channel flow was then applied to study the flow in a sudden expansion pipe flow at a Reynolds number of 30,000 and swirl number of 0.6. As with this case, the validation analysis of the LES WALE results versus RANS k-ω SST results show a large improvement of the LES results over the RANS ones for the mean and RMS of axial and tangential velocities. These improvements were also more evident at the cross sections which were immediately after the expansion. The influence of mesh resolution was also studied and showed a higher sensitivity of the results of this case to the mesh resolution and characteristics. The energy spectrum of the pressure fluctuations also did not exhibit much sensitivity to the mesh resolution, as it was also evident in the channel flow test case.

Author Contributions: Conceptualization, Z.S.M., F.G. and A.G.; methodology, Z.S.M., F.G. and A.G.; validation, Z.S.M., F.G. and A.G.; formal analysis, Z.S.M., F.G. and A.G.; investigation, Z.S.M., F.G. and A.G.; writing—original draft preparation, Z.S.M.; writing—review and editing, Z.S.M., F.G. and A.G.; visualization, Z.S.M.; supervision, F.G. and A.G.; project administration, F.G. and A.G.; funding acquisition, F.G. and A.G. All authors have read and agreed to the published version of the manuscript.

Funding: This research was supported by the Natural Sciences and Engineering research Council of Canada through Discovery Grants (RGPIN-04935-2015) and a Grant from the Collaborative research and Training Experience (CREATE-481695-2016) program in *Simulation-based Engineering Science* (Génie Par la Simulation).

Acknowledgments: This research was supported by the Natural Sciences and Engineering research Council of Canada through Discovery Grants (RGPIN-04935-2015) and a Grant from the Collaborative research and Training Experience (CREATE-481695-2016) program in *Simulation-based Engineering Science* (Génie Par la Simulation).

Conflicts of Interest: The authors declare no conflict of interest.

References

1. Gentner, C.; Sallaberger, M.; Widmer, C.; Bobach, B.J.; Jaberg, H.; Schiffer, J.; Senn, F.; Guggenberger, M. Comprehensive experimental and numerical analysis of instability phenomena in pump turbines. *IOP Conf. Ser. Earth Environ. Sci.* **2014**, *22*, 032046. [CrossRef]
2. Nicolle, J.; Giroux, A.M.; Morissette, J.F. CFD configurations for hydraulic turbine startup. *IOP Conf. Ser. Earth Environ. Sci.* **2014**, *22*, 032021. [CrossRef]
3. Hosseinimanesh, H.; Devals, C.; Nennemann, B.; Reggio, M.; Guibault, F. A Numerical Study of Francis Turbine Operation at No-Load Condition. *J. Fluids Eng.* **2016**, *139*, 011104. [CrossRef]
4. Foroutan, H.; Yavuzkurt, S. Simulation of flow in a simplified draft tube: turbulence closure considerations. *IOP Conf. Ser. Earth Environ. Sci.* **2012**, *15*, 022020. [CrossRef]
5. Krappel, T.; Riedelbauch, S.; Jester-Zuerker, R.; Jung, A.; Flurl, B.; Unger, F.; Galpin, P. Turbulence Resolving Flow Simulations of a Francis Turbine in Part Load using Highly Parallel CFD Simulations. *IOP Conf. Ser. Earth Environ. Sci.* **2016**, *49*, 062014. [CrossRef]
6. Pasche, S.; Avellan, F.; Gallaire, F. Part Load Vortex Rope as a Global Unstable Mode. *J. Fluids Eng.* **2017**, *139*, 051102. [CrossRef]
7. Neto, A.D.A.; Jester-Zuerker, R.; Jung, A.; Maiwald, M. Evaluation of a Francis turbine draft tube flow at part load using hybrid RANS-LES turbulence modelling. *IOP Conf. Ser. Earth Environ. Sci.* **2012**, *15*, 062010. [CrossRef]

8. Sentyabov, A.V.; Gavrilov, A.A.; Dekterev, A.A.; Minakov, A.V. Numerical investigation of the vortex core precession in a model hydro turbine with the aid of hybrid methods for computation of turbulent flows. *Thermophys. Aeromech.* **2014**, *21*, 707–718. [CrossRef]
9. Paik, J.; Sotiropoulos, F.; Sale, M.J. Numerical Simulation of Swirling Flow in Complex Hydroturbine Draft Tube Using Unsteady Statistical Turbulence Models. *J. Hydraul. Eng.* **2005**, *131*, 441–456. [CrossRef]
10. Minakov, A.; Platonov, D.; Dekterev, A.; Sentyabov, A.; Zakharov, A. The analysis of unsteady flow structure and low frequency pressure pulsations in the high-head Francis turbines. *Int. J. Heat Fluid Flow* **2015**, *53*, 183–194. [CrossRef]
11. Javadi, A.; Nilsson, H. A comparative study of scale-adaptive and large-eddy simulations of highly swirling turbulent flow through an abrupt expansion. *IOP Conf. Ser. Earth Environ. Sci.* **2014**, *22*, 022017. [CrossRef]
12. Pacot, O.; Kato, C.; Avellan, F. High-resolution LES of the rotating stall in a reduced scale model pump-turbine. *IOP Conf. Ser. Earth Environ. Sci.* **2014**, *22*, 022018. [CrossRef]
13. Morgut, M.; Jošt, D.; Nobile, E.; Škerlavaj, A. Numerical investigation of the flow in axial water turbines and marine propellers with scale-resolving simulations. *J. Phys. Conf. Ser.* **2015**, *655*, 012052. [CrossRef]
14. Yang, J.; Zhou, L.; Wang, Z. The numerical simulation of draft tube cavitation in Francis turbine at off-design conditions. *Eng. Comput.* **2016**, *33*, 139–155. [CrossRef]
15. Minakov, A.; Platonov, D.; Sentyabov, A.; Gavrilov, A. Francis-99 turbine numerical flow simulation of steady state operation using RANS and RANS/LES turbulence model. *J. Phys. Conf. Ser.* **2017**, *782*, 012005. [CrossRef]
16. Trivedi, C. Compressible Large Eddy Simulation of a Francis Turbine During Speed-No-Load: Rotor Stator Interaction and Inception of a Vortical Flow. *J. Eng. Gas Turbines Power* **2018**, *140*, 112601. [CrossRef]
17. Trivedi, C.; Dahlhaug, O.G. Interaction between trailing edge wake and vortex rings in a Francis turbine at runaway condition: Compressible large eddy simulation. *Phys. Fluids* **2018**, *30*, 075101. [CrossRef]
18. Gant, S.E. Reliability Issues of LES-Related Approaches in an Industrial Context. *Flow Turbul. Combust.* **2010**, *84*, 325–335. [CrossRef]
19. Celik, I.; Klein, M.; Freitag, M.; Janicka, J. Assessment measures for URANS/DES/LES: an overview with applications. *J. Turbul.* **2006**, *7*, N48. [CrossRef]
20. Davidson, L. Large Eddy Simulations: How to evaluate resolution. *Int. J. Heat Fluid Flow* **2009**, *30*, 1016–1025. [CrossRef]
21. Salvetti, M.V. (Ed.) *Quality and Reliability of Large-Eddy Simulations II*; Number 16 in ERCOFTAC Series; OCLC: 837896204; Springer: Dordrecht, The Netherlands, 2011.
22. Poletto, R.; Craft, T.; Revell, A. A New Divergence Free Synthetic Eddy Method for the Reproduction of Inlet Flow Conditions for LES. *Flow Turbul. Combust.* **2013**, *91*, 519–539. [CrossRef]
23. Jarrin, N.; Prosser, R.; Uribe, J.C.; Benhamadouche, S.; Laurence, D. Reconstruction of turbulent fluctuations for hybrid RANS/LES simulations using a Synthetic-Eddy Method. *Int. J. Heat Fluid Flow* **2009**, *30*, 435–442. [CrossRef]
24. Concli, F.; Schaefer, C.T.; Bohnert, C. Innovative Meshing Strategies for Bearing Lubrication Simulations. *Lubricants* **2020**, *8*, 46. [CrossRef]
25. Dellenback, P.A.; Metzger, D.E.; Neitzel, G.P. Measurements in turbulent swirling flow through an abrupt axisymmetric expansion. *AIAA J.* **1988**, *26*, 669–681. [CrossRef]
26. Schlüter, J.U.; Pitsch, H.; Moin, P. Large-Eddy Simulation Inflow Conditions for Coupling with Reynolds-Averaged Flow Solvers. *AIAA J.* **2004**, *42*, 478–484. [CrossRef]

 *fluids*

MDPI

Article

On the Use of a Domain Decomposition Strategy in Obtaining Response Statistics in Non-Gaussian Seas

Griet Decorte [1,*,†], Alessandro Toffoli [2,†], Geert Lombaert [1] and Jaak Monbaliu [1]

[1] Department of Civil Engineering, Faculty of Engineering Science, KU Leuven, Kasteelpark Arenberg 40, B-3001 Leuven, Belgium; geert.lombaert@kuleuven.be (G.L.); jaak.monbaliu@kuleuven.be (J.M.)
[2] Department of Infrastructure Engineering, Melbourne School of Engineering, The University of Melbourne, Parkville, VIC 3010, Australia; toffoli.alessandro@gmail.com
* Correspondence: griet.decorte@kuleuven.be
† These authors contributed equally to this work.

Abstract: During recent years, thorough experimental and numerical investigations have led to an improved understanding of dynamic phenomena affecting the fatigue life and survivability of offshore structures, e.g., ringing and springing and extreme wave impacts. However, most of these efforts have focused on modeling either selected extreme events or sequences of highly nonlinear waves impacting offshore structures, possibly overestimating the actual load to be experienced by the structure. Overall, not much has been done regarding short-term statistics. Although clear non-Gaussian statistics and therefore higher probabilities of extreme waves have been observed in random seas due to wave–wave interaction phenomena, which can impact short-term statistics for the structural load, they have not been studied extensively regarding the assessment of the dynamic behavior of offshore structures. Computational fluid dynamics (CFD) models have shown their viability for studying wave–structure interaction phenomena. Despite the continuously increasing computational resources, these models remain too computationally demanding for applications to the large spatial domains and long periods of time necessary for studying short-term statistics of non-Gaussian seas. Higher-order spectral (HOS) models, on the other hand, have been proven to be efficient and adequate in studying non-Gaussian seas. We therefore propose a one-way domain decomposition strategy, which takes full advantage of the recent advances in CFD and of the computational benefits of HOS. When applying this domain decomposition strategy, it appeared to be possible to deduce response statistics regarding the impact of nonlinear wave–wave interactions.

Keywords: non-Gaussian seas; wave–structure interaction; HOS; OpenFOAM

Citation: Decorte, G.; Toffoli, A.; Lombaert, G.; Monbaliu, J. On the Use of a Domain Decomposition Strategy in Obtaining Response Statistics in Non-Gaussian Seas. *Fluids* **2021**, *6*, 28. https://doi.org/10.3390/fluids6010028

Received: 28 November 2020
Accepted: 5 January 2021
Published: 7 January 2021

Publisher's Note: MDPI stays neutral with regard to jurisdictional claims in published maps and institutional affiliations.

1. Introduction

Due to the world's ever-growing energy demand, the offshore wind industry has known an exponential growth during the last two decades. As a result, there is a strong tendency towards increasingly larger wind turbines being constructed offshore and cost-reduction per kilowatt hour, which poses new challenges regarding design and efficiency [1,2]. XL-monopiles are one such example. Offshore wind turbines have grown increasingly taller and are installed in deeper water depths than ever before, which jeopardizes their fatigue life due to a higher susceptibility to ringing and springing as their natural frequencies decrease further towards typical wave frequencies [3]. Moreover, innovative structures combining wave energy take-off and harvesting wind energy, e.g., combined offshore wind and offshore airborne prototypes, confront designers with substantial differences in dynamic behavior compared to traditional structures stemming from the oil and gas industry. Especially in terms of prevailing fluid phenomena, viscous effects and turbulence gain considerable importance when assessing such structures.

Traditionally, design practices adopt Gaussian seas to represent operational conditions, while real seas are non-Gaussian [4]. In modeling moderately severe wave climates, second-

145

order wave theory is adopted, which gives rise to increased dynamic response for both stiff and compliant structures due to bound second-order wave components [5]. However, in addition to bound components, free wave components are known to interact at third order of nonlinearity for intermediate to deep water wave conditions, which results in spectral energy downshifts and a higher probability of extremes compared to second-order wave theory [6,7]. Due to these spectral energy downshifts, the dynamic response of stiff structures is reduced [8], while at the same time increased probability for extreme events, due to focussing, could lead to a higher occurrence of transient dynamic phenomena, such as ringing and slamming. An improved understanding of the influence of these nonlinear wave–wave interactions on the response statistics could therefore lead to a more appropriate assessment of offshore structures in operational conditions.

Expensive physical experiments in large-scale facilities and simplified engineering models originating from the offshore industry are still the norm. Most numerical models resolve the Euler equations, assuming incompressibility, irrotationality and inviscosity. Moreover, a Taylor expansion around the water level is used to approximate the kinematic and dynamic atmospheric boundary conditions imposed for the wave dynamics problem. As a result, linear potential theory based models, e.g., NEMOH, and expansions up to second-order, e.g., WAMIT, are still of main use, which is a clear consequence of offshore activities being situated in deep water, where second-order theory is generally assumed to capture most phenomena. However, as bottom-founded structures are situated closer to shore and higher-order wave effects, e.g., ringing, originate from higher-order wave harmonics, these models do not suffice in assessing state-of-the-art innovative offshore structures. Therefore, extensive research has been done in recent years to resolve the wave dynamics boundary value problem with fully nonlinear boundary conditions. Two such efforts are the nonlinear potential flow solvers OceanWave3D and formulations of the higher-order spectral equations (HOS). While OceanWave3D is essentially a fully nonlinear potential flow model [9], HOS solves the Euler equations up to a desired level of nonlinearity in the (spectral) wave number domain [10,11]. By truncating up to third-order nonlinearity, HOS has shown to accurately capture the effect of nonlinear four-wave interactions on extreme wave statistics [12]. Nonetheless, all of the aforementioned models still lack in the sense that they do not model viscosity nor turbulence.

As computational fluid dynamics (CFD) solves the full Reynolds-averaged Navier–Stokes (RANS) equations, it overcomes the problems previously posed by the Euler based models, therefore allowing for viscosity and turbulent effects to be taken along [13]. CFD accurately captures the wave–structure interaction, even in highly nonlinear waves. A notable code is the open-source CFD solver OpenFOAM, of which different branches exist and to which researchers worldwide contribute. Notwithstanding the continuously growing computational resources, these models remain computationally demanding, which makes for a real challenge in applying them to large, high resolution domains and long computational times. Consequently, CFD is often used in conjunction with faster models as part of a domain decomposition approach.

In Paulsen et al. [14,15], such a domain decomposition approach was adopted to model respectively wave impact and ringing for monopiles, where the previously mentioned fully nonlinear potential flow solver OceanWave3D as far-field was used in conjunction with OpenFOAM modeling the near-field through coupling by the wave generation and (passive) absorption toolbox waves2Foam. At the near-field boundaries, numerical relaxation zones, which gradually force the computated wave field to comply with the theoretical wave field calculated through the fully nonlinear far-field solver, were applied [16]. A similar strategy was applied in studying breaking of rogue waves arising from modulational instability in a wave train consisting of a carrier wave and side-band perturbations, where a domain decomposition was adopted between the higher-order spectral model (HOSM) and OpenFOAM. First, an initial wave train was simulated in HOSM until a certain threshold for wave height was reached. Subsequently, this wave surface was fed to the wave maker in the wave generation and (active) absorption toolbox olaFlow to closely study the breaking, for which HOSM does not account [17]. More recently, adopting CFD for both the far field and near field, Di Paolo et al. [18] presented a coupling approach with a 2D CFD model representing the far field and a 3D CFD model for the near field, foregoing the limitations associated with potential flow models. However, adopting this approach would still be too time-consuming in studying long time series, as is the case in this work, and CFD also invokes inaccuracies in its long-distance progressive wave modeling [19].

Alternatively, in Luquet et al. [20], field decomposition through the Spectral Wave Explicit Navier-Stokes Equations (SWENSE) approach has been proposed as a way of coupling the computationally efficient HOS solver to fully nonlinear wave tanks. First, the flow field potential is decomposed into the HOS potential and the near field solution in the numerical wave tank. Next, after evaluating the HOS potential, the nonlinear wave tank potential is solved for with the HOS potential as free surface boundary condition. Finally, adding both potentials produces the full flow field potential. SWENSE therefore allows for fast parallellized computations for both the far field solutions in HOS and the near field solutions in numerical wave tanks. In the OpenFOAM branch, foam-extend, this very same strategy has been successfully implemented together with similar passive absorption boundary conditions as in Jacobsen et al. [16] to prevent waves from reflecting into the near field. In doing so, both extreme wave simulations and the accurate assessment of near field effects on the structure have become possible [21,22]. However, although this approach is very effective in obtaining realistic extreme waves arising from an initial wave field and in studying their impacts on offshore structures, HOS essentially resolves the time evolution of an initial wave field. Therefore, this approach is as such less suited for deriving structural response statistics regarding the spatial evolution of random seas and the nonlinear wave–wave interactions involved, which have shown to lead to a departure from Gaussian statistics [7,23].

This work presents an efficient numerical strategy in generating sufficient data to assess the effect of the interacting free wave components on the response of a stiff monopile. The response statistics for a simple offshore structure in non-Gaussian seas are studied in the time domain. To this end, an efficient domain decomposition strategy, where the near field is modeled through CFD and the far field through a fast HOS numerical wave tank model, is adopted. The purpose of this work is to show that such a domain decomposition strategy is an effective and efficient way to study the response characteristics of a structure under wave loads. In the following, first, the methodology of the domain decomposition strategy is discussed. Next, the set-up and numerical input of far field, near field and structural test case are laid out. Based on the thus established models, a benchmark study for the domain decomposition strategy with emphasis on convergence is presented, after which, the simplified structural model's validity is assessed. Finally, a discussion takes place considering the ability of the developed time domain approach regarding response statistics in non-Gaussian seas, proving that the model is able to live up to expectations, but, however efficient, still becomes too expensive in view of the multiple long Monte Carlo simulations needed for the response statistics to be representative.

2. Methodology

2.1. Domain Decomposition

Previous work by Toffoli et al. [23] showed that a large amount of randomly generated (deep) water waves are needed to obtain reliable response statistics regarding nonlinear wave–wave interactions. Moreover, these waves have to travel over long distance to attain their nonlinear properties, when starting from an initial wave field composed of a random superposition of linear waves. If and when maximum non-Gaussianity is reached is primarily related to kd, with k the wave number and d the water depth. For deep water waves this typically occurs after about 10–15 wavelengths, while for intermediate water depths ($kd = 2$), where monopiles are typically built, convergence is not reached even after 20 wavelengths, as will be illustrated further on. As the computational cost of using a full CFD model would be excessive, the domain decomposition strategy, demonstrated in Figure 1 is applied. As shown in Figure 1, the fluid domain is decomposed into a far field modeled by HOS and a near field accounted for by the CFD solver, OpenFOAM [24].

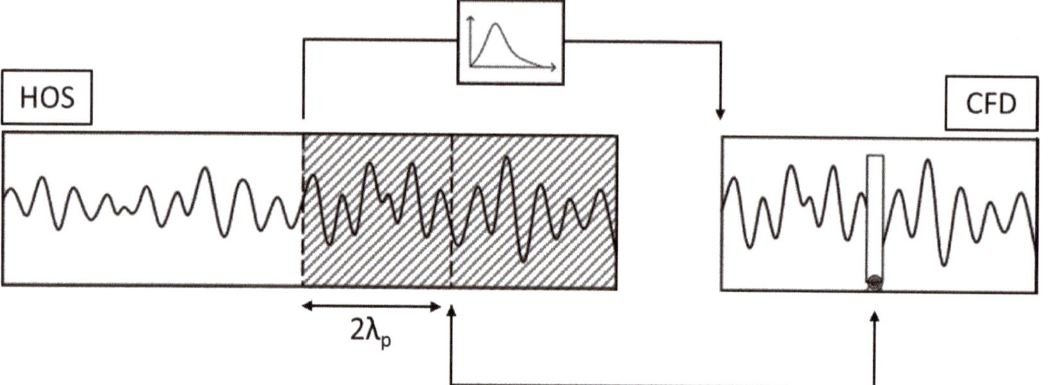

Figure 1. Time domain decomposition strategy.

In the domain decomposition strategy, HOS is used to randomly generate a time series by a wave maker boundary first. In previous work by Toffoli et al. [7,23], the higher-order spectral method (HOSM), has shown to yield statistically significant results regarding the influence of third-order nonlinearities on the occurrence of extreme waves in random wave fields. The higher-order spectral implementation typically resolves the Laplace equations in cyclic wave number space with nonlinear free surface boundary conditions formulated in Equations (1) and (2) with $W(x,y,t)$ and $\phi_s(x,y,\eta,t)$ respectively the vertical velocity and the velocity potential at the free surface, η [25].

$$\frac{\partial \phi_s}{\partial t} + g\eta + \frac{1}{2}\left[\left(\frac{\partial \phi_s}{\partial x}\right)^2 + \left(\frac{\partial \phi_s}{\partial y}\right)^2\right] - \frac{1}{2}W^2\left[1 + \left(\frac{\partial \eta}{\partial x}\right)^2 + \left(\frac{\partial \eta}{\partial y}\right)^2\right] = 0 \tag{1}$$

$$\frac{\partial \eta}{\partial t} + \frac{\partial \phi_s}{\partial x}\frac{\partial \eta}{\partial x} + \frac{\partial \phi_s}{\partial y}\frac{\partial \eta}{\partial y} - W\left[1 + \left(\frac{\partial \eta}{\partial x}\right)^2 + \left(\frac{\partial \eta}{\partial y}\right)^2\right] = 0 \tag{2}$$

Starting from a randomly generated initial wave field, HOSM then uses a series expansion to resolve the wave surface up to desired order of nonlinearity [11].

The higher-order spectral numerical wave tank, HOS-NWT, by Ducrozet et al. [26], is selected to obtain the spatial evolution of a wave field generated by a wave-making boundary. As a numerical wave tank domain is finite by definition and necessitates the presence of wave generation and absorption boundary conditions, HOS-NWT, adopts an additional velocity potential in an approach similar to what was done in Luquet et al. [20] to account for these boundary conditions. Similarly, as is the case for traditional HOS codes, no wave breaking models or dissipation have been implemented into the publicly available code [27], so that when breaking should occur, i.e., if the steepness exceeds a limit, the computation breaks down.

After running HOS-NWT, the input time series for the near field numerical wave basin in OpenFOAM is determined by taking the wave gauge corresponding to maximum kurtosis and assuming this probe to be the equivalent position of the structure in the near field OpenFOAM model. Subsequently, the time series measured at a wave gauge two peak wavelengths back, i.e., the distance between the wave generation boundary and the structure in the near field numerical modeling is then reconstructed by the olaFlow wave generation and absorption toolbox from the measured spectrum and used as boundary for the near field model.

Assuming incompressibility and irrotationality, the continuity and momentum equations in respectively Equations (3) and (4) together with the two-phase mixture continuity in Equation (5) are then solved by OpenFOAM's interFoam solver.

$$\frac{\partial u_i}{\partial x_i} = 0 \tag{3}$$

$$\frac{\partial \rho u_i}{\partial t} + \frac{\partial \rho u_i u_j}{\partial x_j} = -\frac{\partial p^*}{\partial x_i} - g_i \frac{\partial \rho}{\partial x_i} + \frac{\partial}{\partial x_j} \mu \frac{\partial u_i}{\partial x_j} \tag{4}$$

$$\frac{\partial \alpha}{\partial t} + \frac{\partial (\alpha u_i)}{\partial x_i} + \frac{\partial}{\partial x_i} \left(u_{r,i} \alpha (1 - \alpha) \right) = 0 \tag{5}$$

In Equation (4), the hydrostatic pressure p^* relates to the total pressure p as $p^* = p - \rho g_i$, and $u_{r,i}$ depicts a vector coefficient used for volume-of-fluid compression in Equation (5). μ represents the dynamic viscosity, ρ the water density and α the water volume fraction bounded between 0 and 1.

Before applying the measured wave spectrum to the near field boundary, the high frequencies in the HOS-NWT solution need to be filtered. The reason is two-fold. On the one hand, these high-frequencies are unphysical in the sense that they do not lie within the frequency range typically associated with wind-generated waves. On the other hand, when applying an unfiltered wave field including these high-frequency components as boundary condition to the near field, it jeopardizes numerical stability at the inlet. Moreover, including very low-frequency wave components might lead to an increasing water level in the CFD model. A low-frequency cut-off is therefore adopted as well. Filtering only retains energy in the frequency range from $0.33 f_p$ to $3 f_p$, with f_p the peak wave frequency of the applied spectrum, which still includes the phenomena of interest for the simplified monopile herein considered. However, in case of compliant structures, a smaller low-frequent cut-off wave frequency might be considered to trigger the slow structural responses, while stiffer structures will define the high-frequent cut-off wave frequency.

As a first assessment regarding the applicability of the proposed domain decomposition model in obtaining response statistics for offshore structures in non-Gaussian seas, a simplified structural model is added to the wave flume after benchmarking the coupled approach's ability to reproduce the HOS wave field. A monopile-founded offshore wind turbine modeled as a 1 DoF inverted pendulum with a spring-damper hinge aimed at approaching the first fore-aft bending frequency, is chosen. Due to the small excursions of the monopile in the water fraction, a one-way coupled approach, where waves impact on a fixed monopile to then apply the thus obtained wave loads in a structural solver, would also be acceptable. However, the inverted pendulum model also served the purpose of verifying mesh motion for a simple structure before transgressing to more complex multi-degrees-of-freedom motion.

2.2. A Simplified Structural Model

A simplified monopile, modeled as an inverted pendulum, is used to assess the applicability of the domain decomposition in deriving response statistics in non-Gaussian seas. This equivalent monopile is modeled through the capabilities already existing in OpenFOAM's rigid body motion solver, which is frequently used for floating structures. As illustrated in Figure 2, the monopile is gradually simplified to a rigid inverted pendulum with a spring-damper hinge connection at the bottom of the wave flume, which adequately emulates the original monopile-tower assembly first fore-aft bending mode. Although doing so implies that the motion of the structure in the water is overestimated due to stiff modeling of the pile, this should give a first idea considering the possible impact of random wave loading on monopiles.

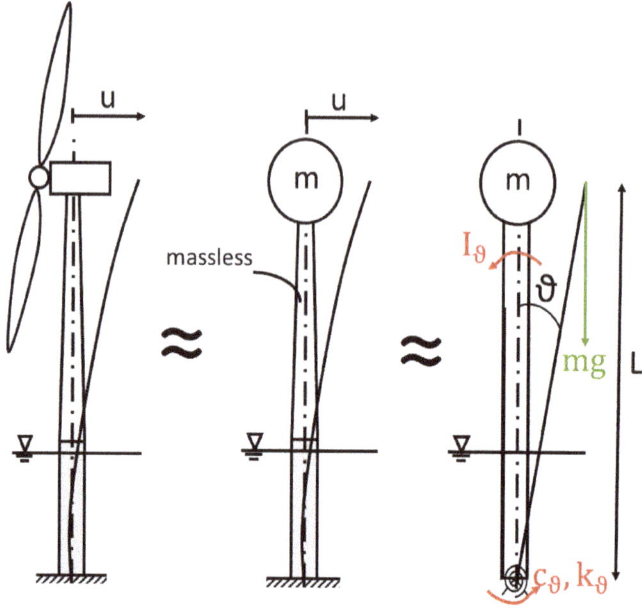

Figure 2. Equivalent monopile model.

First, the monopile is reduced to a one degree-of-freedom system, for which the free vibration has been written down in Equation (6).

$$m\ddot{x} + c\dot{x} + kx = 0 \tag{6}$$

In Equation (6), m is the sum of the lumped mass of the tower and monopile assembly and the mass of the rotor and hub, c is the viscous damping based on the modal damping factor, ζ, of 0.01 and k in this context the stiffness derived for the cantilevered beam.

Assuming the tower-monopile assembly to be stiff and massless, the equation of motion for the first fore-aft bending mode in Equation (6) then reduced to the momentum equilibrium, in terms of ϑ, for an inverted pendulum freely rotating about a spring-damper hinge, as illustrated in the free body diagram on the right-hand side of Figure 2.

$$I_\vartheta \ddot{\vartheta} + c_\vartheta \dot{\vartheta} + (k_\vartheta - mgL)\vartheta = 0 \tag{7}$$

In Equation (7), I_ϑ signifies the inertia about the spring-damper hinge, L the length of the tower-monopile assembly and g the gravitational constant. Enforcing the natural frequency and the modal damping, the equivalent spring-damper hinge constants c_ϑ and k_ϑ can be readily derived. As can be clearly noted, it has to be borne in mind that contrary to the original structural bending model with its inherent small angle approximations, the pendulum model in OpenFOAM takes into account the second-order effect associated with gravity. Hence, the moment resulting from gravity has to be taken along when determining the spring hinge stiffness, k_ϑ.

In Equation (8), the full expression for the momentum equilibrium of the equivalent monopile in water is shown, with a the added mass, c_{hydr} the hydrodynamic damping, k_{hydr} the hydrodynamic restoring coefficient and M the moment resulting from the hydrodynamic forcing.

$$I_\vartheta \ddot{\vartheta} + c_\vartheta \dot{\vartheta} + k_\vartheta \vartheta = -a\ddot{\vartheta} - c_{hydr}\dot{\vartheta} - k_{hydr}\vartheta + mgL + M \tag{8}$$

As such, the monopile is reduced to a one-degree-of-freedom rigid body system with appropriate restraints and structural parameters. Its hydrodynamic properties are implicitly taken along by adopting several loops over the fluid solver and the rigid body motion solver.

3. Numerical Set-Up

3.1. Far Field

As a far field domain, a numerical wave basin of 25 peak wavelengths and 40 m water depth is modeled in HOS-NWT at full-scale, nihilating any possible scale effects, which should be sufficient for the considered sea state to develop its nonlinear characteristics. However, instead of the deep water waves considered in Toffoli et al. [12], intermediate conditions are studied here, which reduces the wave growth rate. An equilibrium state regarding non-Gaussianity might therefore not be reached by the end of the wave basin. At the inlet of the basin, waves are generated by a numerical flap-type wave maker from a JONSWAP spectrum applying linear wave maker theory. Input parameters to the wave maker are peak enhancement factors, γ, of 1, 3.3 and 6, a steepness, $k_p H_s$, of 0.1 and a relative water depth, $k_p h$, equal to 2, i.e., a moderately steep sea state in intermediate water depth, which corresponds to an environment where a monopile would be typically built. k_p depicts the peak wave number and H_s the significant wave height. A grid of $512 \times 1 \times 128$ points, i.e., a spatial resolution of 11.29 m and a maximum resolvable frequency of 0.372 Hz, is used, which provides about ten points to model the peak wave length and two points for the smallest wave length. Wave data are sampled at 50 Hz. Furthermore, the HOS solutions are truncated at third order of nonlinearity, allowing for the nonlinear wave–wave interactions to occur [12]. Wave gauges are equidistantly placed along the full length of the wave flume at an interval of one peak wavelength. To prevent wave reflection at the end of the basin, a numerical relaxation zone of two peak wavelengths is provided at the outlet of the basin.

For each peak enhancement factor, seven runs of 1000 s were run, which amounts to approximately 7000 waves per input spectrum. One serial run for the far field model takes a CPU time of about 12 h for three hours of wave data on one Intel Xeon CPU E5-2630 v3 at 2.40 GHz × 16.

3.2. Near Field

3.2.1. Numerical Wave Flume

For the near field model, the latest version of the wave generation and absorption toolbox olaFlow is used together with OpenFOAM-v1906+. The near field is represented by a wave flume of total length five peak wavelenghts with the monopile being located at two peak wavelengths from the inlet. A schematic view of the near field domain is shown in Figure 3.

The waves obtained from HOS-NWT at two wavelengths before the location of interest are, after filtering, are input to olaFlow's irregular wave generation boundary. To reconstruct the exact same time series and retain the non-Gaussianity developed over the course of the HOS-NWT model, the original phases are input to the irregular wave generation as well. In doing so, the result of the nonlinear wave–wave interactions developed up to the equivalent position of the near field wave generation boundary in the far field is retained. The remaining interactions to occur in between the wave generation boundary and the monopile are then taken care of by the near field CFD model. The benchmark will provide more insight in the validity of these assumptions and the near field resolution needed to guarantee correct reproduction of the original time-series.

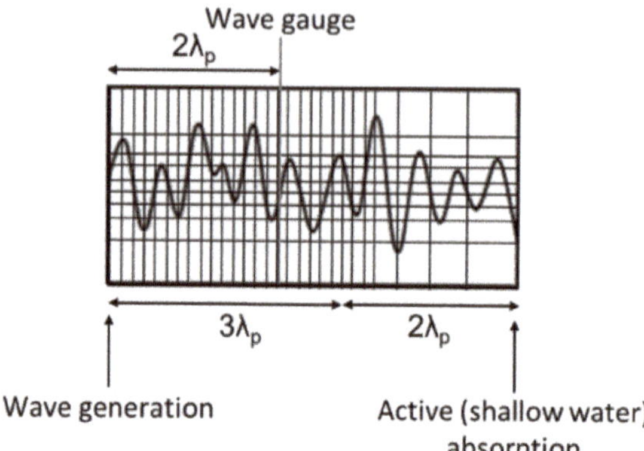

Figure 3. Schematic view of the near field time domain.

At the outlet, active (shallow water) wave absorption is applied. As mentioned before, although olaFlow makes for fast evaluation of wave absorption profiles, it is meant to actively compensate shallow water velocity profiles [28]. Therefore, the (exponential) velocity profiles associated with deeper water waves cannot be adequately absorbed. In Higuera [29], several options for improvement to olaFlow's active wave absorption, i.e., by slightly reformulating the original compensating velocity field and by combining it with passive wave absorption, are explored, generally arriving at reduced reflection at higher computational cost. Instead of applying these proposed improvements, acceptable and computationally efficient wave absorption will be obtained by including a domain of two peak wavelengths consisting of cells of increasing dimensions towards the active wave absorbing boundary, taking full advantage of numerical dissipation. Further on in this work, a reflection analysis along the lines Mansard and Funke [30] will assess the validity of this approach.

In benchmarking the hybrid HOS-CFD approach, the near-field numerical wave flume is modeled as two-dimensional. One run for the herein presented near field 2-D CFD model takes about 72 h on one Intel Xeon CPU E5-2630 v3 at 2.40 GHz × 16 for 3600 s.

3.2.2. Structural Model

As input to the structural model, the monopile presented by Jonkman and Musial [31] with the 77.6 m high National Renewable Energy Laboratory (NREL) 5 MW reference wind turbine by Jonkman et al. [32] on top is used. For nonlinear wave–wave interactions to occur, $kh > 1.36$ must be valid [7]. Therefore, an intermediate water depth of 40 m was chosen and the original monopile of 30 m length in a water depth of 20 m is lengthened up to 50 m. The monopile diameter of 6m and wall thickness of 0.06 m are kept the same. As mentioned earlier, the equivalent monopile is modeled by using OpenFOAM's rigid body motion solver applying appropriate translational restraints and a spring-damper hinge constraint. The mass at the tower top is taken to be the sum of the actual mass of the rotor-hub-nacelle assembly and the lumped tower-monopile mass under the assumption of a clamped connection at the sea floor. The moment of inertia is then simply determined as this mass times total height squared. Knowing the dry natural frequency and the modql damping ratio, the stiffness and damping coefficients can then be readily derived. The original monopile properties and the equivalent monopile input to the structural model are listed in Table 1.

Table 1. Structural parameters for the original monopile and the simplified monopile.

Original Monopile		Equivalent Monopile	
Mass rotor-nacelle	350,000 kg	(Lumped) mass	615,000 kg
Mass tower-monopile	725,575 kg	Moment of inertia	10,013,282,400 kg·m^2
Center of mass	71,946 m	Location of lumped mass	127.6 m
Stiffness	1,711,570 N/m	Spring stiffness	28,637,234,480 N·m/rad
Modal damping factor	0.01	Spring damping	334,080,290 N·m·s/rad
Natural frequency (dry)	0.26550 Hz	Natural frequency (wet)	0.25385 Hz

For the structural computations, the numerical wave flume is modeled by expanding the near field domain in Figure 3 up to 3D and by including the equivalent monopile model presented in Figure 2. As $k_p D$ = 0.3, with D the structural diameter, viscosity is not negligible, but as the Keulegan-Carpenter (KC) number corresponding to the peak of the input spectrum is still small (KC = ±4) no vortex shedding is to be expected. Taking into account the absence of asymmetric vortex shedding and the structural symmetry, it is appropriate to model only half of the domain and the structure, applying symmetry boundary conditions at the symmetry surface, and thus saving considerate computational resources. The resulting three-dimensional mesh is shown in Figure 4. As can be seen in Figure 4, throughout the mesh in the x-direction the target refinement is withheld. For the y- and z-direction, gradual refinement is adopted towards this target refinement in the vicinity of the monopile and the wave surface. As part of the benchmark, this target refinement about the monopile and the free surface will be varied.

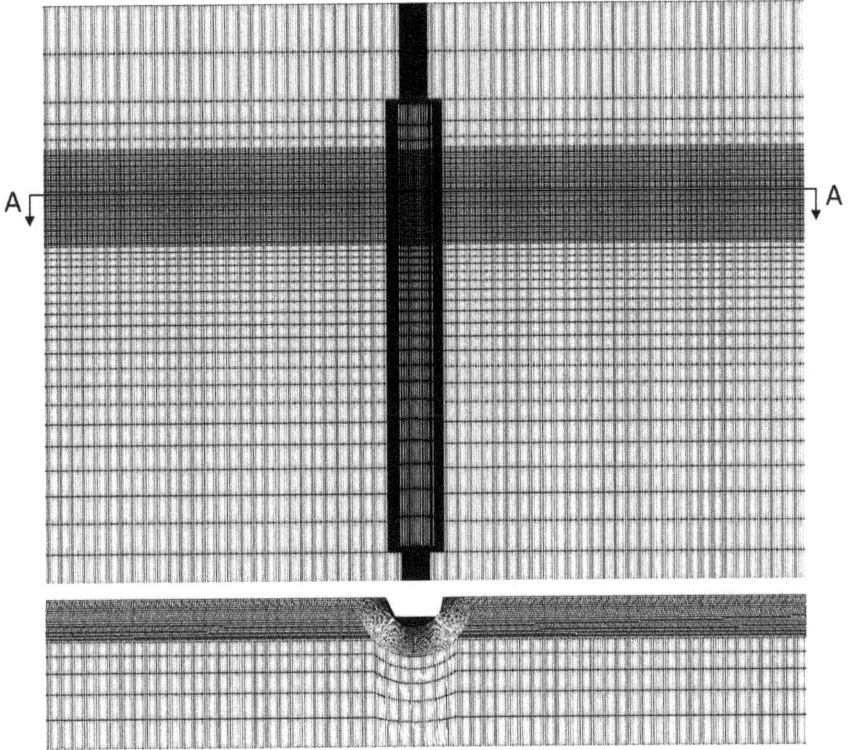

Figure 4. Three-dimensional wave flume mesh with a simplified monopile model.

Simulations are run with an adaptive time step defined by a maximum Courant number (Co) of 0.5. Furthermore, OpenFOAM's interFoam solver adopts the two inner loops for the pressure correction typically of PISO and three outer PIMPLE loop iterations to force convergence between the rigid body and the spring-damper hinge on the one hand and the wave–structure interaction on the other hand during each time step. As the motion in the water is limited, this small number of outer loops suffices. Relaxation and damping factors of, respectively, 0.95 and 0.90 are enforced in the rigid body solver using the implicit second-order Crank–Nicolson scheme. The terms in the Navier–Stokes are discretized using Gaussian integration and using Euler explicit time stepping. Preconditioned conjugate gradient and stabilized bi-conjugate gradient solvers are used for respectively solving for the initial guess for the velocities and the pressure correction equation.

Waves are presented to the near field's numerical wave basin. Both the surface elevation and the structural response are sampled at 50 Hz. For each wave case seven realizations were fed to the near field and run for three weeks on five Intel Xeon CPU's E5-2680 v2 at 2.8 GHz × 20, resulting in time series of approximately 1000 s, i.e., 160 peak waves. The actual length of the obtained time series depended on the spectra simulated, because the irregular wave surface associated with broad-bandedness gives rise to more spurious air velocities at the air-water surface, which significantly reduced the adaptive time steps. Simulating for the Pierson–Moskovitz spectrum took longer than the narrow-banded JONSWAP spectrum generally.

4. Results

4.1. Benchmark for the Wave Field

Before running the fully coupled HOS-CFD model, a benchmark study was conducted to verify the validity of the proposed domain decomposition strategy regarding the wave fields involved and to identify probable discrepancies between the HOS solution and the coupled HOS-CFD solution and their causes. In doing so, first, the HOS model has been used to compute the wave series at 23 peak wavelengths from the models' wave maker. Next, the time series obtained two peak wavelengths back, was applied to the near field boundary and the CFD model was used to let the wave further evolve. Finally, the time series logged at the location of the would-be structure in the CFD model was compared to the original time series obtained at that very same position in the HOS model. The result is presented in Figure 5 for three different near field mesh resolutions, i.e., Hs/5, Hs/10 and Hs/20.

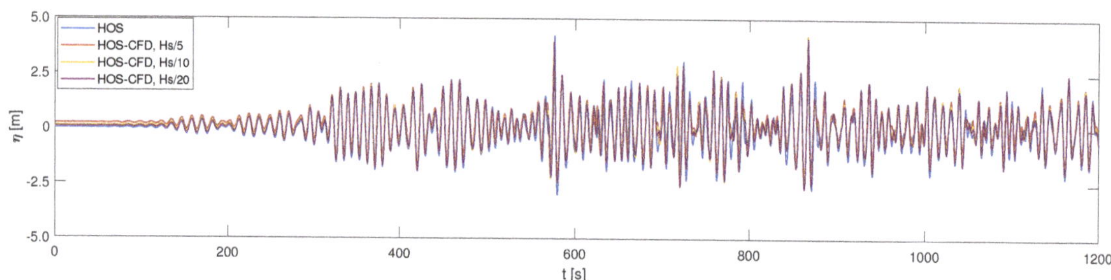

Figure 5. Benchmark with surface elevation plotted against time for the higher-order spectral (HOS) model (blue) and the coupled model for mesh resolutions Hs/5 (red), Hs/10 (yellow) and Hs/20 (purple).

From Figure 5, it can be clearly seen that the signals of the coupled HOS-CFD and of HOS-NWT correspond closely both in terms of phases and peaks for all considered mesh resolutions. As mesh resolution increases, the reconstructed signal in HOS-CFD converges to the target signal obtained from HOS-NWT. Good agreement is found for Hs/20, although the signal corresponding to a resolution Hs/10 is a very close second. As optimal reconstruction of the target signal is key in this work, Hs/20 will be withheld in further considerations. Occasionally, slight underestimations are observed for the waves compared to their HOS counterparts. These underestimations are rather small at the beginning of the time series, but grow more pronounced towards the end. Two explanations for these observed differences in peak values seem plausible.

Firstly, no breaking model is included in HOS-NWT, while waves in OpenFOAM might prematurely break, i.e., before reaching their breaking limit, or lose energy to dissipation. Although publicly available HOS-NWT does not incorporate breaking models and dissipation, it does include a limit to the steepness above which the computation breaks down. Still, large peaks that overshoot the physics might be encountered. On the contrary, OpenFOAM's interFoam solver does include viscosity and two-phase fluid interaction through diffusive transport equations for cell water fractions. The evolution of wave fields can therefore be expected to be more realistic in the sense that they are allowed to break. However, due to the dissipative nature of these transport equations, waves might lose energy to numerical dissipation. Therefore, a breaking detection was performed on the HOS output signal for $a\omega^2/g > \gamma$ with a the wave amplitude, ω the wave frequency and γ the wave limiting steepness. As this condition is only applicable for linear waves, the assessment of the wave breaking is based on a characteristic wave amplitude and characteristic wave frequency computed for each time instance by applying wavelet analysis along the lines of Liu and Babanin [33] with a limiting value, γ, of 0.2. As can be seen from Figure 6, the considered wave field from HOS is in no danger of breaking.

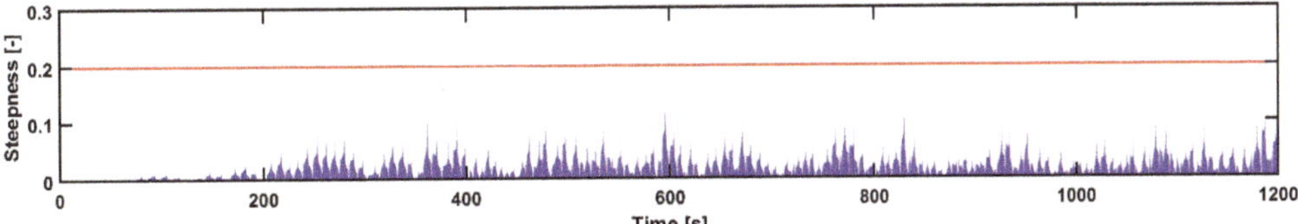

Figure 6. Localized steepness, $a\omega^2/g$, plotted against time for the HOS result.

On the other hand, the differences in peak values could be attributed to the different wave absorption strategies used in HOS-NWT and the CFD model; in HOS-NWT a passive absorption based on numerical relaxation is applied, while in olaFlow active (shallow water) absorption combined with additional numerical dissipation is adopted. Therefore, a reflection analysis according to Mansard and Funke [30], has been applied to both HOS-NWT and CFD at the same respective positions. When comparing the reflection coefficients in Figure 7, it is clear that HOS-NWT shows larger reflection at the high frequencies.

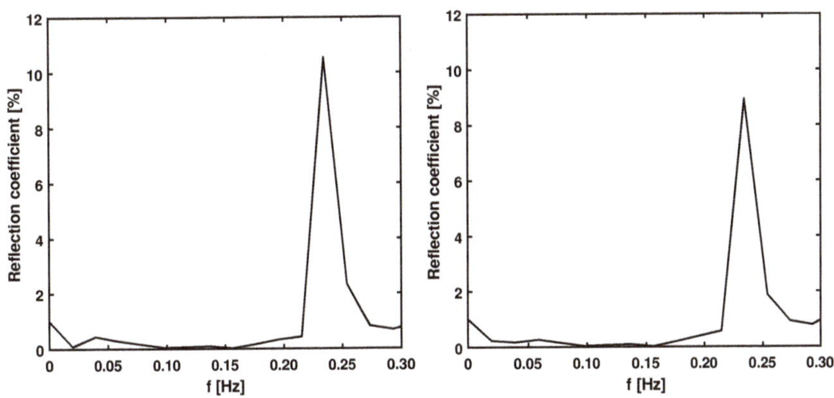

Figure 7. Reflection coefficients for higher-order spectral numerical wave tank (HOS-NWT) **(left)** and OpenFOAM **(right)**.

Notwithstanding, the known inaccuracies of the reflection analysis with respect to the high-frequency tail, about 9% reflection can be noted in the high-frequency tail for CFD, while approximately 11% reflection is shown for HOS-NWT. Both models display small values for the reflection coefficient in the low-frequency tail as well. The values for the reflection coefficient for the HOS-NWT model are generally higher than the ones for the CFD model. This difference in absorbing long waves can be explained by the inherent differences between the active and passive wave absorption applied. As olaFlow is aimed at absorbing shallow water waves, it simply absorbs long waves best. Furthermore, in HOS-NWT, a numerical relaxation zone of two peak wavelengths is used, resulting in the increased reflection of wave components slightly longer than this zone.

The waves obtained by both the CFD and the coupled HOS-CFD model collide remarkably well. Discrepancies can most certainly be attributed to the differences in absorption strategy.

4.2. Verification of the Equivalent Monopile

In order to verify the applied simplified inverted pendulum model and to improve the understanding of the wave–structure interaction results further on, a free decay test for the model's pitch motion has been conducted in the numerical wave flume. Figure 8 shows the pitch motion, ϑ, as it decays in time. Applying modal analysis through the MATLAB toolbox MACEC 3.3 [34], the natural frequency and the damping ratio are defined as respectively 0.253Hz and 1.4%, which lies close to the target values earlier defined in Table 1.

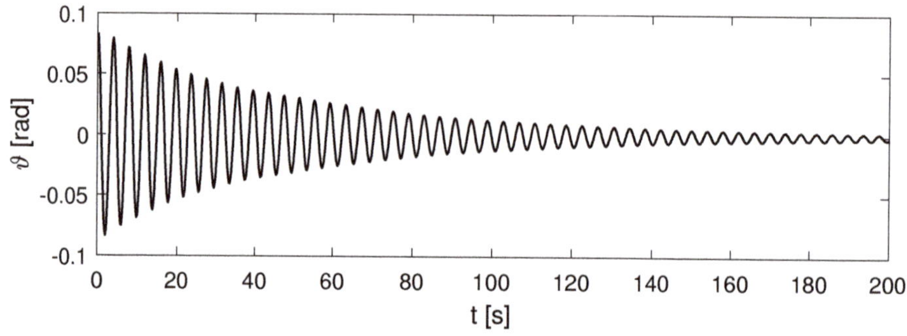

Figure 8. Free decay test for the inverted spring-damper pendulum.

Furthermore, as can be seen from Figure 9, the equivalent monopile response closely mimics the wave elevation, which is expected as the natural frequency is situated on the right-hand side of the wave peak frequency. The high-frequency stiff monopile response is superposed onto the quasi-static response to the waves. Even for the short window presented, multiple high-frequency events can be noted. The most notable one is situated in between 650 and 700 s and corresponds most plausibly to a typical high-frequency ringing event; the response appears to build up over several periods and subsequently decays.

The equivalent monopile is therefore not only able to reproduce the monopile's structural properties, it captures its typical high-frequency behavior as well. Having verified the validity of both the domain decomposition and the equivalent monopile, in the remainder of this work, some statistics regarding the waves and the responses will be drawn and their reliability will be assessed.

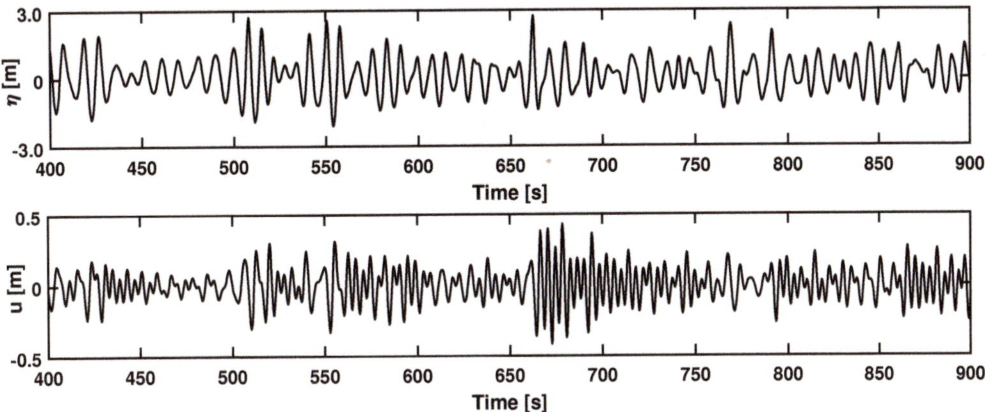

Figure 9. Results for wave elevation (**top**) and structural response (**bottom**).

4.3. Application to Obtaining Non-Gaussian Response Statistics

While adopting the same wave parameters as previously used for the benchmark study, the applicability of the proposed domain decomposition strategy to obtaining response statistics for structures in non-Gaussian seas is hereafter studied. Seven Monte Carlo realizations of about 500 peak waves have been simulated for three wave spectra with different peak wave factors; i.e., 1, 3.3 and 6. The input spectra were chosen to contain similar energy content, i.e., same zero-order moment, m_0, or standard deviation, σ. Using different spectral peak factors changes the shape of the wave spectrum with the spectrum corresponding to $\gamma = 1$ classified as broad-banded and the one with $\gamma = 6$ being narrow-banded.

Narrow-banded spectra are known to be especially prone to a special case of resonant wave–wave interactions, the so-called Benjamin–Feir instability, which modulates a wave train originally narrow-banded by exchanging energy with its close side-bands. As the Benjamin–Feir instability modulates narrow-banded wave trains, it is a considered to be a plausible candidate in explaining the formation of freak waves [35]. The Benjamin–Feir instability manifests itself to differing degrees in long-crested deep to intermediate water wave trains ($kd \leq 1.36$) [7,23]. Onorato et al. [36] summarized the influence of bandwidth and water depth on the occurrence of the Benjamin–Feir instability in the Benjamin–Feir index (BFI) as

$$\text{BFI} = \frac{2k_p a}{(\Delta k/k_p)} \sqrt{\frac{|\beta|}{\sigma}} \tag{9}$$

with factors σ and β for finite water depths defined as

$$\sigma = 2 - v^2 + \frac{8\left(k_p h\right)^2 \cosh\left(2k_p h\right)}{\sinh^2\left(2k_p h\right)} \tag{10}$$

$$\beta = \frac{8 + \cosh\left(4k_p h\right) - 2\tanh^2\left(k_p h\right)}{8\sinh^4\left(k_p h\right)} - \frac{\left(2\cosh^2\left(k_p h\right) + 0.5v\right)^2}{\sinh^2\left(2k_p h\right)\left(\frac{k_p h}{\tanh\left(k_p h\right)} - v^2/4\right)} \tag{11}$$

$$\text{with} \quad v = 1 + \frac{2k_p h}{\sinh\left(2k_p h\right)} \tag{12}$$

The Benjamin–Feir indices for the herein considered wave spectra then become 0.09, 0.58 and 0.73 for respectively $\gamma = 1, 3.3$ and 6, the highest BFI incorporating the fact that instabilities will most likely occur for narrow-banded spectra.

Figure 10 shows the probabilities of the wave elevations captured in the near field for the three wave spectra considered. These wave elevations correspond to the sea state captured at 23 peak wavelengths from the wave maker in the far field. For the sake of comparison, all wave elevations are normalized by their standard deviation $\sigma = Hs/4 = \sqrt{m_0}$. In addition, the normal probability density function and its second-order correction, computed along the lines of Tayfun [37], are shown. From Figure 10, all realizations appear to be skewed with respect to the Gaussian. The positive surface elevations evidently conform closest to the Tayfun distribution. However, the negative values overestimate the values expected for typically second-order waves. Furthermore, not much can be said regarding the differences between the spectra.

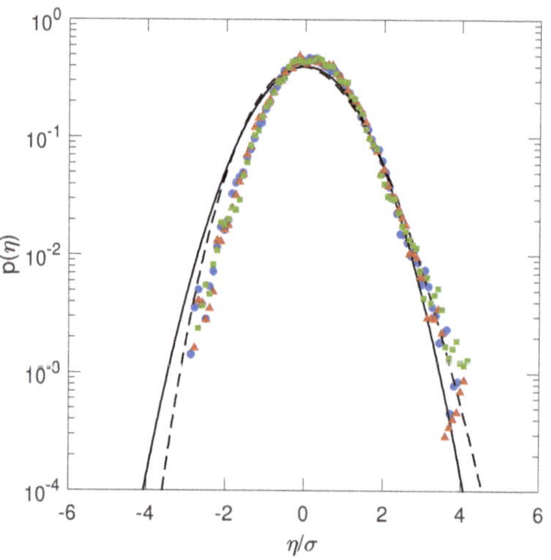

Figure 10. Probability density function of the surface elevations with respect to the surface elevations normalized by their standard deviation, as computed by near-field time domain for BFI = 0.09 (•), BFI = 0.58 (▲) and BFI = 0.73 (■) plotted against the normal pdf (full line) and its second-order correction (dashed line).

In Figure 11, the probability distributions for the wave crests are shown. The thresholds have been normalized by their wave height, i.e., Hs = 4σ. For comparison, the Rayleigh distribution, to which wave crests for linear waves tend, are drawn together with the Tayfun distribution for second-order wave crests. As expected, the wave crests resulting from all three wave specra conform best to the Tayfun distribution. Depending on the bandedness of the spectra, differences in wave crest distribution can be noted as well. For the largest BFI, corresponding to the narrow-banded case, strongest deviations from the Tayfun distribution can be noted. Overall, the statistics presented in Figure 11 are in strong agreement with the works by Onorato et al. [36] and Toffoli et al. [7].

Figure 12 shows the tower top excursions with respect to the Gaussian. All displacements have been normalized by their respective standard deviation. The skewness of the responses compared to the Gaussian appears to reflect the skewness originally found for the probability density function of the wave elevations in Figure 10. Compared to Figure 10, the lower and upper tails seem to be heavier and clear differences are noted depending on wave spectral bandedness. The tails seem to be heaviest for the broad-banded specrum, corresponding to lowest BFI, while the narrow-banded wave spectrum results in slightly heavier tails than does the response spectrum corresponding to $\gamma = 3.3$.

Moreover, Figure 13, depicting the probability distribution for the tower top displacements, appears to confirm the more extreme responses already found in the upper tail in Figure 12 for the broad-banded spectrum. Similarly, the narrow-banded spectrum results in more extreme responses than does the wave spectrum lying in between.

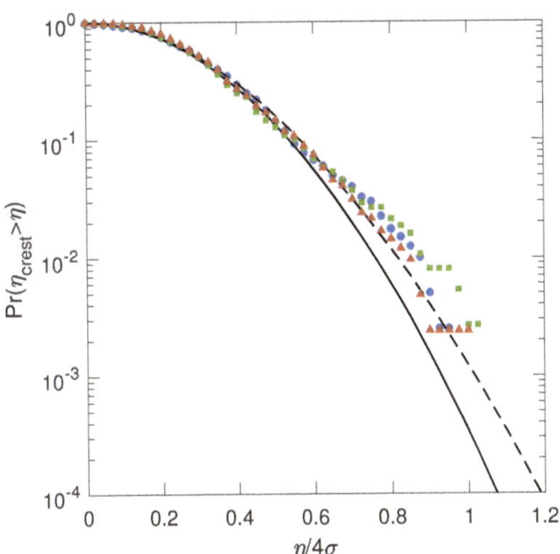

Figure 11. Probability distribution of the wave crests with respect to the wave elevations normalized by their standard deviations, as computed by near-field time domain for BFI = 0.09 (•), BFI = 0.58 (▲) and BFI = 0.73 (■) plotted against the Rayleigh distribution (full line) and the Tayfun distribution (dashed line).

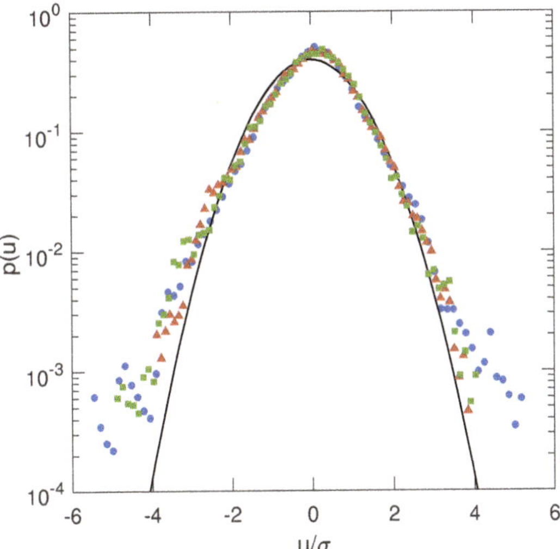

Figure 12. Probability density function tower top displacements with respect to the displacements normalized by their standard deviations, as computed by near-field time domain for BFI = 0.09 (•), BFI = 0.58 (▲) and BFI = 0.73 (■) plotted against the normal pdf (–).

The largest response noted for the broad-banded spectrum could be attributed to its heavier high-frequency spectral tail, which continuously excites the tower's first fore-aft bending resonant frequency. The remaining narrower-banded wave spectra have similar energy in their tails, with the narrowest-banded wave spectrum having obviously the least. Interestingly though, the narrowest-banded wave spectrum results in larger response than does the wave spectrum corresponding to $\gamma = 3.3$. Therefore, energy considerations probably not tell the full story.

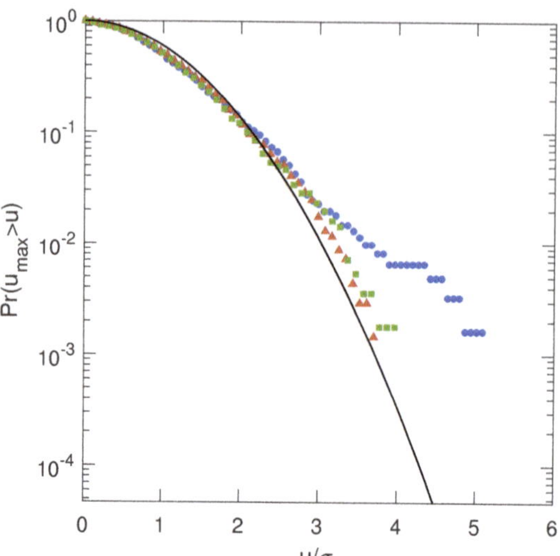

Figure 13. Probability distribution maximum tower top displacements with respect to the displacements normalized by their standard deviation, as computed by near-field time domain for BFI = 0.09 (•), BFI = 0.58 (▲) and BFI = 0.73 (■) plotted against the Rayleigh distribution (–).

Figure 14, depicting the response spectra for the considered wave spectra smoothed using a Hann window of 512 samples (sampled at 50 Hz) and averaged over the seven realizations, shows that indeed the broad-banded wave spectrum results in highest response in its tail. The second broadest wave spectrum appears to do so as well. The narrow-banded spectrum shows overall the least energy. Therefore, it is most plausible that the larger response for the narrowest-banded spectrum as noted in Figure 13 originates from the larger waves associated with the higher probability of occurrence of the Benjamin–Feir instability. Moreover, looking at the response spectra, one would expect the response at the wave spectral peak to be largest for the narrow-banded wave spectrum and smallest for the broad-banded wave spectrum.

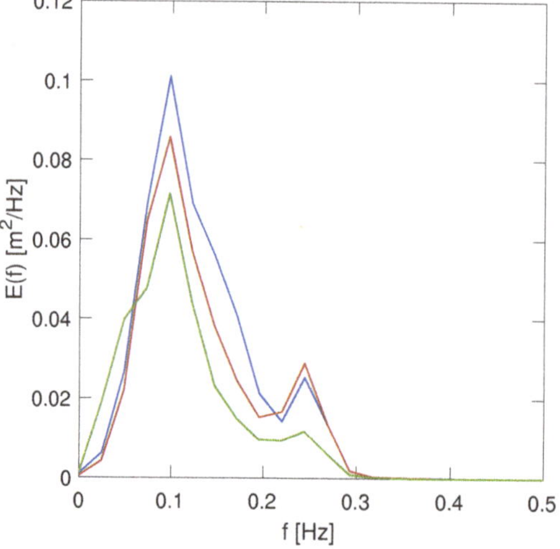

Figure 14. Response density spectra for the tower top displacements for BFI = 0.09 (blue), BFI = 0.58 (orange) and BFI = 0.73 (green).

5. Discussion

From the above presented results, the proposed domain decomposition strategy is clearly able to both reproduce the non-Gaussianity of the wave field obtained from HOS-NWT with remarkable accuracy and capture some interesting dynamic response phenomena. Furthermore, a link appears to exist between the non-Gaussian wave statistics obtained through the decomposed model and the perceived simplified monopile responses, indicating that the model could be applied to studying the impact of nonlinear wave–wave interactions in offshore structural response. During this study, the question was raised of whether such a model would be able to generate enough data for converged response statistics. In answering this question, three points have to be considered: the extent of the data gathered and the sampling rate; the availability of computational resources; and most importantly, the response characteristics of the structure under consideration.

Regarding the first point, reducing the sampling frequency by five could deliver data at one-third of the above-mentioned computational costs as a large overhead exists in CFD in writing data to disk space, still leaving us at 7 × 1 week of computational time on five Intel Xeon CPU's E5-2680 v2 at 2.8 GHz × 20. per wave spectral case. Having sufficient computational resources, simulating for two months would deliver enough data, i.e., ±10,000 waves, for a simple stiff structures as is the monopile. This is in line with the amount of waves previously used in studying modulational instability as one of the mechanisms underlying extreme wave formation [7].

For the monopile, the results indicate that nonlinear wave–wave interactions have their influence on its response statistics. Obviously, the Pierson–Moskowitz spectrum continuously excites the first fore-aft bending natural frequency due to its relatively high energy content in its high-frequency tail. However, results also indicate that although its limited energy content in the upper spectral tail, narrow-banded wave spectra might lead to larger response maxima than does a traditional JONSWAP spectrum characterized by $\gamma = 3.3$. Based on the premise that narrow-banded waves lead to a higher occurrence of extreme waves and taking into account the ringing observed in Figure 9, a plausible guess could be that ringing more often occurs for these narrow-banded spectra. The above presented results agree with what would be expected from literature and therefore confirm the applicability of the proposed model in studying response statistics in non-Gaussian seas. Nonetheless, it has to be borne in mind that these preliminary observations are based on statistics that are not converged yet.

When considering compliant floating structures, eigenfrequencies are about an order of magnitude lower than, e.g., the monopile's first natural frequency. Many more data would therefore be needed to arrive at converged response statistics for compliant structures. The bottleneck lies in the number of realizations of the slow difference frequency response realizations rather than the amount of waves. Computations are expected to become truly demanding in cases of 3D motion and directional wave fields, rendering the previously applied symmetry plane invalid. Therefore, although the model has shown to be able to capture the response phenomena related to nonlinear wave–wave interactions in non-Gaussian seas, and the estimate of the cost required for converged results for stiff offshore structures seems reasonable, the proposed methodology is not expected to be feasible for 3D motion, directional wave fields and compliant structures.

6. Conclusions

In this work, a time-domain approach for the response statistics in non-Gaussian seas based on domain decomposition through coupling between HOS-NWT and OpenFOAM has been proposed to overcome the combined cost of detailled near field CFD and the long domains needed for waves to develop the required non-Gaussian properties. A benchmark case for the coupled HOS-CFD model showed that the peaks and phases of the wave signal obtained by HOS-NWT are well reproduced. Furthermore, an equivalence between the monopile and its simplified structural model adopted herein exists, which allows one to capture typical hydrodynamic phenomena, such as ringing. It appears to be possible to deduce response statistics regarding the impacts of nonlinear wave–wave interactions, and a major implication for further research was found, as the wave extremes for the narrow-banded spectrum were mirrored in the corresponding response. However, not enough data were obtained for these observations to be conclusive.

Due to computational cost, the proposed methodology seems to be limited to stiff structures and unidirectional wave trains.

Author Contributions: Conceptualization, G.D. and A.T.; methodology, G.D. and A.T.; software, G.D.; validation, G.D..; formal analysis, G.D.; investigation, G.D.; resources, J.M. and A.T.; data curation, G.D.; writing—original draft preparation, G.D.; writing—review and editing, G.D., A.T., J.M and G.L.; visualization, G.D.; supervision, A.T., J.M. and G.L.; project administration, J.M. and G.L.; funding acquisition, G.D. and J.M. All authors have read and agreed to the published version of the manuscript.

Funding: This research was funded by the Research Foundation—Flanders grant numbers 11A1217N and V431019N.

Acknowledgments: The first author is a Ph.D. fellow of the Research Foundation—Flanders (FWO) (Ph.D. fellowship 11A1217N). This specific collaborative research came to fruition thanks to the travel grant awarded to the first author (Long stay travel grant V31019N) by the Research Foundation—Flanders (FWO).

Conflicts of Interest: The authors declare no conflict of interest.

References

1. Breton, S.P.; Moe, G. Status, plans and technologies for offshore wind turbines in Europe and North America. *Renew. Energy* **2009**, *34*, 646–654. [CrossRef]
2. Bilgili, M.; Yasar, A.; Simsek, E. Offshore wind power development in Europe and its comparison with onshore counterpart. *Renew. Sustain. Energy Rev.* **2011**, *15*, 905–915. [CrossRef]
3. Bhattacharya, S. Challenges in design of foundations for offshore wind turbines. *Eng. Technol. Ref.* **2014**, *1*, 922. [CrossRef]
4. DNV GL. DNV-OS-J103—Design of Floating Wind Turbine Structures. Available online: https://rules.dnvgl.com/docs/pdf/DNV/codes/docs/2013-06/OS-J103.pdf (accessed on 6 January 2021).
5. Moan, T.; Zheng, X.Y.; Quek, S.T. Frequency-domain analysis of non-linear wave effects on offshore platform responses. *Int. J. Non-Linear Mech.* **2007**, *42*, 555–565. [CrossRef]
6. Onorato, M.; Osborne, A.R.; Serio, M.; Cavaleri, L.; Brandini, C.; Stansberg, C. Extreme waves, modulational instability and second order theory: Wave flume experiments on irregular waves. *Eur. J. Mech.-B/Fluids* **2006**, *25*, 586–601. [CrossRef]
7. Toffoli, A.; Onorato, M.; Bitner-Gregersen, E.; Osborne, A.R.; Babanin, A. Surface gravity waves from direct numerical simulations of the Euler equations: A comparison with second-order theory. *Ocean Eng.* **2008**, *35*, 367–379. [CrossRef]
8. Schløer, S.; Bredmose, H.; Bingham, H.B. The influence of fully nonlinear wave forces on aero-hydro-elastic calculations of monopile wind turbines. *Mar. Struct.* **2016**, *50*, 162–188. [CrossRef]
9. Engsig-Karup, A.P.; Bingham, H.B.; Lindberg, O. An efficient flexible-order model for 3D nonlinear water waves. *J. Comput. Phys.* **2009**, *228*, 2100–2118. [CrossRef]
10. Dommermuth, D.G.; Yue, D.K. A high-order spectral method for the study of nonlinear gravity waves. *J. Fluid Mech.* **1987**, *184*, 267–288. [CrossRef]
11. West, B.J.; Brueckner, K.A.; Janda, R.S.; Milder, D.M.; Milton, R.L. A new numerical method for surface hydrodynamics. *J. Geophys. Res. Ocean.* **1987**, *92*, 11803–11824. [CrossRef]
12. Toffoli, A.; Gramstad, O.; Trulsen, K.; Monbaliu, J.; Bitner-Gregersen, E.; Onorato, M. Evolution of weakly nonlinear random directional waves: Laboratory experiments and numerical simulations. *J. Fluid Mech.* **2010**, *664*, 313–336. [CrossRef]
13. Weller, H.G.; Tabor, G.; Jasak, H.; Fureby, C. A tensorial approach to computational continuum mechanics using object-oriented techniques. *Comput. Phys.* **1998**, *12*, 620–631. [CrossRef]
14. Paulsen, B.T.; Bredmose, H.; Bingham, H.B. An efficient domain decomposition strategy for wave loads on surface piercing circular cylinders. *Coast. Eng.* **2014**, *86*, 57–76. [CrossRef]
15. Paulsen, B.T.; Bredmose, H.; Bingham, H.B.; Jacobsen, N.G. Forcing of a bottom-mounted circular cylinder by steep regular water waves at finite depth. *J. Fluid Mech.* **2014**, *755*, 1–34. [CrossRef]
16. Jacobsen, N.G.; Fuhrman, D.R.; Fredsøe, J. A wave generation toolbox for the open-source CFD library: OpenFoam®. *Int. J. Numer. Methods Fluids* **2012**, *70*, 1073–1088. [CrossRef]
17. Alberello, A.; Pakodzi, C.; Nelli, F.; Bitner-Gregersen, E.M.; Toffoli, A. Three dimensional velocity field underneath a breaking rogue wave. In Proceedings of the ASME 2017 36th International Conference on Ocean, Offshore and Arctic Engineering, Trondheim, Norway, 25–30 June 2017.

18. Di Paolo, B.; Lara, J.L.; Barajas, G.; Losada, Í.J. Wave and structure interaction using multi-domain couplings for Navier-Stokes solvers in OpenFOAM®. Part I: Implementation and validation. *Coast. Eng.* **2020**, 103799. [CrossRef]
19. Larsen, B.E.; Fuhrman, D.R.; Roenby, J. Performance of interFoam on the simulation of progressive waves. *Coast. Eng. J.* **2019**, *61*, 380–400. [CrossRef]
20. Luquet, R.; Ducrozet, G.; Gentaz, L.; Ferrant, P.; Alessandrini, B. Applications of the SWENSE Method to seakeeping simulations in irregular waves. In Proceedings of the 9th International Conference on Numerical Ship Hydrodynamics, Ann Arbor, MI, USA, 5–8 August 2007.
21. Gatin, I.; Vukčević, V.; Jasak, H. A framework for efficient irregular wave simulations using Higher Order Spectral method coupled with viscous two phase model. *J. Ocean Eng. Sci.* **2017**, *2*, 253–267. [CrossRef]
22. Gatin, I.; Vladimir, N.; Malenica, Š.; Jasak, H. Green sea loads in irregular waves with Finite Volume method. *Ocean Eng.* **2019**, *171*, 554–564. [CrossRef]
23. Toffoli, A.; Benoit, M.; Onorato, M.; Bitner-Gregersen, E. The effect of third-order nonlinearity on statistical properties of random directional waves in finite depth. *Nonlinear Process. Geophys.* **2009**, *16*, 131. [CrossRef]
24. Weller, Henry G and Tabor, Gavin and Jasak, Hrvoje and Fureby, Christer A tensorial approach to computational continuum mechanics using object-oriented techniques *Comput. Phys.* **1998**, *12*, 620-631.
25. Zakharov, V.E. Stability of periodic waves of finite amplitude on the surface of a deep fluid. *J. Appl. Mech. Tech. Phys.* **1968**, *9*, 190–194. [CrossRef]
26. Ducrozet, G.; Bonnefoy, F.; Le Touzé, D.; Ferrant, P. A modified high-order spectral method for wavemaker modeling in a numerical wave tank. *Eur. J. Mech.-B/Fluids* **2012**, *34*, 19–34. [CrossRef]
27. Ducrozet, G.; Bonnefoy, F.; Perignon, Y. Applicability and limitations of highly non-linear potential flow solvers in the context of water waves. *Ocean Eng.* **2017**, *142*, 233–244. [CrossRef]
28. Higuera, P.; Lara, J.L.; Losada, I.J. Realistic wave generation and active wave absorption for Navier–Stokes models: Application to OpenFOAM®. *Coast. Eng.* **2013**, *71*, 102–118. [CrossRef]
29. Higuera, P. Enhancing active wave absorption in RANS models. *Appl. Ocean Res.* **2020**, *94*, 102000. [CrossRef]
30. Mansard, E.; Funke, E. The Measurement of Incident and Reflected Spectra Using a Least squares Method. In Proceedings of the 17th International Conference on Coastal Engineering (ICCE), Sydney, Australia, 23–28 March 1980.
31. Jonkman, J.; Musial, W. *Offshore Code Comparison Collaboration (OC3) for IEA Wind Task 23 Offshore Wind Technology and Deployment*; Technical Report; National Renewable Energy Lab. (NREL): Golden, CO, USA, 2010.
32. Jonkman, J.; Butterfield, S.; Musial, W.; Scott, G. *Definition of a 5-MW Reference Wind Turbine for Offshore System Development*; Technical Report, National Renewable Energy Lab. (NREL): Golden, CO, USA, 2009.
33. Liu, P.C.; Babanin, A.V. Using wavelet spectrum analysis to resolve breaking events in the wind wave time series. *Ann. Geophys.* **2004**, *22*, 3335–3345. [CrossRef]
34. Reynders, E.; Schevenels, M.; De Roeck, G. MACEC 3.3: A Matlab Toolbox for Experimental and Operational Modal Analysis. 2014. Available online: https://bwk.kuleuven.be/bwm/macec/macec.pdf (accessed on 6 January 2021).
35. Onorato, M.; Waseda, T.; Toffoli, A.; Cavaleri, L.; Gramstad, O.; Janssen, P.; Kinoshita, T.; Monbaliu, J.; Mori, N.; Osborne, A.R.; et al. Statistical properties of directional ocean waves: The role of the modulational instability in the formation of extreme events. *Phys. Rev. Lett.* **2009**, *102*, 114502. [CrossRef]
36. Onorato, M.; Osborne, A.R.; Serio, M.; Cavaleri, L. Modulational instability and non-Gaussian statistics in experimental random water-wave trains. *Phys. Fluids* **2005**, *17*, 078101. [CrossRef]
37. Tayfun, M.A. Narrow-band nonlinear sea waves. *J. Geophys. Res. Ocean.* **1980**, *85*, 1548–1552. [CrossRef]

Article

Numerical Flow Characterization around a Type 209 Submarine Using OpenFOAM [†]

Ruben J. Paredes [1,*], Maria T. Quintuña [1], Mijail Arias-Hidalgo [1] and Raju Datla [2]

[1] Escuela Superior Politécnica del Litoral, ESPOL, ESPOL Polytechnic University, Campus Gustavo Galindo Km. 30.5 Vía Perimetral, Guayaquil 09-01-5863, Ecuador; mtquintu@espol.edu.ec (M.T.Q.); mijedari@espol.edu.ec (M.A.-H.)

[2] Davidson Laboratory, Stevens Institute of Technology, Hoboken, NJ 07030, USA; rdatla@stevens.edu

[*] Correspondence: rparedes@espol.edu.ec

[†] This paper is an extended version of our paper published in the 15th OpenFOAM Workshop (OFW15).

Abstract: The safety of underwater operation depends on the accuracy of its speed logs which depends on the location of its probe and the calibration thoroughness. Thus, probes are placed in areas where the flow of water is smooth, continuous, without high velocity gradients, air bubbles, or vortical structures. In the present work, the flow around two different submarines is numerically described in deep-water and near-surface conditions to identify hull zones where probes could be installed. First, the numerical setup of a multiphase solver supplied with OpenFOAM v7 was verified and validated using the DARPA SUBOFF-5470 submarine at scaled model including the hull and sail configuration at $H/D = 5.4$ and $Fr = 0.466$. Later, the grid sensitivity of the resistance was assessed for the full-scale Type 209/1300 submarine at $H/D = 0.347$ and $Fr = 0.194$. Free-surface effect on resistance and flow characteristics was evaluated by comparing different operational conditions. Results shows that the bow and near free-surface regions should be avoided due to high flow velocity gradient, pressure fluctuations, and large turbulent vortical structures. Moreover, free-surface effect is stronger close to the bow nose. In conclusion, the probe could be installed in the acceleration region where the local flow velocity is 15% higher than the navigation speed at surface condition. A 4% correction factor should be applied to the probe readings to compensate free-surface effect.

Keywords: submarine; flow characterization; vortex identification; full-scale simulation; type 209 class; DARPA SUBOFF

Citation: Paredes, R.J.; Quintuna, M.T.; Arias-Hidalgo, M.; Datla, R. Numerical Flow Characterization around a Type 209 Submarine Using OpenFOAM. *Fluids* **2021**, *6*, 66. https://doi.org/10.3390/fluids6020066

Academic Editor: Eric G. Paterson
Received: 30 November 2020
Accepted: 26 January 2021
Published: 3 February 2021

Publisher's Note: MDPI stays neutral with regard to jurisdictional claims in published maps and institutional affiliations.

1. Introduction

Although submarines are the most powerful underwater marine vehicles used to guarantee the maritime sovereignty of nations, their operational safety depends on several mechanical and electronic equipment. For instance, the warship underwater position estimation relies on the accuracy of its speed log, which is affected by the location of its probe on the hull and the calibration procedure to compensate the hydrodynamic non-linearities on the flow velocity caused by the hull curvature. The accuracy of the measured speed depends on technology used by the log, such as electromagnetic (EM), pitometer, Doppler, impeller, and Global Positioning Satellite (GPS). Although latter is the preferred alternative for modern surface vessels due to its high precision, it is useless for submarines operating completely submerged. One of the viable options for submarines navigating completely submerged are EM logs with flush or protruding probes that estimate the local speed of water by measuring the amplitude of the electrical signal and whose accuracy relays on the probe protruding distance from the hull, flow linearity, turbulence intensity at the probe location, and sensor calibration thoroughness.

The procedure to determinate the probe installation location requires to have an insight into the physics of the water flowing around the full-scale submarine's light hull considering every hydrodynamic variable available in deep-water and near-to-surface conditions. These flow characteristics depend on fluid properties, hull geometry, appendage distribution, vessel motion orientation, and free-surface proximity. Typically, towing tank tests are used to measure the total resistance force, local pressure and velocity, and streamline patterns around underwater vehicles at model scale. Later, full-scale extrapolation is performed by parts: (i) skin friction and pressure components are estimated assuming an axisymmetric body moving through an unbounded fluid, and (ii) wake-making component is estimated considering wave train generation and non-linear effects from free-surface proximity that modifies the pressure distribution around the light hull. For the last step, it is possible to perform full-scale CFD simulations to characterize this free-surface effect on the flow pattern around submarines through detailed analysis of streamline distribution, boundary layer development, vortex core identification, velocity and pressure distributions and turbulence levels. The probe calibration procedure could include free-surface effects using CFD results after a proper verification and validation procedure [1,2] is performed using experimental data from towing tank experiments or sea trials measurements .

One of the experimental databases used for CFD validation was published by the SUBOFF program sponsored by the Defense Advanced Research Projects Agency (DARPA), where a submarine with polynomial geometry was developed and tested including typical appendages at several operational conditions [3,4]. Thereafter, many studies have been performed to understand the flow field around the SUBOFF submarine through numerical prediction of resistance, maneuverability, self-propulsion, both in deep-water and near-surface conditions. Regarding deep-water analysis, Gross et al. [5] numerically studied the velocity, eddy viscosity, and skin friction at different angles of attack of the bare hull configuration. They confirmed the presence of streamwise vortices on the leeward side at large angles of attack and identified a laminar separation near the submarine stern at low Reynolds number. A thorough investigation of a more complex geometry configuration was performed by Lungu [6], who analyzed the hydrodynamic performance of SUBOFF hull including the sail and four stern appendages using a modified Detached Eddy Simulation (DES) approach. In this work, the streamwise velocity, the pressure distribution and turbulent structures of the nose of the SUBOFF bow and sail confirmed the expected flow physics and showed that the wake at the propeller plane has significant velocity defects, which suggest that special attention is required when modeling the propeller performance.

Regarding free-surface effects, Vasileva and Kyulevcheliev [7] numerically predicted the submarine resistance at different depths for a wide range of velocities and demonstrated that the submarine resistance increases as its free-surface distance decreases due to the wave train generated around submarine. However, this effect decreases at higher speeds in agreement with the different evolution of wave making and viscous resistance coefficients as a function of Froude number. Another alternative used to assess near-surface conditions was proposed by Gourlay and Dawson who used a Havelock Source Panel Method [8] to take advantage of its low computational requirements. Wave-making resistance was successfully compared with model tests data results except in extreme near-surface condition.

Additionally, several self-propulsive simulations, where the DARPA SUBOFF submarine is fitted with a seven bladed INSEAN E1619 propeller, have been performed to assess the propeller thrust, torque, advance coefficient and wake fraction. Di Felice et al. [9] investigated the performance and fluid dynamics of the INSEAN E1619 propeller running in open water conditions through experiments and simulations. It was suggested that submarine propellers require more cell elements than conventional propellers to capture the weaker flow structures generated by the propeller motion. For this reason, different methods such as body force and sliding mesh method [10] or LES method are used to simulate the flow [11] and thrust identity method [12]. For example, Drogul et al. [13] demonstrate the robustness of the actuator disc theory on the self-propulsion point estimation by coupling open water propeller results with a RANSE solver in a single phase analysis. Here, two methods of self-propulsion were analyzed; (i) where the propeller was modeled as an actuator disc based on a body force method, and (ii) where the propeller behind the hull is included in the simulation. It was shown that the former underpredicts the propulsive efficiency and overpredicts the delivered power. In a complementary study, Delen et al. [14] compared the hydrodynamic performance predictions of DARPA SUBOFF with the previous actuator disk method and empirical method [15]. The CFD resistance predictions are more accurate than those of the empirical method when compared with the experimental data. Finally, Sezen et al. [16] successfully extended the resistance and self-propulsion analysis to bare hull, sail, and four stern appendages configuration, modeling the propeller action with a Moving Reference Frame (MRF) method. It was found that the hull efficiency is higher in self-propelled case suggesting that this could be due to positive propeller-hull interaction.

Regarding maneuvering analysis, Duman et al. [17] investigated the propeller effects on maneuvering forces, moments, and its derivatives in six oblique towing conditions of the bare hull configuration approximating the propeller effect through an actuator disk. It was found that using this actuator disk model, the propeller does not affect the sway force and yaw moment, but produce a considerable difference in longitudinal forces at small drift angles. Moreover, Chase et al. [18] performed self-propulsion simulations of the fully appendaged DARPA Suboff traveling straight-ahead and turning with several drift angles using RANS and DDES approaches implemented in the CFDShip-Iowa software efficiently coupled with a modified version of a vortex-lattice potential flow propeller code PUF-14 with and a discretized propeller. The most complex case considered was a surfacing maneuver simulation of the DARPA and a discretized E1619 propeller including a sea state of 4. It was showed that the submarine adopts a bow-down attitude as it reaches the surface, that is accompanied with large fluctuations in thrust and torque because of the highly unsteady inflow to the propeller. Furthermore, this configuration was used to analyze a methodology used in surface ships, concluding that this could be applied to compute self-propulsion of submarines [19]. The comparison of four different turbulence models (RANS, DES, DDES, and no turbulence model) showed that RANS overly dissipates the wake and no turbulence model tip vortices become unphysically unstable. The results confirmed that the approach applied to surface ships is applicable to self-propulsion performance prediction of submarines. However, to the best of the author knowledge, there is no hydrodynamic study of the flow around a different class type of submarine, other than SUBOFF geometry, considering its full-scale length.

Additional CFD studies on the free surface effect on underwater vehicles with different submarines types, other than SUB-OFF, have been performed. For example, the total drag and wake of an axisymmetric UWV model were studied at different depths and speeds using ANSYS CFX [20]. Numerical predictions showed good agreement with the experimental data, where drag coefficient is inversely proportional to Reynolds number for all submergence depths and it is inversely proportional to submergence depth form constant Reynolds number due to the wave-making resistance. In addition, Moonesun and Korol [21] analyzed two torpedo-like models with different L/D ratios using Flow Vision and found that the free surface effects banished for high Froude numbers at an immersion depth of 4.5 D. For regular and irregular waves, a torpedo-like model was analyzed using Flow 3D and Panel method respectively [22,23]. It was recommended an immersion depth of 10% of the incident wavelength to operate safely for AUVs, research submersibles and submarines. Furthermore, a benchmark geometry model, proposed by the CSSRC, was used to assess the hull/propeller interactions in submergence and near surface conditions [24]. Numerical results of thrust, torque, and self-propulsion factors showed a fairly good agreement when compared with experimental data with a small variation of the self-propulsion factors in the near-surface condition. An additional experimental and numerical study on hull/propeller interaction of a realistic modern geometry and a five-blade Wageningen B-series propeller was done by Vali et al. [25]. It was found that when the propeller is operating near the free-surface, the thrust deduction and wake factors decrease and that the thrust coefficient is reduced by 9% when compared to the open water condition. Moreover, Carrica et al. [26] simulated the near-surface self-propulsion of the generic Joubert BB2 submarine considering two conditions, even keel and in controlled free running considering calm water and regular waves representative of sea state 2 to 7. Here, the top of the sail distance to the calm water surface is used as a reference depth. It was shown that near surface operation causes an increase in the propeller thrust and in waves presented a decay of the wave influence with depth and with decreasing wave amplitude. On the other hand, a compensation through trim tanks is required in controlled free running in response to the vertical forces and pitching moments. Another study was carried out by Zhang et al. [27] and included experimental and numerical simulations of the emergency rising of a Russia's Kilo-class diesel-electric submarine . The effect of course keeping on the roll stability was assessed and it was demonstrated a coupling effect between roll and yaw. It was concluded that the excessive yaw deviation is the cause of emergency rising roll instability at moderate incidence angles.

Among the few available studies on full-scale submarines operating in fully submerged condition, Moonesun et al. [28] analyzed a torpedo-like and an IHSS model through CFD, and a Tango nose model in a towing tank. In all three cases, it was shown that the total resistance coefficient becomes Reynolds number independent after 5×10^6, suggesting that there is a critical Reynolds number that is independent of the submarine geometry. Finally, full-scale numerical simulations of a modern submarine type were done to estimate the hydrodynamic derivatives required to evaluate its maneuvering characteristic [29]. Simulations considering the submarine moving in a straight line and rotating showed that asymmetric forces are generated by a vortex flow developed behind the conning tower that affects diving motion. Hence, all the previously described works focused on global variables estimation using experimental or numerical methods, including hull-propeller interaction without describing the flow structure on the near bow region where speed sensors are usually fitted.

In this work, the flow around a Type 209 submarine is simulated at full-scale using OpenFOAM to identify the suitable hull regions for installing speed logs to guarantee sensor precision. First, the numerical setup of a multiphase solver supplied with OpenFOAM v7 was verified and validated using the DARPA SUBOFF-5470 submarine at scaled model including the hull and sail configuration navigating at $H/D = 5.4$ and $Fr = 0.466$. Later, the grid sensitivity analysis of the resistance force is assessed following the Validation and Verification guidelines recommended by the ITTC. Afterward, several fluid characteristics such as: dynamic pressure, velocity on boundary layer limit, skin friction, and vortex cores are used to understand the flow evolution around the submarine light hull at the emerged condition. Later, the influence of the free surface is evaluated using the validated mesh density to identify the flow variations, especially on velocity distribution, including a fully submerged condition. Finally, the linear variation of local flow speed at the probe installation location is assessed to confirm measurement usefulness. It is expected that the methodology used in this work to determine appropriate areas on the hull for probe installation could be generalized to existing or new submarines type designs.

2. Numerical Method

The air and water flow around the submarine light hull was modeled as a steady-state multiphase flow using OpenFOAM v7 [30]. Following current numerical hydrodynamics practices used for surface vessels, this study adopted a Volume of Fluid (VOF) method, implemented in the interFoam solver, to capture free-surface evolution when the submarine is navigating at near-surface and deep-water depth conditions. Also, the $\kappa - \omega$ SST eddy viscosity model including wall functions were used to capture turbulent structures. The interFoam solver approximates in each cell the Continuity and the Navier-Stokes equations for two immiscible, incompressible, and viscous fluids. An additional transport equation is included to capture free-surface evolution as follows

$$\frac{\partial \rho}{\partial t} + \frac{\partial (\rho u_i)}{\partial x_i} = 0 \tag{1}$$

$$\frac{\partial(\rho u_i)}{\partial t} + \frac{\partial[\rho u_i u_j]}{\partial x_j} = -\frac{\partial p}{\partial x_i} + \frac{\partial \tau_{ij}}{\partial x_j} + \rho f_i \tag{2}$$

$$\frac{\partial \alpha}{\partial t} + \frac{\partial(\alpha u_j)}{\partial x_j} = 0 \tag{3}$$

where ρ is a mixture density defined as $\rho = \alpha \rho_1 + (1 - \alpha)\rho_2$, being $\alpha = 1$ if cell is filled with water, and 0 otherwise. Hence, the free-surface interface is located when $0 < \alpha < 1$, and the available MULES correction algorithm is used for the free-surface reconstruction guaranteeing its sharpness. In general, equations are integrated in time using the localEuler scheme with a global maximum CFL number of 4 even at the free surface interface. At post-processing, the local flow velocity (V_{flow}) over the hull and outside boundary layer is estimated from dynamic pressure (p_rgh) and nondimensionalized using submarine navigation velocity (V_{sub}), as follows:

$$V^*_{hull} = \frac{V_{flow}}{V_{sub}} = \alpha \sqrt{1 - \frac{2 \cdot p\_rgh}{\rho \cdot V^2_{sub}}} \tag{4}$$

2.1. DARPA SUBOFF and Type 209/1300 Submarine Geometry

As described above, the geometry of the DARPA SUBOFF model is defined by polynomial equations [31] and consists of an axi-symmetrical hull tested in eight different arrangement configurations (AFF) depending on the appendages included [3]. In the present study, only the geometry of the bare hull with the sail is modelled because log probes are typically fitted on the forward half of the hull to avoid stern appendages or propeller interaction. Hence, the experimental resistance force data measured in the DTMB Tow Tank that is used in the validation is estimated from AFF-1 (bare hull), AFF-3 (hull with four stern appendages), and AFF-8 (hull with sail and four appendages).

On the other hand, Type 209 submarine is a diesel-electric warcraft built by HDW of Germany since 1971, with 61 units still operating in 13 countries worldwide. There are five variants depending on the surface displacement, from 1100 to 1500. In the last decade, most of the oldest submarines have being refitted including electronic equipment upgrades. In particular, the 209/1300 submarine analyzed in this study upgraded to two EM speed logs, located at starboard and upper deck, see Figure 1, that must rely on experimental or numerical hydrodynamics studies to determine its light hull location and further calibration curve for near-surface and deep-water navigation.

(a) Navigating at surface condition   (b) EM speed probe at starboard   (c) EM speed probe at upper deck

Figure 1. Type 209/1300 class submarine with Electromagnetic (EM) speed logs. Subfigure (**a**) was provided by the Submarine Squadron of Ecuador.

This calibration curve is estimated from sea trials to guarantee the accuracy of the readings reported by the probes. This curve compares the submarine navigation speed and local flow velocity at probe location to compensate non-linearities for hydrodynamic variations produced by the curvature of the hull. In the submarine case, two calibration curves, near-surface and deep-water operation conditions, are used. Typically, the calibration of the deep-water curve uses the near-surface one as the reference navigation speed. However, this heuristic methodology wrongly assumes that flow at deep-water and near-surface conditions are similar, despite experimental evidence proving otherwise, as previously discussed. One alternative to quantify free surface effects at the full-length scale at regions near probe locations is to use an open-access experimental database, such as DARPA SUB-OFF. However, there are significant differences in displacement and length proportions with the 209/1300 class, as shown in Table 1, that prevent a straight forward extrapolation. Hence, the only feasible option is to perform a numerical study at full-scale to characterize flow conditions around the Type 209/1300 submarine light hull.

Table 1. Main characteristics comparison between DARPA–SUBOFF and Type 209/1300 submarines.

Parameter	Symbol	SUB-OFF Model Scale	SUB-OFF Equivalent Prototype Scale	Type 209/1300
Length between perpendiculars [m]	L	4.356	58.167	58.167
Length-Diameter ratio	L/D	8.575	8.575	9.950
Draft-Diameter ratio	T/D	0.863	0.863	0.863
Length percentage of fore body	L_{FB}/L	0.233	0.233	0.189
Length percentage of parallel middle body	L_{PMB}/L	0.512	0.512	0.496
Length percentage of aft body	L_{AB}/L	0.255	0.255	0.314
Relative sail location	L_{FC}/L	0.21	0.21	0.40
Wetted area of hull at surface [m^2]	S_{surf}	4.760	848.759	775.220
Wetted area of hull + sail [m^2]	$S_{hull-Sail}$	6.160	1098.394	1182.420
Wetted surface area of sail [m^2]	S_{sail}	0.184	32.855	83.258
Displacement at surface condition [tons]	Δ_{surf}	0.650	1547.675	1309.950
Displacement submerged hull + sail [tons]	Δ_{sub}	0.703	1674.560	1578.00

Finally, Table 2 shown the coordinates of four points, P1 to P4, that corresponds to the base and measurement location of fixed protruding probes at starboard (P1, P3) and upper deck (P2, P4), as shown in Figure 1. The distance between each pair of points is 40 cm, corresponding to distance between the probe base and its measurement position. This control points are used to assess the lineal variation of the local flow speed with navigation speed.

Table 2. Control points locations measured from baseline used to probe flow velocity on the hull of a Type 209/1300 submarine.

Control Point	X (m)	Y (m)	Z (m)
P1	53.82	−2.34	1.34
P2	50.64	−0.53	7.02
P3	53.82	−2.68	1.14
P4	50.64	−0.53	7.41

2.2. Simulation Setup for Submarines

For the DARPA SUBOFF-5470 submarine, the bare hull and sail configuration at model scale was considered, as shown in Figure 2. The hull is Axis-symmetric and is defined by polynomials as described in Groves et al. [31]. The 3D CAD geometry was generated using the Grasshopper plugin in Rhinoceros and the surface was exported to OpenFOAM as an STL file. Additionally, after transforming the geometry to mesh type, the water tightness of the geometry was verified with the software Blender.

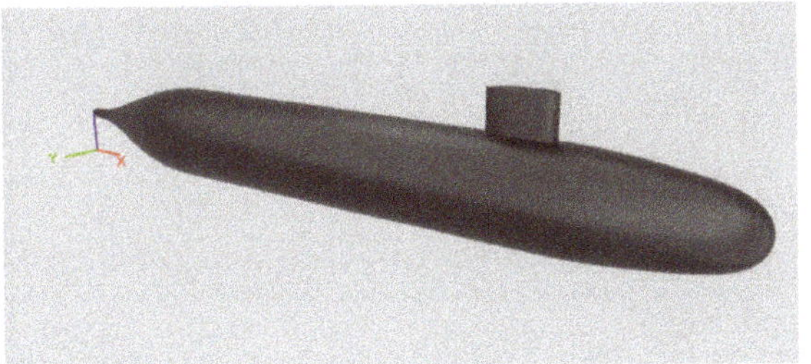

Figure 2. 3D CAD model of DARPA SUB-OFF submarine used in the present study.

For the Type 209/1300 submarine, a complete CAD model was generated by combining original blueprints provided by shipowner and measurements taken during a scheduled dry-dock maintenance. However, appendages with dimensions smaller than 10 cm such as: hydroplanes, passive sonars, antennas, and collision bumpers, were not included in the CAD version used in the current work. This simplified CAD version, shown in Figure 3, was used in the numerical simulations to decrease the required computational time per simulation, where neither propeller geometry nor its hull interaction is included.

Figure 3. Simplified 3D CAD model of Type 209/1300 submarine used in the present study.

Although the flow around the submarine may be vortex dominated, the dominant vortical structures are generated by the sail [32], that produce asymmetrical flow structures about the central plane that are shed downstream along the hull, hence requiring a full-domain approach. Nevertheless, there is evidence that when a submarine is moving straight-ahead, it is possible to use the half-domain approach [5,7,10,21]. On the other hand, if the submarine propeller [9,17], self-propulsion problem [12–14,18,26], or maneuvering characteristics [27,29] are numerically assessed; then a full-domain approach must be used. The present work focus on the forward hull region below the free surface interface, where the flow is expected to be axisymmetric when fully submerged as reported by Lungu [6], that is slightly perturbed by the occurrence of breaking wave in front of the bow due to vertical, cylindrical bow that is symmetric around centerline when operating at surface depth, as reported by Jasak et al. [33]. Thus, half-domain approach is chosen in this work and decreasing the required time per simulation.

Hence, the computational domain size was defined following ITTC guidelines [34], where the inlet, side, bottom, atmosphere, and outlet boundaries are placed far enough from the submarine to avoid spurious flow interaction, see Figure 4. The boundary conditions are shown in Table 3 for all patches of the computational domain, where the no-slip wall condition is imposed on the submarine boundary. Regarding initial conditions, the flow velocity is initialized with the navigation speed in the inlet, bottom, and internal field.

In addition, turbulent kinetic energy (κ) and its specific rate of dissipation (ω) are set assuming a turbulence intensity level of 1.5% considering wall functions for the hull patch.

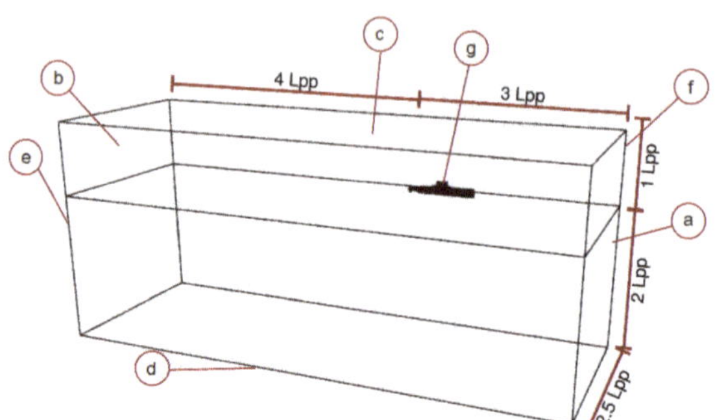

Figure 4. Computational domain and boundaries used to model DARPA SUBOFF and Type 209/1300 submarines.

Table 3. Boundaries names and conditions.

Label	Name	U	*p_rgh*
a	Inlet	fixedValue	fixedFluxPressure
b	Side	symmetryPlane	symmetryPlane
c	Atmosphere	pressureInletOutletVelocity	totalPressure
d	bottom	fixedValue	fixedFluxPressure
e	outlet	outletPhaseMeanVelocity	zeroGradient
f	midPlane	symmetryPlane	symmetryPlane
g	hull	movingWallVelocity	fixedFluxPressure

The computational grids are generated in stages: (i) an hexahedral background grid using blockMesh with refinement in the region where free-surface interface is expected, (ii) six refinement levels using topoSet tool to half the element size progressively as the elements get closer to the submarine hull, (iii) the surface is inserted with additional (1 or 2) refinement levels if the elements is less than 0.1 D from the submarine hull, (iv) eight layers of prismatic elements parallel to submarine surface are included with a target yPlus, that is estimated considering the laminar boundary layer thickness as a reference length to capture boundary layer evolution efficiently. Figure 5 shows slices of centerline showing one of the generated grids for SUBOFF and S209 submarines. The progressive refinement in regions close to the submarine is evident, aiming to capture high gradients of velocity in the boundary layer and pressure distribution around the submarine's light hull.

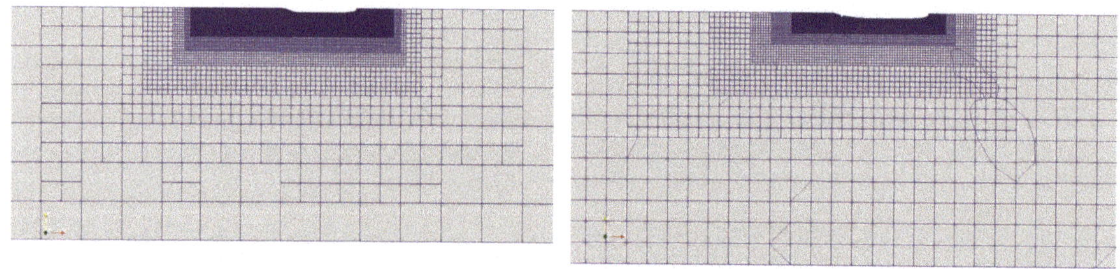

Figure 5. Slices showing mesh refinement for fine grid around SUBOFF (left hand side) and Type 209/1300 (right hand side) submarine with 4,500,437 and 9,145,512 cells respectively.

Table 4 shows the number of cells used for modelling the computational domain to estimate grid errors and uncertainties. The element size in the inner *topoSet* region is used as the reference length (Δx) and it is uniform in all three directions to improve the reliability of free surface evolution and turbulence prediction. The Medium grid was generated using the ITTC recommendations and the additional grids were generated using a refinement ratio of $rG = \sqrt{2}$ in both directions. For the Medium grid: (i) DARPA SUBOFF, an element size close to the submarine of 0.05 m ($L/\Delta x = 87.12$) and a target yPLus of 50 were used, (ii) Type 209/1300, an element size of 0.20 m ($L/\Delta x = 291.45$) and a target yPLus of 300 were used. For the second geometry, the element size is of the same order of magnitude than the probe dimension and it will capture the relevant flow characteristics for speed log accuracy.

Table 4. Number of cells for uncertainty analysis.

#	Grid Type	DARPA SUBOFF	Type 209/1300
1	Finer	11,986,756	22,832,329
2	Fine	4,500,437	9,145,512
3	Medium	1,576,885	3,892,550
4	Coarse	600,071	1,348,794
5	Coarser	215,582	514,008

Figure 6 shows the distribution of the number of surface layers obtained for each of the considered meshes, despite having requested eight layers, suggesting that most of the hull has the necessary elements to capture the viscous effects within the boundary layer. This difference is due to the mesh quality parameters and curvature of the light hull at the bow and stern sections. Despite the above, the prism surface layer reaches the requested thickness in most of the outer hull, not shown here. Although the yPlus distribution near the surface at the bow, sail, and aft regions has values greater than 300, it is expected that it will not affect the main flow characteristics near the probe location due to its small influence area.

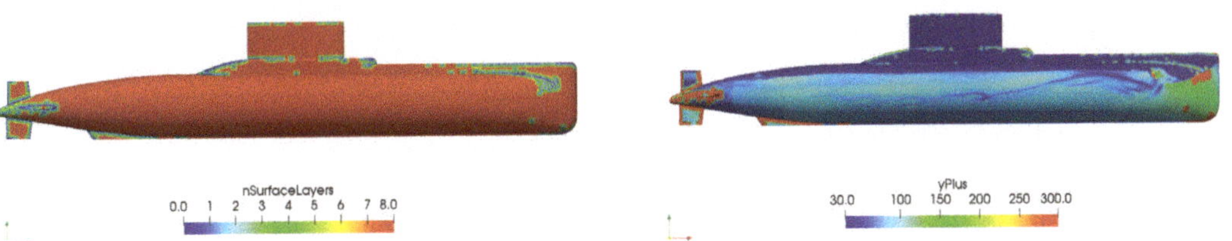

Figure 6. Actual number of surface prism layers generated around submarine by *snappyHexMesh* tool.

2.3. Test Matrix

The DARPA SUBOFF submarine was used for the Verification and Validation considering $H/D = 5.4$ and $Fr = 0.46$ condition, where H is the distance from the free-surface to the hull diameter centerline. The numerical prediction is compared to the experimental data provided by Liu and Huang [3]. Additional simulations predictions using the verified grid were compared with the experimental data provided by Neulist [35] for different immersion depth H/D ratios and Froude number, as shown in Table 5:

For the Type 209/1300, the navigation conditions shown in Figure 7 were used to assess the flow characteristics, free-surface effects, and potential zones on the light hull of the submarine for probe installation to guarantee measurement accuracy fulfilling the manufacturer requirements. Thus, three different drafts measured at the aft perpendicular were considered without initial trim, namely: (i) surface depth at 5.18 m, (ii) periscope depth at 12 m, and (iii) deep-water at 40 m corresponding to a non-dimensional H/D of 0.33, 1.47, and 6.14 respectively.

Table 5. Experimental data used for Verification and Validation of DARPA–SUBOFF submarine simulations.

H/D	Fr	$C_T \times 10^3$	Source
1.1	0.132	4.730	
	0.310	5.980	Neulist [35]
	0.463	7.440	
2.2	0.132	4.460	
	0.310	3.690	Neulist [35]
	0.463	4.110	
5.4	0.466	3.310	Liu and Huang [3]

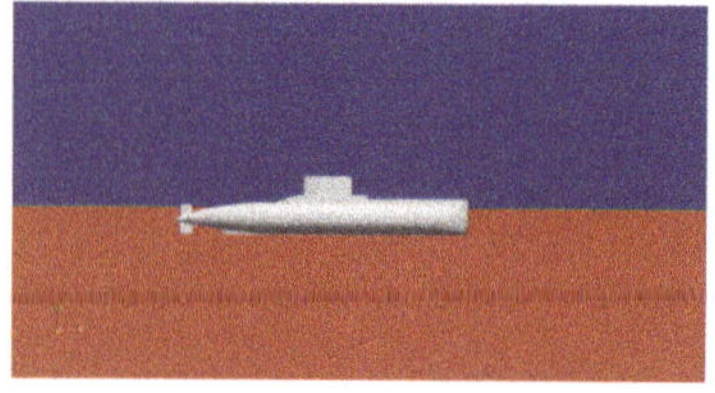

(**a**) Condition 01: Surface depth and no trim, $H/D = 0.33$

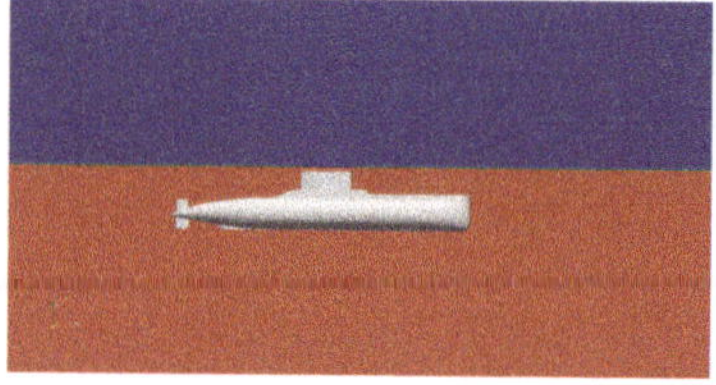

(**b**) Condition 02: Periscope depth and no trim, $H/D = 1.47$

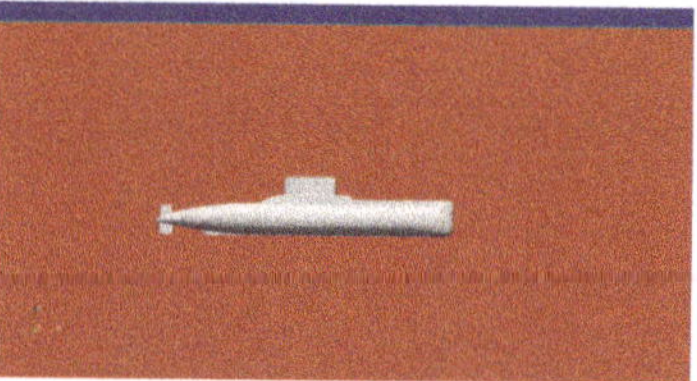

(**c**) Condition 03: Deep water depth and a no trim, $H/D = 6.14$

Figure 7. Navigation conditions considered for numerical simulations. Initial condition of alpha water distribution where red is water and blue is air.

Table 6 shows the navigation details for each Type 209/1300 simulation performed in the present study, such as velocity, draft, and grid density considering operational limits reported by the submarine crew. First, simulations 1–3 were used to assess grid convergence independence with the submarine navigating at 9 knots ($Fr = 0.194$) at surface depth and no trim. Later, simulations 7–9 and 12–14 are completed to confirm grid convergence study navigating at 6 knots ($Fr = 0.129$) at periscope and deep-water depths respectively. Additional simulations with the submarine at conditions 1, 2, and 3 were used to verify the linear variation of the local speed measured by the probe including a speed of 3 knots ($Fr = 0.065$).

Table 6. Parameters used in each of the simulations in the present study of the Type 209/1300 submarine.

Simulation	Navigation Condition	Velocity [Knots]	Coarse Grid	Intermediate Grid	Fine Grid
1–3		9.0	X	X	X
4	C1: Surface depth	6.0			X
5		3.0			X
6		9.0			X
7–9	C2: Periscope depth	6.0	X	X	X
10		3.0			X
11		9.0			X
12–14	C3: Deep-water	6.0	X	X	X
15		3.0			X

3. Results

The numerical results obtained with the settings described in the previous section are summarized here. First, a Verification and Validation procedure is performed using the DARPA SUBOFF submarine geometry at model scale. Next, the grid sensitivity at full-scale was assessed using the total resistance predictions acting on the Type 209/1300 submarine sailing at three navigation conditions: surface depth, periscope depth, and deep-water immersion depth, considering the recommendations described by [1,33,36,37]. Later, the flow was characterized considering the fine grid by assessing the flow around the external hull, wave train, dynamic pressure, yPlus, and wall Shear Stress distribution, and visualization of vortex cores following [38,39].

3.1. Verification and Validation

For this procedure, simulations were performed considering the navigation conditions, $H/D = 5.4$ and $Fr = 0.466$, previously described in Table 5. Figure 8 shows: the evolution of: (i) the residual for dynamic pressure, the fraction of water, κ, and ω variables and (ii) viscous and pressure force components acting on the hull, both for the previously described navigation condition using the fine grid with 4,500,437 cells. All residuals smoothly decreased several orders of magnitude, but the dynamic pressure oscillates around 10^5. Also, both force components achieved convergence after 2000 time steps, which could be related to the absence of free-surface effects. The others grids show the same behaviour.

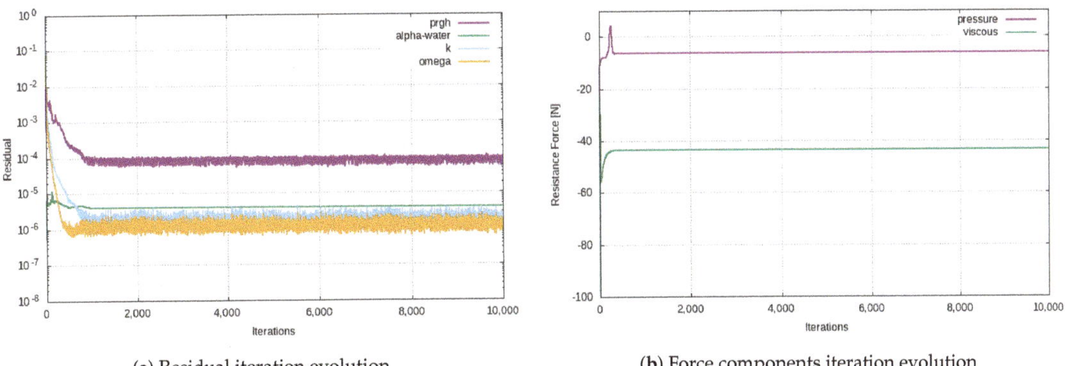

(a) Residual iteration evolution   (b) Force components iteration evolution

Figure 8. Residual and force components evolution acting on the DARPA SUB-OFF submarine navigating at $Fr = 0.466$ and $H/D = 5.4$ using grid 02 with 4,500,437 cells.

Figure 9 shows the evolution of the resistance force for all grids. It is evident that convergence is achieved after 2000 iterations. Additionally, all curves are smooth, but Grid 03 that has spurious oscillations. Grid 01 and 02 are close, suggesting that grid convergence have been achieved. Table 7 shows the average values of pressure and viscous force components acting on the whole hull area, and it is estimated as two times the average of the last 1000 iterations for each grid considered in this study. Although the numerical prediction approximates to the experimental value and the error percentage decreased considerably from Grid 05 to 01, it is still at 4.7% for the latter. It is believed that a possible explanation of the current simulations is due to the existing limitations of snappyHexMesh to add enough prism layers in about 1.1% of the hull area, especially on the hull-sail intersection and the aft end of the hull. Another source of error is from the procedure estimation of the experimental data, since the hull-sail configuration is obtained from three different configurations (AFF1 + AFF8 - AFF3) as described in [3]. Further simulations of SUBOFF use the fine grid due to current hardware limitations.

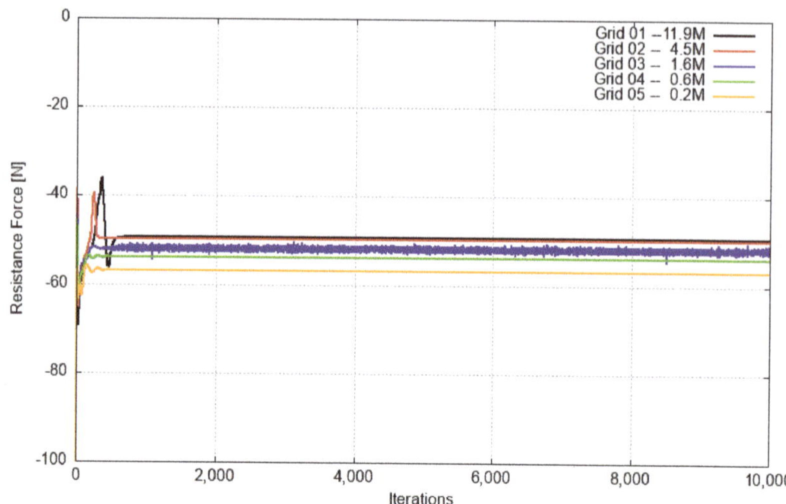

Figure 9. Convergence of forces on SUBOFF in CFD simulations for $H/D = 5.4$ and $Fr = 0.466$.

Table 7. Average and standard deviation of force components vs grid density for SUB-OFF submarine at $Fr = 0.466$ and $H/D = 5.4$.

Grid Density	Number of Cells	Pressure	Viscous	Total	% SD	Exp Value	% Error	Avg. $y+$
Coarser	215,582	30.840	82.840	113.685	0.004		20.6%	103.24
Coarse	600,071	21.643	86.270	107.913	0.009		14.4%	58.29
Medium	1,576,885	17.631	86.271	103.902	0.798	94.29	10.2%	43.39
Fine	4,500,437	12.508	86.954	99.461	0.002		5.5%	23.06
Finer	11,986,756	10.884	87.813	98.696	0.097		4.7%	16.48

Tables 8 and 9 show the results of the Verification calculations obtained using the Total Resistance coefficient from the numerical results previously shown in Table 7. The convergence ratio R_G of 0.172 and 0.605 means that the numerical results for these grid sets (1–3, and 3–5) are in the convergence region and Richardson extrapolation can be used to estimate the numerical force coefficient corresponding to a grid with infinity number of cells. For the grid 1–3 set, the low values of $\delta_G = 0.8\%$ and $U_{GC} = 0.6\%$ suggest that numerical results achieved grid convergence and it could be considered as verified.

Table 8. Total resistance coefficient convergence for SUB-OFF submarine at $Fr = 0.466$ and $H/D = 5.4$.

Grid	Coarser—G5	Coarse—G4	Medium—G3	Fine—G2	Fine—G1	Richardson Extrapolation	Exp Value—S
$C_T \times 10^3$	3.985	3.783	3.642	3.486	3.460	3.433	3.305
$E\%$		5.1%	3.7%	4.3%	0.8%		
$\epsilon_{ij} \times 10^3$		0.202	0.141	0.156	0.027		

Table 9. Verification and Validation parameters considering grids 1–5 for SUB-OFF submarine at $Fr = 0.466$ and $H/D = 5.4$.

Analysis Set	R_G	$\%GCI_{FINE}$	P_G	C_G	$U_G\%S$	$\delta_G\%S_C$	$U_{GC}\%S_C$	S_C
1–3	0.172	0.2%	4.98	4.842	1.4%	0.8%	0.6%	3.433
2–4	1.107							
3–5	0.695	14.3%	0.90	0.433	8.9%	4.0%	2.7%	3.501

Finally, Figure 10 compares numerical Total Resistance Coefficient prediction to the experimental values provided by Neulist [35] and Liu and Huang [3]. Although the numerical prediction seems to slightly over-predicts the Total drag coefficient for all H/D ratios but for the lowest Froude number $Fr = 0.132$, simulations are able to capture the non linear variations at intermediate Froude $Fr = 0.31$ for the near-surface condition $H/D = 1.1$. Furthermore, the difference at the lowest Froude condition could be explained by limitations on computational resources and should decrease further refining the grid.

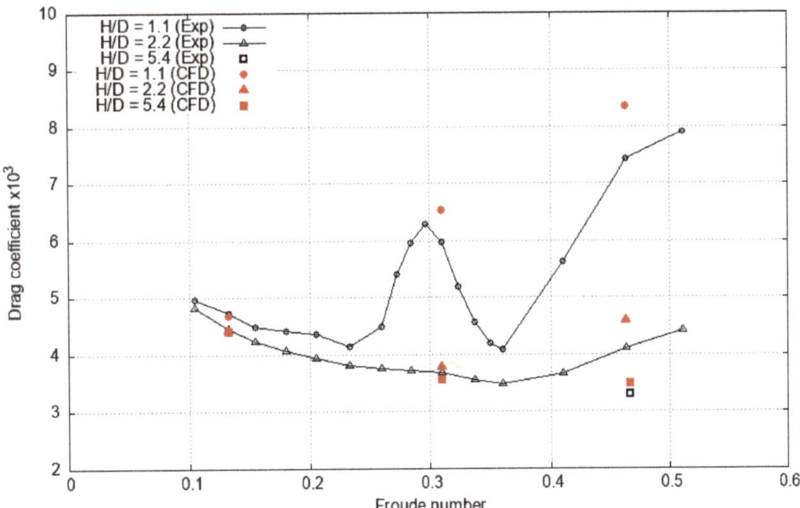

Figure 10. Total Drag coefficient variation with Froude number and immersion depth ratio for the DARPA SUB-OFF submarine.

3.2. Grid Convergence for Type 209 at Full-Scale

Figure 11 shows the evolution of the viscous and pressure force components acting on the submarine light semihull for simulation 03 using $CFL = 4$ in the fluid domain and free-surface interface taking advantage of the implicit MULES scheme. As usual, the viscous component is related to the shear stresses of both fluids and the pressure component is linked to normal stresses due to the interaction of fluid motion and waves generated around the submarine. Both resistance components reach steady values after 10,000 iterations, with a smoother behaviour for the coarser. Furthermore, the viscous curve converges faster, and it is smoother than the pressure curve, as expected. The force evolution predicted using coarse and intermediate grids shows a similar behavior but with an increased dispersion on the pressure curve, that is suspected to be produced by smaller vortices being captured, and spurious instability produced by high CFL number. For this reason, the smaller CFL number of 4.0 was chosen to avoid the occurrence of numerical instability. In summary, the viscous component represents roughly 30% of the total resistance force, which agrees with the expected predictions of cylindrical bluff bodies moving near a free surface at moderate Froude numbers.

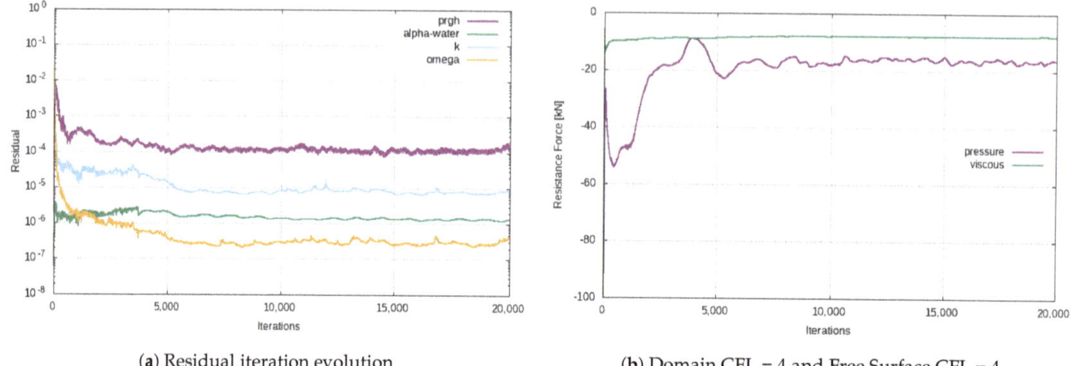

(a) Residual iteration evolution   (b) Domain CFL = 4 and Free Surface CFL = 4

Figure 11. Residual and force evolution of viscous and pressure components acting on the Type 209/1300 submarine navigating at $H/D = 0.33$ and $Fr = 0.194$ using the fine grid.

The viscous and pressure force predictions shown in Table 10 correspond to average values of the last 1000 iterations for each of the grids considered in Simulation 03, namely, $H/D = 0.33$ and $Fr = 0.194$. The standard deviation is used as an indicator of the variation between iterations of the total force, having a maximum value of almost 460 N (2% of average value) of the pressure component for the finer grid, that is similar to the behaviour described in Jasak et al. [33] where a transient simulation of a general cargo carrier shows the resistance force oscillating due to breaking waves in front of the bow.

Table 10. Average and standard deviation of force components vs grid for submarine at condition 1 with Domain CFL = 4 and Free surface CFL = 4.

Grid Density	Number of Cells	Pressure (N)	Viscous (N)	Total (N)	SD (N)
Coarser	524,008	17,560.51	7390.06	24,950.57	446.85
Coarse	1,348,794	15,602.31	7927.64	23,529.95	336.51
Medium	3,892,550	16,467.29	7807.49	24,274.78	292.47
Fine	9,145,512	16,640.02	7768.89	24,408.91	450.07
Finer	22,832,329	15,709.69	8113.64	23,823.22	459.44

Figure 12 shows the iteration evolution of the total resistance for grids 01 to 05. All curves seem to achieve a quasi-steady state value with oscillations that could be related to the bow breaking wave shown in Figure 13. In this study, convergence condition (R_G) is calculated as 0.18 for the Grid set 2–4 which means that the solution is converged monotonically. For Grid sets 1–3 and 3–5, R_G is calculated as −4.366 and −0.524 respectively, which means that the solution has an oscillating convergence. The Grid Convergence Index (GCI) for the set 2–4 is 1.0%, 0.3%, and 3.9% for sets 1–3, 2–4, and 3–5, where the latter has the higher uncertainty. Furthermore, the total resistance is extrapolated from set 2–4 using the Richardson extrapolation procedure [37] to 24,458.84 N for an infinite number of elements. Hence, the fine grid is considered as acceptable considering its balance between the resolution of the flow characteristics and the required computational resources. All remaining simulations will use this fine grid with 9,145,512 elements.

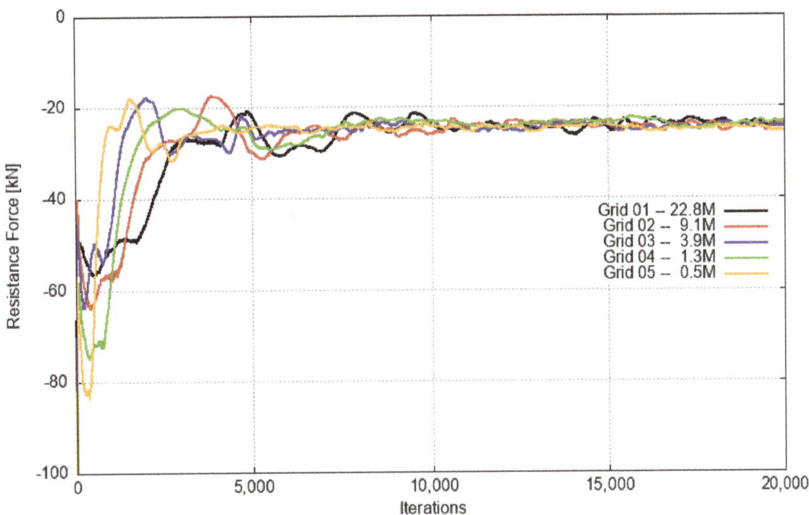

Figure 12. Convergence of forces in CFD simulations for $H/D = 0.347$ and $Fr = 0.194$.

3.3. Flow Characterization at Condition 01: Surface Depth — H/D = 0.33

Figure 13 shows the wave train generated around the Type 209/1300 submarine at condition 1, namely, $H/D = 0$ and $Fr = 0.194$. On the left-hand side, it is evident that two Kelvin wave patterns are obtained, regardless of a spurious green water on the submarine upper deck produced by the instantaneous acceleration imposed as initial condition. Regarding both wave systems: (i) the system beginning at the submarine's bow with a 0.4 m wave amplitude forward stagnation point, and (ii) the small system beginning at the higher aft foil with a 0.2 m wave amplitude. On the right-hand side, the water fraction distribution profile shows that the first wave crest is almost high enough to wet the upper deck area. However, the next wave trough is modified by a small wave system generated at the upper deck and light hull intersection. The wave profile seems to be canceled after this area as the flow moves downstream, in agreement with the theoretical prediction of the first wave resistance hollow.

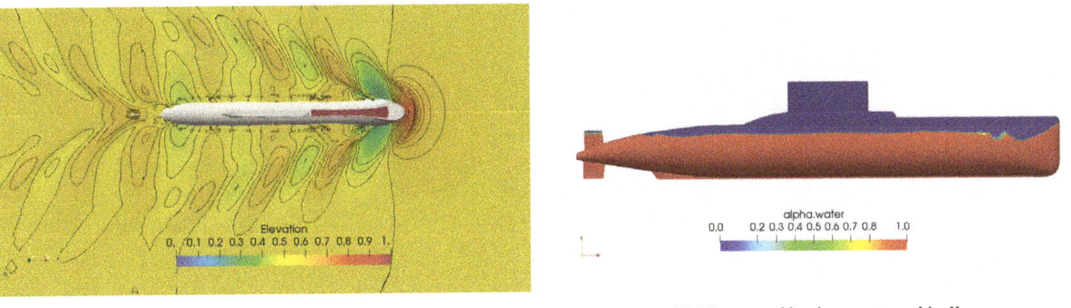

(a) Wave train

(b) Wave profile along external hull

Figure 13. Distribution of wave generated around a Type 209/1300 submarine at condition 1 $H/D = 0.33$ and $Fr = 0.194$. This distribution includes contour lines between 0.45–0.80 with steps of 0.025 [m].

Figure 14 shows the dynamic pressure and wall shear stress distribution around the Type 209/1300 submarine at condition 1 and $Fr = 0.194$. On the right-hand side, it is possible to distinguish a stagnation area at the bow of the forward body section as predicted by a Pitot-tube like device and several pairs of high-low pressure areas produced by wave crest-trough propagation. Moreover, there are stagnation areas at the nose of each aft foil. On the left-hand side, the wall shear stress distribution seems smooth, but in the near-surface bow region at the bow wave trough, where high gradients are observed and the presence of turbulent eddies at the free surface level is suspected.

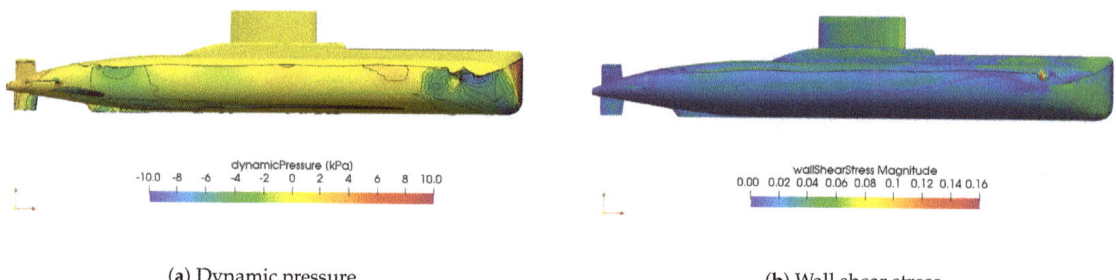

(**a**) Dynamic pressure   (**b**) Wall shear stress

Figure 14. Numerical prediction of dynamic pressure and wall shear stress distribution around a Type 209/1300 submarine at condition 1 $H/D = 0.33$ and $Fr = 0.194$. Dynamic pressure distribution includes contour lines between range limits with steps of 1000 [Pa]. Wall shear stress distribution includes the free surface profile.

Figure 15 shows the flow velocity distribution relative to the navigation speed of the submarine including contour lines between 0.5 and 1.3. The stagnation area at the submarine bow described earlier is followed by a wide high-velocity gradient area where, in principle, the speed probe should not be fitted. Moreover, a small high velocity region is revealed at the first wave trough described before that could be related to vortices being shed from this location. Although P1, corresponding to the side probe base location (marked by a grey dot), is at the end of the acceleration region on a wide velocity region where the local flow speed is between 1.1 and 1.2 times higher than the navigation speed. Velocity probe could be installed here, because of appendage restrictions, if the local flow velocity remains inside this range for the whole range of operational speeds. Hence, it is required to include the results from Simulations 04 and 05.

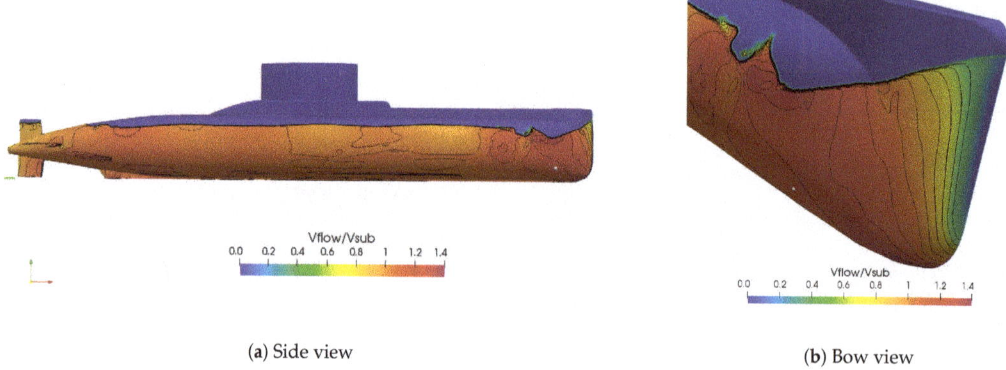

(**a**) Side view   (**b**) Bow view

Figure 15. Non-dimensional flow velocity distribution, outside boundary layer, including contour lines between 0.5–1.3 with steps of 0.1, around a Type 209/1300 submarine at condition 1 $H/D = 0.33$ and $Fr = 0.194$.

Figure 16 shows the local flow velocity as a function of navigation speed at two locations: P1 and P3, corresponding to the measurement position of a protruding probe that is 40 cm away from P1. Both trend lines for condition 01 can be approximated by a lineal function with a slope of 1.15, which suggests that the flow velocity remains in the

1.1–1.2 range described before and that there is not a significant difference in local flow if measured by a flush or protruding probe when the boundary layer effect is neglected.

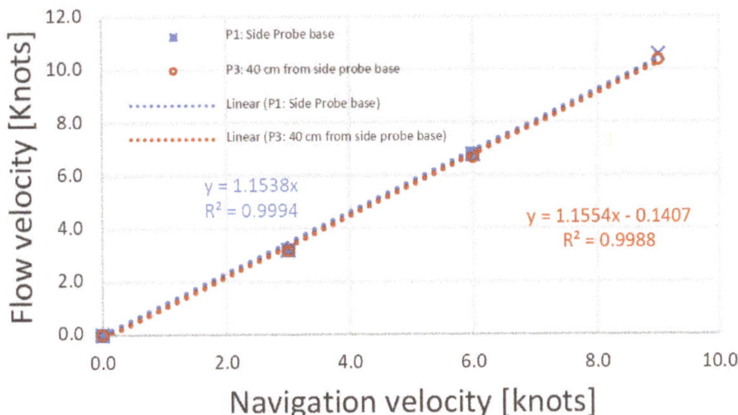

Figure 16. Local flow velocity as a function of navigation speed of a Type 209/1300 submarine at condition 1 at probe base location P1 and measurement protruding probe location P3.

Figure 17 shows vortex cores at the submarine bow identified by using Q and λ_2 parameters. Both methods detect a dominant vortex structure at the bow wave and its downstream propagation along the light hull. Small vortex structures with low dynamic pressure (high flow velocity) are detected at the free surface level near the first trough wave, as suggested by the wall shear stress distribution shown above. The latter method is able to detect smaller structures at the free surface and upper deck level.

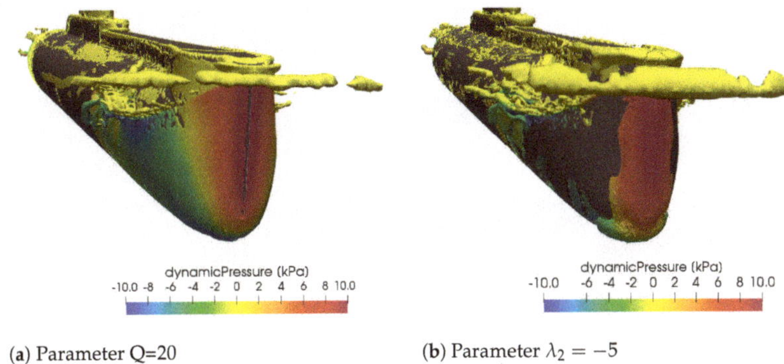

(a) Parameter Q=20   **(b)** Parameter $\lambda_2 = -5$

Figure 17. Vortex cores identification around the bow of a Type 209/1300 submarine at condition 1 and $Fr = 0.194$.

Finally, Figure 18 shows the numerical prediction of drag coefficients for total, viscous, and pressure force components as a function of Froude number, including ITTC'57 curve for the submarine at condition 01 and Total Drag Coefficient for SUBOFF at $H/D = 1.1$. The total coefficient prediction of the Type 209/1300 submarine has a different behaviour than the SUBOFF, and it is closer to the variation expected for surface vessels. Viscous coefficient data compares well with ITTC'57 curve despite its light hull has fuller form than standard surface vessels. Furthermore, the pressure component, which includes wave-making and form factor effects, increases with Froude number as expected, and becomes dominant for higher Froude number. The viscous coefficient difference could be related to insufficient elements being used to describe the hull surface for lower Froude numbers.

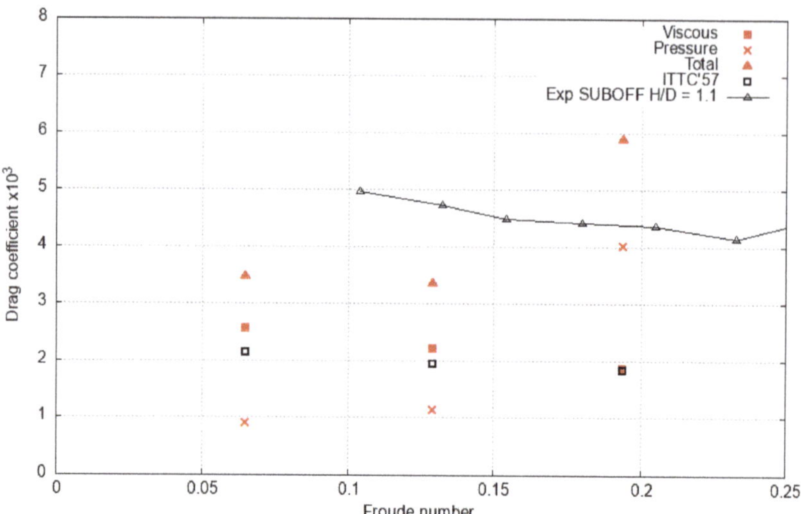

Figure 18. Total Drag coefficient vs Froude number for a Type 209/1300 submarine at condition 1 ($H/D = 0.33$).

3.4. Free-Surface Depth Influence on Flow Characteristics

Figure 19 shows the wave profile at midplane generated by the Type 209/1300 submarine navigating at $H/D = 1.47$ and $Fr = 0.194$. On the left-hand side, the sail deforms the free surface above it due to a Venturi effect and produces a short length wave system beginning at the sail end regardless of no free surface piercing. On the right-hand side, the wave elevation demonstrates a weak free-surface interaction with the submarine light hull, where a small wave amplitude system is generated.

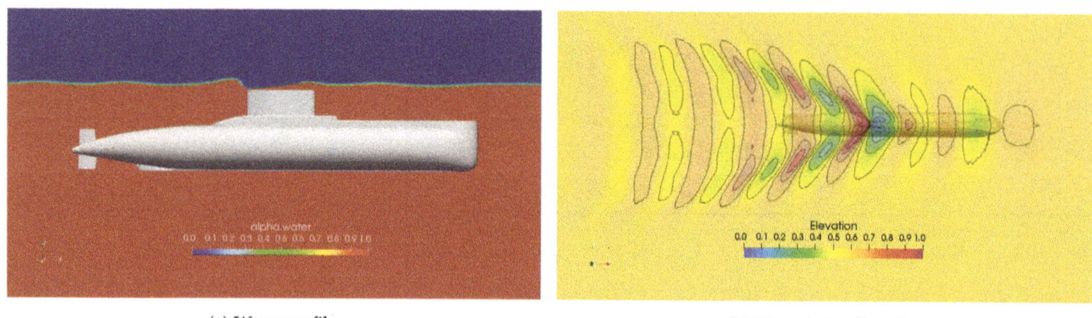

(a) Wave profile　　　　　　　　　　　　　　(b) Wave train elevation

Figure 19. Wave train generated by Type 209/1300 submarine at $H/D = 1.47$ (condition 2—periscope depth) for $Fr = 0.194$.

Figure 20 shows the flow velocity distribution on light hull relative to the submarine speed including contour lines between 0.5 and 1.3 for $H/D = 1.47$ and 6.14. In both conditions, the stagnation area at the submarine's bow, sail, and aft foils are followed by a high-velocity gradient where the probe should not be installed. On the other hand, P1—side probe base, marked by a grey dot, is located on a wide velocity region between 1.1–1.2 times the navigation speed and P2—upper deck probe base in on a slower region between 1.0–1.1 times the navigation speed is used when the submarine is completely submerged. Hence, speed probes could be mounted on both locations given the local flow velocity remains this range for the whole range of operational navigation speeds regardless of its difference with the condition 01 $H/D = 0.33$. Moreover, the calibration procedure used for the upper deck probe is discouraged because it is located on a different velocity range. Finally, the flow velocity distribution is similar for $H/D = 1.47$ and 6.14 but at the sail produced by its free surface proximity where changes are evident at the sail side.

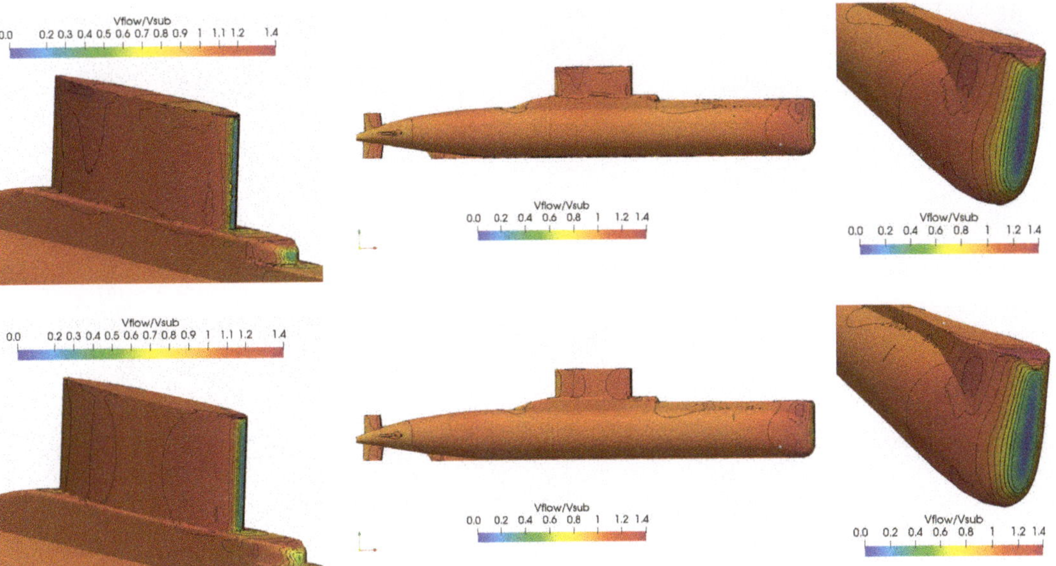

Figure 20. Non-dimensional flow velocity distribution, outside boundary layer, including contour lines between 0.5–1.3 with steps of 0.1, around a Type 209/1300 submarine at condition 2 (above) and 3 (below)—$Fr = 0.194$.

On the left-hand side, Figure 21 shows local flow velocity as a function of navigation speed at two locations: P1 and P2 for $H/D = 1.47$. Both trend lines can be approximated by linear functions with a slope of 1.12 and 1.03 respectively, suggesting that the local flow velocity around the side probe at P1 remains in the 1.1–1.2 range predicted for the surface depth condition and that it is safe to use this probe when the submarine is operating at near-surface condition. On the other hand, the upper deck probe at P2 is within a different speed range 1.0–1.1, suggesting that the calibration procedure based on the side probe is not appropriate producing an error of 0.25 knots when the submarine is navigating to 9 knots ($Fr = 0.194$). On the right-hand side, the local flow velocity at P1 is shown as a fraction of the navigation speed. The local flow velocity decreased almost 4% when the submarine move from emerged ($H/D = 0.33$) to the deep-water condition ($H/D = 6.14$) slightly increasing the error of the measured velocity by side probe sensor.

Finally, Figure 22 shows drag components coefficients as a function of Froude number, including the ITTC'57 curve for submarine at $H/D = 1.47$, and $H/D = 6.14$ conditions and the Total Drag Coefficient for SUBOFF at $H/D = 1.1$ and $H/D = 3.3$. On the right-hand side, viscous coefficient data compares well with the ITTC'57 curve, despite the wetted area includes the sail contribution. The pressure component is almost parallel to the viscous one, suggesting that the form factor could be estimated. Total drag coefficients have the same trend as the experimental values for SUBOFF geometry at $H/D = 3.3$, where some free surface effects are still present. On the left-hand side, viscous data have the same behavior as before, and pressure data seems to increase as the Froude number increases. The Wave-making coefficient could be estimated by deducting the form factor previously estimated.

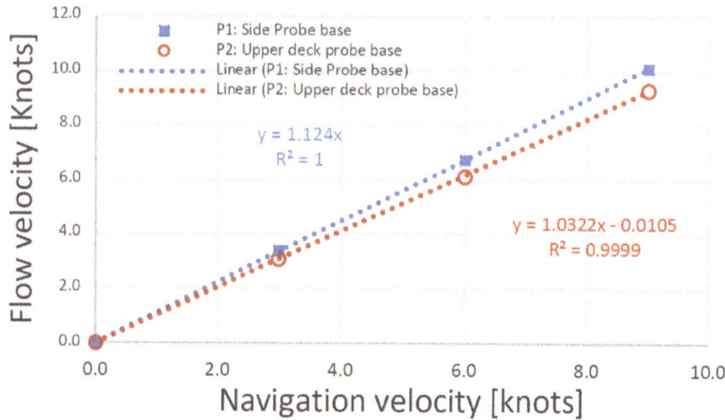

Figure 21. Local flow velocity at condition 2 at side probe base location P1 and upper deck probe base location P2.

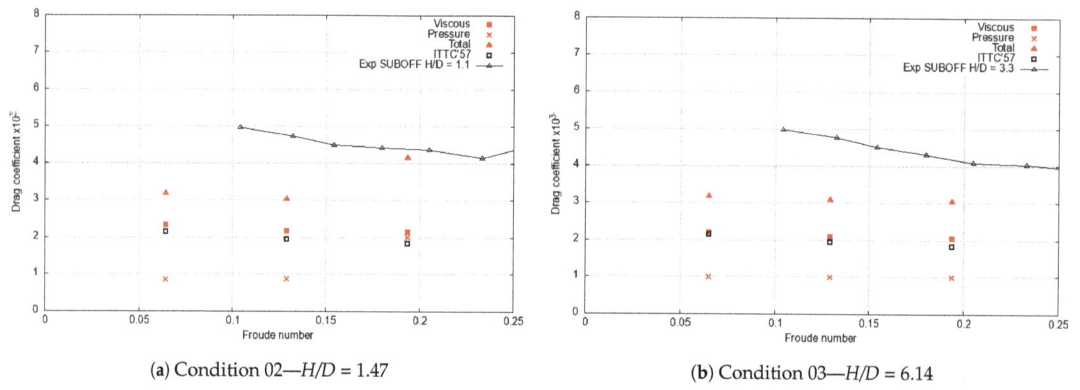

(**a**) Condition 02—*H/D* = 1.47

(**b**) Condition 03—*H/D* = 6.14

Figure 22. Drag coefficients vs Froude number for a Type 209/1300 submarine at condition 2 - periscope depth and 3 - deep Water depth.

4. Conclusions

Results for DARPA SUB-OFF geometry demonstrate that the numerical methodology used in this work is reliable at the engineering level when 4,500,000 cells are used in the simulations. The Verification and Validation procedure was used to estimate the numerical uncertainty of three grid sets combining 5 different densities. Also, strong nonlinear effects due to proximity to free-surface can be captured for Froude number of $Fr = 0.3$ at $H/D = 1.1$.

Considering the results and discussion shown in the previous section, it is concluded that the side probe sensor could be installed at P1 location considering that this area presents a continuous flow without recirculating zones or vortex cores for all the navigation conditions considered. However, turbulent structures are concentrated in the near-surface regions. This probe will be located close to the end of the acceleration region at bow, where the local flow speed is still 15% higher than actual submarine speed. This location is recommended to be used when submarine is at surface condition because the flow is considered linear because its velocity remains at the same level for navigation velocities between 0 and 9 knots. Moreover, speed measurements could be useful for near-surface and deep-water conditions if a 4% correction is included to compensate for free-surface effect. On the other hand, the current calibration procedure used for the upper deck probe at P2 using the side probe measurement at P1 as a reference should be further modified to take into account the difference in the flow velocity range of both locations P1 and P2.

Finally, it is recommended to verify the effect of boundary layer thickness on speed measurements of the flush probe type to guarantee the expected accuracy of the flush probe and to generate validation data during sea trials, including at least 8 calibration points for the flush probe type to compensate the manufacturer declared difference in precision (±0.2 knots) between flush and protruding probe types.

This research could follow several interesting paths: (i) to improve results reliability, authors could perform transient simulations, include ramping initial conditions, improve the prism layer generation to cover 100% of the submarine hull, include a wave damping region at the outlet boundary to avoid reflections; (ii) to assess the non-linear free surface effect for Froude numbers between 0.2 and 0.4, to assess the boundary layer thickness effect on speed measurements or assessing the operational trim effect on flow characteristics. Finally, it will be challenging to assess the Type 209 submarine in self propulsion or maneuvering conditions.

Author Contributions: Conceptualization, R.J.P., M.T.Q., M.A.-H., R.D.; methodology, R.J.P.; software, R.J.P.; validation, R.J.P., M.T.Q. and R.D.; formal analysis, R.J.P. and R.D.; investigation, R.J.P. and M.T.Q.; resources, R.J.P.; data curation, R.J.P. and M.T.Q.; writing—original draft preparation, R.J.P.; writing—review and editing, M.T.Q., M.A.-H., and R.D.; visualization, R.J.P. and M.T.Q.; supervision, R.J.P.; project administration, R.J.P.; funding acquisition, R.J.P. All authors have read and agreed to the published version of the manuscript.

Funding: This research was partially funded by The Ecuadorian Navy through ASTINAVE EP Shipyard grant number REGNMIN-ASTEP-177-18.

Institutional Review Board Statement: Not applicable.

Informed Consent Statement: Not applicable.

Data Availability Statement: The data presented in this study are available on request from the corresponding author. The data are not publicly available due to its potential impact on submarine operational safety.

Acknowledgments: This research was carried out using the research computing facilities and/or advisory services offered by Scientific Computing Laboratory of the Research Center on Mathematical Modeling: MODEMAT, Escuela Politecnica Nacional—Quito. Special thanks to Mercy Anchundia for her support and the help provided to solve remote access issues. In addition, Authors are deeply thankful with the Submarine Squadron of Ecuador and the submarine crew for their predisposition to share their know-how.

Conflicts of Interest: The authors declare no conflict of interest. The funders had no role in the design of the study; in the collection, analyses, or interpretation of data; in the writing of the manuscript, or in the decision to publish the results.

Abbreviations

The following abbreviations are used in this manuscript:

AUV	Autonomous Underwater Vehicle
CFD	Computational Fluid Dynamics
CSSRC	China Ship Scientific Research Centre
DARPA	Defense Advanced Research Projects Agency
DDES	Delayed Detached Eddy Simulation
DES	Detached eddy simulation
HDW	Howaldtswerke-Deutsche Werft
IHSSS	Iranian Hydrodynamic Series of Submarines
ITTC	International Towing Tank Conference
MRF	Moving Reference Frame
RANSE	Reynolds-averaged Navier–Stokes equations
VOF	Volume of Fluid

References

1. ITTC. Uncertainty Analysis in CFD Verification and Validation Methodology and Procedures. In *ITTC—Recommended Procedures and Guidelines*; 28th ITTC Executive Committee: Wuxi, China, 2017; Chapter 7.5-03-01-01.
2. Stern, F.; Wilson, R.; Shao, J. Quantitative V&V of CFD simulations and certification of CFD codes. *Int. J. Numer. Meth. Fluids* **2006**, *50*, 1335–1355. [CrossRef]
3. Liu, H.L.; Huang, T.T. *Summary of DARPA SUBOFF Experimental Program Data*; Technical Report CRDKNSWC/HD-1298-11; Naval Surface Warfare Center, Carderock Division (NSWCCD): West Bethesda, MD, USA, 1998.
4. Huang, T.T.; Liu, H.L.; Groves, N.C. *Experiments of the DARPA SUBOFF Program*; Technical Report DTRC/SHD-1298-02; Davidson Taylor Research Center: Bethesda, MD, USA, 1989.
5. Gross, A.; Kremheller, A.; Fasel, H. Simulation of Flow over SUBOFF Bare Hull Model. In Proceedings of the 49th AIAA Aerospace Sciences Meeting including the New Horizons Forum and Aerospace Exposition, Orlando, FL, USA, 4–7 January 2011; American Institute of Aeronautics and Astronautics: Orlando, FL, USA, 2011. [CrossRef]
6. Lungu, A. DES-based computation of the flow around the DARPA suboff. *IOP Conf. Ser. Mater. Sci. Eng.* **2019**, *591*, 12053. [CrossRef]
7. Vasileva, A.; Kyulevcheliev, S. Numerical Investigation of the Free Surface Effects on underwater vehicle resistance. In Proceedings of the Fourteenth International Conference on Marine Sciences and Technologies, Varna, Bulgaria, 10–12 October 2018; Black Sea: Varna, Bulgaria, 2018; pp. 100–104.
8. Gourlay, T.; Dawson, E. A Havelock source panel method for Near-surface Submarines. *J. Mar. Sci. Appl.* **2015**, *14*, 215–224. [CrossRef]
9. Di Felice, F.; Felli, M.; Liefvendahl, M.; Svennberg, U. Numerical and experimental analysis of the wake behavior of a generic submarine propeller. In Proceedings of the First Internationnal Symposium on Marine Propulsors, Trondheim, Norway, 22–24 June 2009; p. 7.
10. Takahashi, K. Numerical Simulations of Comprehensive Hydrodynamics Performance of DARPA SUBOFF Submarine. Master's Thesis, Florida Institute of Technology, Melbourne, FL, USA, 2019.
11. Liefvendahl, M.; Troeng, C. Computation of Cycle-to-Cycle Variation in Blade Load for a Submarine Propeller, using LES. In Proceedings of the Second International Symposium on Marine Propulsors, Hamburg, Germany, 15–17 June 2011; p. 7.
12. Chase, N. Simulations of the DARPA Suboff Submarine Including Self-Propulsion with the E1619 Propeller. Master's Thesis, University of Iowa, Iowa City, IA, USA, 2012. [CrossRef]
13. Dogrul, A.; Sezen, S.; Delen, C.; Bal, S. Self-propulsion simulation of DARPA suboff. In Proceedings of the 17th International Congress of the International Maritime Association of the Mediterranean (IMAM 2017), Lisbon, Portugal, 9–11 October 2017; Maritime Transportation and Harvesting of Sea Resources, Tecnico Lisboa: Lisbon, Portugal, 2017; Volume 1, pp. 503–511.
14. Delen, C.; Sezen, S.; Bal, S. Computational Investigation of Self Propulsion Performance of Darpa Suboff Vehicle. *Tamap J. Eng.* **2017**, *2017*, 4.
15. Moonesun, M.; Javadi, M.; Charmdooz, P.; Mikhailovich, K. Evaluation of submarine model test in towing tank and comparison with CFD and experimental formulas for fully submerged resistance. *Indian J. Mar. Sci.* **2013**, *42*, 1049–1056.
16. Sezen, S.; Dogrul, A.; Delen, C.; Bal, S. Investigation of self-propulsion of DARPA Suboff by RANS method. *Ocean. Eng.* **2018**, *150*, 258–271. [CrossRef]
17. Duman, S.; Bal, S. Propeller effects on maneuvering of a submerged body. In Proceedings of the 3rd International Meeting - Progress in Propeller Cavitation and its Consequences: Experimental and Computational Methods for Predictions, Istanbul, Turkey, 15–16 November 2018.
18. Chase, N.; Thad, M.; Carrica, P.M. Overset simulation of a submarine and propeller in towed, self-propelled and maneuvering conditions. *Int. Shipbuild. Prog.* **2013**, *60*, 171–205. [CrossRef]
19. Chase, N.; Carrica, P.M. Submarine propeller computations and application to self-propulsion of DARPA Suboff. *Ocean Eng.* **2013**, *60*, 68–80. [CrossRef]
20. Nematollahi, A.; Dadvand, A.; Dawoodian, M. An axisymmetric underwater vehicle-free surface interaction: A numerical study. *Ocean Eng.* **2015**, *96*, 205–214. [CrossRef]
21. Moonesun, M.; Mikhailovich, Y. Minimum immersion depth for eliminating free surface effect on submerged submarine resistance. *Turk. J. Eng. Sci. Technol.* **2015**, *1*, 36–46.
22. Moonesun, M.; Ghasemzadeh, F.; Korneliuk, O.; Korol, Y.; Valeri, N.; Yastreba, A.; Ursalov, A. Effective depth of regular wave on submerged submarine. *Indian J. Geomar. Sci.* **2019**, *48*, 1476–1484. [CrossRef]
23. Moonesun, M.; Mahdian, A.; Korneliuk, O.; Korol, Y.; Bandarinko, A.; Valeri, N. Evaluation of submarine motions under irregular ocean waves by panel method. *Indian J. Geomar. Sci.* **2019**, *48*, 1485–1495.
24. Zhang, N.; Zhang, S.L. Numerical simulation of hull/propeller interaction of submarine in submergence and near surface conditions. *J. Hydrodyn.* **2014**, *26*, 50–56. [CrossRef]
25. Vali, A.; Saranjam, B.; Kamali, R. Experimental and numerical study of a submarine and propeller behaviors in submergence and surface conditions. *J. Appl. Fluid Mech.* **2018**, *11*, 1297–1308. [CrossRef]
26. Carrica, P.M.; Kim, Y.; Martin, J.E. Near-surface self propulsion of a generic submarine in calm water and waves. *Ocean Eng.* **2019**, *183*, 87–105. [CrossRef]

27. Zhang, S.; Li, H.; Zhang, T.; Pang, Y.; Chen, Q. Numerical simulation study on the effects of course keeping on the roll stability of submarine emergency rising. *Appl. Sci.* **2019**, *9*, 3285. [CrossRef]
28. Moonesun, M.; Mikhailovich, K.Y.; Tahvildarzade, D.; Javadi, M. Practical scaling method for underwater hydrodynamic model test of submarine. *J. Korean Soc. Mar. Eng.* **2014**, *38*, 1217–1224. [CrossRef]
29. Nguyen, T.T.; Yoon, H.K.; Park, Y.; Park, C. Estimation of Hydrodynamic Derivatives of Full-Scale Submarine using RANS Solver. *J. Ocean. Eng. Technol.* **2018**, *32*, 386–392. [CrossRef]
30. The OpenFOAM Foundation. *OpenFOAM v7 User Guide*; The OpenFOAM Foundation: London, UK, 2019; p. 237.
31. Groves, N.C.; Huang, T.T.; Chang, M.S. *Geometric Characteristics of DARPA SUBOFF Models (DTRC MODEL Nos. 5470 and 5471)*; Technical Report DTRC/SHD-1298-01; David Taylor Research Center: Bethesda, MD, USA, 1989.
32. Liu, Z.H.; Xiong, Y.; Wang, Z.Z.; Wang, S.; Tu, C.X. Numerical simulation and experimental study of the new method of horseshoe vortex control. *J. Hydrodyn.* **2010**, *22*, 572–581. [CrossRef]
33. Jasak, H.; Vukčević, V.; Gatin, I.; Lalović, I. CFD validation and grid sensitivity studies of full scale ship self propulsion. *Int. J. Nav. Archit. Ocean. Eng.* **2019**, *11*, 33–43. [CrossRef]
34. ITTC. Practical Guidelines for Ship CFD Applications. In *ITTC—Recommended Procedures and Guidelines*; 26th ITTC Executive Committee: Rio de Janeiro, Brazil, 2011; Chapter 7.5-03-02-03.
35. Neulist, D. *Experimental Investigation into the Hydrodynamic Characteristics of a Submarine Operating Near the Free Surface*; Technical Report; Australian Maritime College: Launceston, Australia, 2011.
36. Islam, H.; Guedes Soares, C. Uncertainty analysis in ship resistance prediction using OpenFOAM. *Ocean Eng.* **2019**, *191*, 1–16. [CrossRef]
37. Celik, I.B.; Ghia, U.; Roache, P.J.; Freitas, C.J.; Coleman, H.; Raad, P.E. Procedure for estimation and reporting of uncertainty due to discretization in CFD applications. *J. Fluids Eng. Trans. ASME* **2008**, *130*, 0780011–0780014. [CrossRef]
38. Jeong, J.; Hussain, F. On the identification of a vortex. *J. Fluid Mech.* **1995**, *285*, 69–94. [CrossRef]
39. Dubief, Y.; Delcayre, F. On coherent-vortex identification in turbulence. *J. Turbul.* **2000**, *1*, 37–41. [CrossRef]

MDPI
St. Alban-Anlage 66
4052 Basel
Switzerland
Tel. +41 61 683 77 34
Fax +41 61 302 89 18
www.mdpi.com

Fluids Editorial Office
E-mail: fluids@mdpi.com
www.mdpi.com/journal/fluids